# Soil Aeration and Its Role for Plants

Authors

**Jan Gliński, Ph.D., Dr.Sc.**

**and**

**Witold Stępniewski, Ph.D., Dr.Cs.**

Institute of Agrophysics
Polish Academy of Sciences
Lublin, Poland

**CRC Press, Inc.**
**Boca Raton, Florida**

**Library of Congress Cataloging in Publication Data**

Glínski, Jan.
Soil aeration and its role for plants.

Bibliography: p.
Includes index.
1. Soil aeration. 2. Plant-soil relationships.
I. Stępniewski, Witold, 1947- . II. Title.
S593.7.G55 1985 631.4'33 83-23144
ISBN 0-8493-5250-9

This book represents information obtained from authentic and highly regarded sources. Reprinted material is quoted with permission, and sources are indicated. A wide variety of references are listed. Every reasonable effort has been made to give reliable data and information, but the author and the publisher cannot assume responsibility for the validity of all materials or for the consequences of their use.



Direct all inquiries to CRC Press, Inc., 2000 Corporate Blvd., N.W., Boca Raton, Florida, 33431.



International Standard Book Number 0-8493-5250-9
Library of Congress Card Number 83-23144
Printed in the United States

# PREFACE

Investigations into soil aeration carried out by the Institute of Agrophysics of the Polish Academy of Sciences and the study of its extensive collection of literature of the subject (more than 800 items) stimulated us to undertake the provision of a systematic and comprehensive review of the present state of knowledge concerning soil aeration and its role for plants.

This book is directed to the interests of our fellow workers on soil aeration. Due to the interdisciplinary nature of the subject, however, it may also be read by research workers, teachers, and students of agronomy, soil physics, soil chemistry, soil biology, ecology, and plant physiology.

SI units are used throughout the text, the necessary conversions and recalculations of data and results having been performed in respect of all reports to which reference is made. The change of units together with the need to compile data from many sources in order to provide representative material has precluded the use of figures, diagrams, and tables as originally published and all such items have been prepared specifically for this edition.

We are indebted to the many colleagues and authors who have sent us reprints of their papers. We are grateful to Prof. Dr. P. Kowalik, Prof. Dr. W. Myśków, and Dr. W. Zygmunt for their reviews of the manuscript and for many helpful suggestions and criticisms they have offered. The assistance of Mr. J. G. W. Munns in correcting the language of the English text is also appreciated.

Finally, we tender our best thanks to our wives for their patient support and encouragement.

**J. Gliński**
**W. Stępniewski**
June 1983

## THE AUTHORS

**Jan Gliński, Ph.D., Dr.Sc.,** Professor of Agricultural Sciences, pedologist, is a director of the Institute of Agrophysics of the Polish Academy of Sciences in Lublin, and a head of the Laboratory of Soil Physicochemistry in this Institute.

Earlier he worked at the Soil Science Department of the Agricultural University in Lublin, where he carried out investigations on soil chemistry and especially on trace elements in soils and plants and instrumental methods of their determination.

In the Institute of Agrophysics, Prof. Gliński leads the group that works on electrical and aeration properties of soils. He has published more than 10 papers on this subject alone from a total of 120 of his scientific publications.

As a member of the International Society of Soil Science, he participated in the Congresses of this Society in Madison (1960), Bucharest (1964), Moscow (1974), and Edmonton (1978), where he presented his papers.

**Witold Stępniewski, Ph.D., Dr.Cs.,** Associate Professor of Agricultural Sciences, studied chemistry at Maria Curie-Sklodowska University in Lublin where he earned his B.Sc. Since 1970, he has been employed in the Institute of Agrophysics of the Polish Academy of Sciences in Lublin as a research worker. He earned a Ph.D. in 1975 and Dr.Sc. in 1980.

The main subjects of his research are problems of soil aeration, particularly field studies of soil air, gas diffusion in soil, and plant response to soil oxygen status. He is the author of about 30 papers on those problems.

# TABLE OF CONTENTS

# INTRODUCTION

One of the most important factors influencing soil fertility, besides water and nutrient content, is soil air.

Clements[117] traced the first notes on the importance of soil air for plants in the works of Mayow in 1668, and Huygens and Papin in 1674. It was not until a century later that Scheele and Priestley discovered oxygen, and soon after that its role in the respiratory processes of living organisms had been explained. During this era another pioneer of aeration research was Ingen Housz.[261]

The first paper on soil aeration appeared in France in 1853 and was published by Boussignault and Léwy.[68] Soon after that, in 1874, Sorauer[261] established deficient soil aeration as a cause of plant diseases and in 1904 Buckingham[87] stated that the main mechanism of gas exchange in soil is diffusion.

At the beginning of the 20th century, a number of papers on the composition of soil air appeared, of which the most important are those of Russell and Appleyard,[621] Clements,[117] and Cannon and Free.[103]

The latter authors, as well as Hutchins,[329] pointed out that the rate of oxygen influx to the plant roots rather than the oxygen content in the soil has a decisive role in supplying plant roots with oxygen.

Hutchins[329] elaborated the first, already historical, method for the evaluation of oxygen availability, the placement of oxygen absorber in the soil. This method was replaced by Lemon and Erickson[419] in 1952 with a method of polarographic reduction of oxygen on a platinum cathode.

Further progress in research on soil air-plant relations was also connected with the development of new measuring techniques such as gas chromatography in 1952 and membrane-covered oxygen electrodes in 1963 (cf. Chapter 6).

In the literature concerning the soil aeration problem there are some terminological inconsistencies. The term "soil aeration" is sometimes inadequately used in a narrow sense to denote the gas exchange between the soil and the atmosphere, or the oxygen content of the soil air, and even as a name for a process of artificially forcing air into the soil.

In this book the term "soil aeration" is used with a wide meaning including all the related aspects such as soil air composition and its role for plants and the processes of absorption, production, and transfer of gases in soil. That part of soil aeration connected with oxygen distribution in soil and its availability for microorganisms and plant roots will be called "soil oxygenation". This term is also used by other authors.[25,381]

Soil aeration is a complex subject. Soil air composition is a resultant effect of biological and chemical processes of uptake and evolution of gases on the one hand and of the physical processes of their transport on the other. Plant response to soil air depends, however, not only on the air composition but also on the other related chemical, physical, and biological soil processes which are significant for plants.

It is true that the problem of soil aeration or its particular aspects have been discussed in some valuable reviews[25,56,261,381,623,624,668,705-707,769,787] but a more precise, factual treatment of this subject is needed. Hence the aim of this book is to provide a comprehensive review of all aspects of the problem and of the methods used in connection with it.

Chapter 1

# OXYGEN UPTAKE AND CARBON DIOXIDE PRODUCTION IN THE SOIL ENVIRONMENT

The processes of absorption and production of gases in soil involve oxygen, carbon dioxide, nitrogen and its oxides, hydrogen, methane, ethylene, and some other volatile organic compounds (Figure 1).*

For the soil air composition, undoubtedly, oxygen and carbon dioxide are of major importance. Oxygen is mainly consumed, in the process of respiration. However, it may also be produced, for example, by autotrophs assimilating carbon dioxide among which algae predominate. The presence of the latter in rice paddy soils constitutes an important source of oxygen for the soil and protects rice from diseases.[464]

Similarly, carbon dioxide, apart from the prevailing process of its production owing to degradation of organic matter, may also be assimilated by soil autotrophs and also by some heterotrophs, e.g., *Rhizobia*.[414,657]

Besides these biological processes, both oxygen and carbon dioxide may be involved in some chemical reactions with mineral and organic soil components (e.g., oxidation of reduced forms of iron and manganese, decomposition of carbonates, etc.).

In this chapter only those topics which concern the metabolic uptake of oxygen and production of carbon dioxide by soil microbes and plant roots are presented. Other biological processes as well as chemical reactions and the respiration of soil microfauna are omitted as being of minor importance for the oxygen and carbon dioxide budget in soil. The processes that involve other components of soil air, i.e., nitrogen and its oxides, ethylene, methane, and hydrogen are described later as they are related to the soil oxygenation status.

## I. RESPIRATION AND FERMENTATION — A GENERAL DESCRIPTION

The process of oxic respiration (i.e., respiration at abundance of oxygen) can be summarized as follows:

$$C + O_2 \rightarrow CO_2 + \text{energy} \quad (1)$$

In the case of glucose[559] it is

$$C_6H_{12}O_6 + 6O_2 \rightarrow 6CO_2 + 6H_2O + 2\,883 \text{ kJ (total energy)} \quad (2)$$
$$\text{or } + 1\,270 \text{ kJ (biologically useful energy, i.e., 38 mol of ATP)}$$

The molar (volumetric) ratio of carbon dioxide evolved to oxygen consumed is called the respiratory quotient (RQ). This quotient depends both on the substrate used and on the type of metabolism. In oxic respiration it is equal to about 1.3 for organic acids, 1.0 for carbohydrates, 0.8 for proteins, and 0.7 for fats.

A typical example of transformations connected with the respiratory process is the oxidation of glucose, which is the basic substrate of this process.

The first step of glucose oxidation does not require oxygen and is identical in the presence of oxygen (oxic conditions) and in its absence (anoxic conditions).[345,559] This step consists of the transformation of glucose to pyruvic acid and may be performed, depending on the organism, along three pathways, i.e., glycolysis, penthosophosphate, or the Entner-Doudoroff.

* Figures and tables for this chapter appear at the end of the text.

Pyruvic acid, which constitutes a central metabolite, is a starting point for two pathways of further transformations: the oxic and the anoxic. Thus at the stage of pyruvic acid, the splitting of the metabolic pathways begins.

On the oxic pathway, pyruvic acid is oxidized in the Krebs cycle to carbon dioxide. The electrons released in the cycle are transported to oxygen molecules through the electron transport chain, the last step of which is a cytochrome oxidase.

In anoxic conditions the metabolism of pyruvic acid may be changed, for instance, towards its decarboxylation to acetaldehyde:

$$CH_3COCOOH \xrightarrow{\text{carboxylase}} CO_2 + CH_3CHO \qquad (3)$$

and to its subsequent reduction to ethanole (alcohol fermentation):

$$CH_3CHO + NADH_2 \xrightarrow{\text{alcohol dehydrogenase}} C_2H_5OH + NAD \qquad (4)$$

Another possibility is the reduction of pyruvic acid to lactic acid (lactic fermentation):

$$CH_3COCOOH + NADH_2 \xrightarrow{\text{lactic dehydrogenase}} CH_3CHOHCOOH + NAD \qquad (5)$$

Summarizing, for the alcohol fermentation[559] we obtain:

$$\begin{aligned} C_6H_{12}O_6 \rightarrow 2C_2H_5OH + 2CO_2 + 226\text{ kJ (total energy)} \\ \text{or } + 67\text{ kJ (biologically useful energy, i.e., 2 mol of ATP)} \end{aligned} \qquad (6)$$

This shows that the energy efficiency of fermentation is much lower than that of normal (oxic) respiration. During fermentation carbon dioxide is produced without oxygen and this means a rise of RQ to infinity.

Alcohol fermentation is one of many types of fermentation (cf. Figure 4, Chapter 5). Like lactic fermentation it occurs not only in microorganisms but also in animals and higher plants, though alcohol fermentation is typical for plants, while lactic fermentation prevails in animal tissues. Other types of fermentation are the domain of microorganisms and of some lower animals, e.g., nematodes.

A relationship between the metabolic intensity of respiration and the oxygen concentration, in plant and animal tissues as well as in the cells of facultative microorganisms, is shown in Figure 2.

A decrease of oxygen content is initially accompanied by a decrease in organic substrate use and of carbon dioxide production but later, at low oxygen concentrations, an increase of both these factors is observed. The phenomenon of increased substrate use in anoxic conditions is called the Pasteur effect. For its evaluation the Pasteur quotient (PQ) is used. It is defined as follows:

$$PQ = \frac{\begin{array}{c}\text{substrate used in}\\ \text{anoxic conditions}\end{array} - \begin{array}{c}\text{substrate used in}\\ \text{oxic conditions}\end{array}}{\text{substrate used in oxic conditions}} \qquad (7)$$

The PQ value is between 2 and 6.5[559] which means that cells using one substrate unit in oxic conditions increase substrate use in anoxic conditions by a factor of 3 to 7.5.

Another expression, which is more convenient to use in experimental conditions, is the Meyerhoff quotient (MQ), defined as follows:

$$MQ = \frac{\text{CO}_2 \text{ volume in anoxic conditions} - \text{CO}_2 \text{ volume in oxic conditions}}{\text{CO}_2 \text{ volume in oxic conditions}} \tag{8}$$

The MQ value is between 1.0 and 2.5.[559]

A very important point, "x" in Figure 2, is called the extinction point. It denotes an oxygen concentration at which anoxic metabolism appears. Below the extinction point, carbon dioxide production exceeds oxygen uptake which implies a rise of the RQ value above 1. Thus the RQ value can be regarded as a sensitive index of soil oxygenation status.

For higher plants oxic metabolism is typical. Anoxic metabolism may also occur in plants but in this condition most plants may survive for only a short period of time. There is a high differentiation among plant species with respect to tolerance to anoxic metabolism. This will be treated in detail in Chapter 5.

Microorganisms, in turn, are divided into three groups according to their oxygen requirements: aerobes, facultative anaerobes, and obligate anaerobes (Table 1). More appropriate names should be oxic, facultative anoxic, and obligate anoxic organisms, but the traditional names commonly used by microbiologists will be retained here, though such traditional terms as "anaerobic respiration", "aerobic respiration", "anaerobiosis", etc. will be avoided here and replaced by "oxic respiration", "anoxic respiration", "anoxibiosis", etc. In soil containing oxygen both aerobes and facultative anaerobes are active. With the exhaustion of oxygen, the activity of aerobic microorganisms ceases while facultative anaerobes change their metabolism into an anoxic one and carry out their life processes simultaneously with obligate anaerobes, which become active in anoxia.

The terminal electron acceptor in the process of oxic respiration is molecular oxygen. During anoxic respiration in soil, the function of electron acceptor is performed by oxidized inorganic compounds, i.e., $NO_3^-$, Fe(III), and Mn(IV) compounds, $SO_4^{--}$, $H^+$, and by some organic compounds. Simultaneously these compounds are reduced. They can be reoxidized by autotrophic bacteria which are able to carry out chemosynthesis of organic compounds from carbon dioxide, utilizing the energy of oxidation of these reduced inorganic forms. Thus sulfuric bacteria oxidize $S^{--}$, $S_2O_3^{--}$, and free sulfur to $SO_4^{--}$, iron bacteria oxidize $Fe^{++}$ to $Fe^{+++}$, nitrifying bacteria transform $NH_4^+$ and $NO_2^-$ into $NO_3^-$, and hydrogen bacteria transform gaseous $H_2$ into the $H^+$ ion. These processes are described in Chapter 4.

Thus the presence of oxygen, besides the kind of substrates undergoing decomposition, is a decisive factor for the composition of the soil microflora, and for the type and direction of the transformations of organic and inorganic substances performed by the microbes.

## II. RESPIRATION IN THE SOIL

Respiration in the soil is a result of the biological activities of all the soil organisms, the most important being the respiratory activity of soil microbes and that of plant roots. The measure of this activity is the amount of carbon dioxide evolved or oxygen taken up in unit of time per volume or mass unit of the soil. Examples of respiratory activities of different organisms are presented in Table 2.

### A. Soil Microbial Respiration and Its Determinants

Table 3 is a compilation of the soil microbial activity values given by various authors. They are normally within the range from 0.2 to 20 $cm^3\ O_2\ kg^{-1}\ hr^{-1}$. This corresponds to 0.05 to 5 $mm^3kg^{-1}s^{-1}$ that is approximately 3 to 300 $\mu mol\ m^{-3}s^{-1}$ which is equivalent to 0.1 to 10 $mg\ m^{-3}s^{-1}$. However, in particular cases, e.g., in some organic soils, in litters or after incorporation of organic substances into the soil, the respiratory activity may be as much as one order of magnitude higher than the upper value for normal soils. Soil respiratory activity is the result of the activity of all the microbial population in the soil. Therefore it depends on the composition, size, and metabolic activity of this population and these are, in turn, influenced by such factors as moisture, oxygen availability, organic matter decomposability, mineral nutrients, and by such parameters as temperature, carbon dioxide content, soil bulk density, pH, presence of heavy metals or pesticides, and soil sterilization. Moreover, in each case a parameter of time is involved. The latter is included while treating each topic below.

*1. Soil Moisture*

Microorganisms are divided into three groups with respect to their moisture requirements: hygrophiles, mesophiles, and xerophiles. These groups exhibit different reactions to water deficiency,[573] namely,

1. Hygrophiles — disappearance of activity at soil moisture tension above 7.1 MPa (pF 4.85)
2. Mesophiles — disappearance of activity within a soil moisture tension interval from 7.1 MPa to 30 MPa (pF 4.85 to 5.48)
3. Xerophiles — disappearance of activity at moisture tension above 30 MPa (pF 5.48). Upper limit of activity of known xerophiles is 48 MPa (pF 5.68)

The majority of bacteria, as well as yeasts and some other fungi, belong to hygrophilic microorganisms. The great majority of fungi and some bacteria belong to mesophiles. Only some *Phycomycetes* belong to the group of xerophilous microorganisms, e.g., *Aspergillus niger, A. glaucus, Penicillium glaucum*.

The influence of soil moisture tension on the number of different microbes is shown in Figure 3.

The relationship between soil microbial activity and soil moisture content is curvelinear with a maximum.[4,76,179,481,573,605,606,727,798] A typical example of this type of dependence is shown in Figure 4. A decrease in the respiration rate is observed at both low and high moisture contents. The maximum of the soil microbial respiration rate is usually within a soil moisture tension interval of 0.01 to 1 MPa (pF 2 to 4) which is shown as an example in Figure 5.

A reduction in the respiration rate at low moisture contents is caused by a low availability of soil water. The cessation of soil respiration is usually observed at a soil moisture tension above 10 MPa,[179,573,803] i.e., after the permanent wilting point for plants has been reached.

In the range of high moisture contents, however, a decrease in the observed respiration rate is caused by limitation of oxygen accessibility in the entire bulk of soil due to the soil pores becoming filled with water. Thus it is not a direct effect of water on soil microorganisms but an indirect one, as a result of physical blocking of gas transport in soil, accompanied by rise of RQ value above 1.

Abrasimova[4] found hysteresis of carbon dioxide evolution as a function of soil moisture content. Respiratory activity at a moisture content obtained by dewatering a saturated sample was lower than in the case of the same moisture content achieved by adding water to the dry sample. This may be explained by different rates of development of the microbes and by different degrees of substrate utilization during the two processes used to reach the desired moisture level; the factor of time is of highest importance.

It should be emphasized that respiratory activity, at constant moisture level, decreases in time[373,614,615,727,740] and this is shown in Figures 6 and 7.

This relationship can be described by the exponential formula:[379,380,532]

$$q = a + b^{kt} \tag{9}$$

or by the hyperbolic equation:[113]

$$q = \frac{a}{t^b} \quad (10)$$

where q = respiration rate, t = time, and a,b,k = constants.

Respiration rate of an air-dry soil immediately after its remoistening is relatively high and then decreases, stabilizing after several days[137,218,341,343,532,702] and sometimes only after several weeks.[149] The stimulating effect which drying the soil has on its oxygen uptake or carbon dioxide production immediately after its remoistening is the higher the higher the temperature and the longer the period of the drying or storing; this is shown, as an example, in Figure 8. The following reasons for this were suggested:[218] physiological youth of the surviving microbes, newly available nutrients, destruction of inhibitory substances, occurrence of free enzymes in the soil, spore germination, and a rapid mycelial growth. It was also found[341] that during the first days after remoistening the air-dried soil, a considerable part of the carbon dioxide evolved is of chemical origin. Nevertheless, remoistening the air-dry soil is also accompanied by an increase in the number of soil microbes (Figure 8) due to the increased decomposability of organic matter after prolonged drying, caused by chemical processes such as decarboxylation or oxidation.

There is also a reverse situation, i.e., excessive soil moistening for a considerable period causes reduction processes in the soil which greatly increases soil oxygen uptake after the excess water has been drained off.[269,768] For example, Grechin and Ignatiev[269] found that soddy-podzolic soil, after long term flooding (189 days in laboratory conditions) accompanied by redox potential decreased down to −260 mV, showed, after removal of an excess of water, a 35-fold increase in oxygen uptake rate and "only" an 8-fold increase in carbon dioxide evolution. A very low RQ value within 0.3 to 0.16 denotes that most of the oxygen consumed was used in chemical oxidation of reduced compounds formed during the previous period of flooding.

In the light of the above reports it is obvious that repeated cycles of moistening and drying,[296,342] as well as oxic and anoxic cycles,[585] increase respiration intensity which implies an increased rate of organic matter depletion compared to conditions with constant moisture content.

### 2. *Temperature*

Microorganisms in general, including those in the soil, are divided (according to their temperature requirements) into three groups: kriophiles, mesophiles, and thermophiles, their optimum temperatures being <20, 20 to 40, and >40°C, respectively.[247]

Normally, soil contains a variety of different microbes, and its respiratory activity as a function of temperature shows a maximum, as in the case of moisture.

As early as 1896, Déherain and Demoussy[151] found that a maximum of carbon dioxide evolution occurred at 65°C (respiratory maximum). At temperatures above 90°C they observed a second increase in carbon dioxide production, connected with intensification of chemical oxidation.

The respiratory maximum in soils usually occurs within the temperature range 40 to 70°C and even for the same soils it is subjected to change with differing moisture contents.[686,727] At temperatures normally found in soils the respiratory maximum is not reached, and an increase in oxygen uptake or carbon dioxide production rate is observed with temperature rise.[4,174,323,614,632,686,727,733,755]

This relationship is described usually by two types of equations, that of Van't Hoff or that of Arrhenius, which are applied for chemical reactions.[322,323,755]

The first of these says that the reaction rate increases by a factor of 2 to 3 with a temperature rise of 10°C.

Thus for the respiration rate q we obtain:

$$\ln q = c + T\frac{\ln Q_{10}}{10} \tag{11}$$

where q = respiration rate at a temperature T, c = constant, and $Q_{10}$ = the so-called Q-ten temperature coefficient, showing by how many times the respiration intensity increases, when the temperature increases by 10°C.

For chemical processes the $Q_{10}$ value is 2 to 3, and for the physical ones (diffusion, conductivity) it is within the range 1.2 to 1.3.

For plant and animal tissues $Q_{10}$ lies in the interval 1.5 to 3.0 at temperatures below 23°C. In soil this interval may be much wider. Howard and Howard[323] found for leaf litter $Q_{10}$ equaled 1.7 to 5.3, Yoneda[818] found 1.70 for wood litter, and for different forest soils Yoneda and Kirita[819] found 1.6 to 2.94. Abrasimova[3,4] found values for several cultivated soils of the U.S.S.R. of 1.6 to 7.6, while Ross and Cairns[614] obtained 1.2 to 9.3 for nine soils. The higher $Q_{10}$ values measured in soil in comparison with plant or animal tissues are due to the double influence of temperature on soil respiration intensity, and through an increase in the vital activity of the microbial cells, and through an increase in their number.

The $Q_{10}$ value for soils is of little utility because even for the same soil it is not constant and decreases with temperature.[3,614]

A second and more precise description of the dependence of chemical reaction rate upon temperature is the equation of Arrhenius:

$$\ln k = \frac{-E}{RT} + a \tag{12}$$

where k = reaction rate constant, a = constant, R = universal gas constant, T = absolute temperature, and E = activation energy of the reaction.

For respiration activities $q_1$ and $q_2$ at temperatures $T_1$ and $T_2$ one obtains:

$$\ln\frac{q_2}{q_1} = \frac{E}{R}\frac{(T_2 - T_1)}{T_1T_2} \tag{13}$$

According to this model, the actual respiration rate should be a linear function of the reciprocal of T. This equation also implies a decrease in $Q_{10}$ value with temperature; at $T_2 - T_1$ equal 10°C the higher the absolute temperature, the higher $T_1T_2$, and the lower $q_2/q_1$ which is in accordance with the experimental data.

Neither of the above models is always satisfactory for describing the influence of temperature on soil respiratory activity because of the impact of some other environmental factors such as moisture, microflora composition, the kind of the organic substrate and its degree of decomposition, time, etc. In many cases the above relationship can be satisfactorily described by a straight line model.[323]

Freezing the soil does not have an unequivocal effect on its respiration[333,341,613] although a tendency to a decrease in the respiration rate seems to prevail, especially at repeated cycles of freezing and thawing.

Franz[214] found that an increase in altitude from 100 to 5200 m above sea level, connected with a decrease in mean annual temperature from 25 to 0°C, was accompanied by a drop (of about 100-fold) in aerobic bacteria counts, as well as by a 10-fold reduction in oxygen uptake per gram of organic matter in mountain soils.

### 3. *Organic Amendments*

The content of organic matter varies widely from below 1% in sandy soils up to almost 100% in peat soils. In natural conditions the organic matter content of a particular soil is generally constant.

Although soil organic matter undergoes permanent transformations, each soil preserves, in particular ecological conditions, a specific equilibrium between an afflux of fresh organic material and its decomposition rate. Due to this it is not easy to increase the organic matter content of the soil, especially in the stable form of humus, even after the incorporation of high amounts of organic material.

Incorporation of organic material is accompanied by an increase in soil microbial activity and in the rate of mineralization. This acceleration may sometimes result in a negative phenomenon of an increased decomposition rate of the native soil organic matter.[732] It is generally assumed that about 3/4 to 4/5 of the organic material incorporated in the soil undergoes mineralization, while the rest is transformed into much more stable specific humic substances. In agricultural practice the amounts of organic amendments introduced to the soil during 1 year are of the order of several thousands of kilograms per hectare and equivalent amounts of oxygen are necessary for their mineralization.

The influence of an organic amendment on the soil microbial respiration depends on the decomposability of that material (kind and age of the plant debris). Organic materials commonly incorporated into soil contain the following components:[305] (1) cellulose, 15 to 60% of the dry weight, (2) hemicellulose, 10 to 30% of the dry weight, (3) lignin, 5 to 30% of the dry weight, (4) compounds soluble in water (simple saccharides, amino acids, aliphatic acids), 5 to 30% of the dry weight, (5) components soluble in ether and in alcohol (fats, waxes, resins, pigments), (6) proteins. The content of proteins and water soluble compounds decreases as the plant ages, while that of components more resistant to microbial decomposition like cellulose, lignin, and hemicelluloses increases.

Beside materials normally reaching the soil, like litter in forest soils and postharvest plant residues in arable soils, the most important organic additions are farmyard manure, animal slurries, composts, sewage water, and green manure. These materials supply soils with organic matter.

Artyushenko,[28] in field experiments with manuring and composting, and Das,[149] applying poultry manure to the soil, found that carbon dioxide evolution from soil was almost doubled as compared to the control, while Grabert and Steinbrenner[260] observed a sevenfold increase in carbon dioxide production in the dunged soil.

In addition, the incorporation of grass,[732] clover,[1] straw,[242,309] and residues of alfalfa, oats, and sorgho[119] stimulated the respiratory activity of the soil by up to 500%.

Gilbert and Griebel[232] treated 12 different soils with volatile compounds obtained from alfalfa which caused as much as an eightfold increase in their respiration rates.

The influence of organic amendment on soil respiration is shown in Figure 9, and also in Figure 14.

Similar effects to those described above were also found while treating soils with many other substances such as glucose,[85,328,614,616] saccharose,[218] starch, cellulose, and casein.[85] The respiratory activity of the soil can also be stimulated by incorporating small amounts of ethanol and acetaldehyde,[282] and such fenolic acids as gallic acid, 3,4-dihydroxybenzoic acid, vanilic acid, ferulic acid, *p*-hydroxycinnamonic acid, *p*-hydroxybenzoic acid, 2,4-dihydroxybenzoic acid.[328]

It should be emphasized, that from the point of view of soil oxygenation large additions of materials such as animal slurries and sewage must be applied with care, because while they increase soil oxygen demand they simultaneously impede gas exchange by clogging the soil pores.

*4. Oxygen*

As already mentioned, microorganisms are divided, according to their oxygen needs, into aerobes, facultative anaerobes, and obligate anaerobes.

Aerobic microorganisms utilize oxygen both as a terminal electron acceptor from the cytochrome oxidase and as a material for biosynthesis of sterols and unsaturated fatty acids, and in the reaction of aromatic ring oxidation.

The cell respiration rate as a function of oxygen concentration is described by the Michaelis-Menten equation:

$$q = \frac{q_{max}}{1 + \frac{K_M}{C_{O_2}}} \tag{14}$$

where q = oxygen uptake rate, $q_{max}$ = maximum oxygen uptake rate, $K_M$ = Michaelis' constant, and $C_{O_2}$ = oxygen concentration.

The $K_M$ value is that of the oxygen concentration in the solution, at which the oxygen uptake rate is equal to $^1/_2\ q_{max}$. The $K_M$ values measured by Longsmuir[437] for several bacterial strains are within $1.1 \cdot 10^{-8}$ *M (Micrococcus candicans)* to $3.57 \cdot 10^{-6}$ *M (Bacillus me gatherium-glicine)*. Similar values from 2.7 to $4.4 \cdot 10^{-6}$ *M* for water suspensions of four soils were obtained by Greenwood.[274]

The basic patterns of the relationship between oxygen uptake and carbon dioxide evolution by different types of microbes and the ambient oxygen concentration is shown in Figure 10. As can be noted, a sharp decrease in oxygen uptake occurs only at low oxygen concentrations in the gas phase below 0.01 to 0.02 $m^3m^{-3}$.

The equation of Michaelis-Menten in the case of aerobic microorganisms (Figure 10A) describes both the oxygen uptake and the corresponding carbon dioxide production, as the RQ value is equal to 1. For facultative anaerobes (Figure 10B) this equation is applicable only to oxygen uptake, because of an increase in RQ value above 1 at low oxygen concentrations.

For obligate anaerobes (Figure 10C), the relationship is quite different and their activity, expressed by carbon dioxide production, takes place only at an oxygen concentration close to zero.

In soils which normally contain all the groups of microbes, the relationship of oxygen uptake and carbon dioxide production to oxygen concentration should be similar to that of facultative anaerobes (Figure 10B). However, the available literature does not provide sufficient data concerning oxygen uptake and carbon dioxide production, throughout the entire range of oxygen concentrations, to enable this hypothesis to be verified.

In Figure 11 a relationship between the oxygen uptake and the oxygen concentration in the soil is presented. This figure, and also other data, shows that a decrease in soil respiration rate to half of its maximum value takes place below 0.02 $m^3m^{-3}$ of oxygen in the soil air.[273,536,724] However, the oxygen concentration at which the RQ value increases above 1 may sometimes be higher than 0.02 $m^3m^{-3}$, especially in aggregated soil. This is because microbial respiration in soil is also dependent on the physical properties of the soil which may limit the oxygen access to the whole bulk of the soil, whereas the pure microbial cells considered above were supplied with oxygen directly from the solutions in which they were contained.

The oxygen transport limitation may often be the cause of a decrease in oxygen uptake in soil below its potential oxygen demand. The potential respiratory activity means the respiration rate under the conditions of unlimited oxygen supply (range of plateau in Figure 10A) and is identical with the $q_{max}$ value in the Michaelis-Menten equation. This implies the applicability of the Michaelis-Menten equation to only a given point in the soil (local applicability) but not to the bulk soil. In the latter case, due to soil heterogeneity, some anoxic microsites may occur although the average oxygen concentration may even be close to the atmospheric level (cf. Chapter 2, Section II.C).

### 5. Carbon Dioxide

It seems obvious that carbon dioxide as a product of respiration should impede microbial activity. However, this effect, has been little studied in soils.

MacFadyen[454] found that the inhibitory effect of carbon dioxide on microbial activity was dependent on soil texture. In acid sandy heath soil 0.0025 $m^3m^{-3}$ carbon dioxide markedly reduced the oxygen consumption and 0.007 $m^3m^{-3}$ suppressed it. The heavier the soil texture, the higher the carbon dioxide concentration that was needed to decrease respiration rate by a given proportion. In the most heavily textured soil investigated (a pasture clay), oxygen uptake suppression appeared at 0.10 $m^3m^{-3}$ of carbon dioxide. MacFadyen[454] suggested that microbes in heavier soils are better adapted to higher carbon dioxide concentrations than in sandy ones.

*6. Soil Air-Filled Porosity, Bulk Density, and Aggregate Size*

These physical parameters of the soil influence its respiratory activity indirectly, through their influence on the transport of oxygen and carbon dioxide. At low air-filled porosities (usually below 0.2 $m^3m^{-3}$) a decrease in soil respiratory activity occurs, accompanied by an RQ rise above 1[76,605,606] which is shown, as an example, in Figure 12. This is caused by unsatisfactory oxygen access to the entire soil bulk and the appearance of anoxic metabolism in some part of the soil volume.

Bulk density influences soil respiration by changing its air-filled porosity.

Soil aggregates are specific microsites in soil with porosity lower than the average of the soil. For this reason, inhibition of the gas exchange takes place first within soil aggregates and the anoxic processes appear there first. Due to this, a reduced oxygen uptake by bigger soil aggregates is observed at higher moisture contents (Figure 13).

*7. Soil Reaction*

The activity of microorganisms is strongly dependent on the pH of the environment. The following four groups of microorganisms are distinguished according to their sensitivity to pH:[247]

1. Microorganisms insensitive to pH, active in a wide pH range with optimum close to 7
2. Microorganisms requiring neutral reaction and sensitive to acidification or alkalinization
3. Acidophylic microorganisms with optimum pH 2.0 to 3.5
4. Basidophylic microorganisms active at pH above 8

Significant changes in soil reaction due to, for example, uptake or leaching of nutrients or due to industrial emissions influencing the composition and number of soil microflora[7] may also influence its respiratory activity. This effect, however, has been little studied.

Bryant et al.[85] in their studies of soil acidification in field conditions (long-term effect of sulfur deposits) and in the laboratory (acidification with $H_2SO_4$) found that a lowering of pH may sometimes, but not always, cause a decrease in soil respiratory activity, shown in Figure 14. An inhibitory effect on respiration, caused by acidification due to the incorporation of sulfur in the soil was also confirmed by Salonius and Mahendrappa.[633]

*8. Soil Minerals*

Soil minerals affect respiration of microorganisms by sorptive interactions which can be divided into three groups:[461]

1. Sorption of microorganisms on the surfaces of large mineral particles
2. Sorptive interactions between cells and mineral particles of approximately bacterial size (formation of aggregates)
3. Sorption of very small mineral particles onto the surfaces of microbial cells

Moreover, the effect of soil minerals on microorganisms involves adsorption of substrates and products of vital activity of microorganisms, hydrogen ions, antibiotics, etc. on the mineral particles.

The problems of interactions between microorganisms and soil mineral particles have been studied by many authors. More information on this subject can be found in reviews of Burns,[94] Dąbek-Szreniawska and Drązkiewicz,[147] Filip,[202,203] Filip et al.,[204] Hattori,[305] Hattori and Furusaka,[306] Marshall,[461] and Zvyagintsev.[828]

There is not, however, an agreement with respect to the influence of these interactions on respiration rate. Both stimulating and inhibitory effects have been reported. These effects have not yet been fully explained as they are connected with many factors such as mineral type, kind of sorbed microorganisms, carbon source, pH, cation exchange capacity, the kind of cations saturating the mineral, and the specific surface area of the colloids.[290,306,373,394,406,534,716-719,796]

Among soil minerals, montmorillonite-type minerals exhibit the greatest effect on microbial respiration.[716-719] Montmorillonite, generally, increases respiration of bacteria but hampers that of the fungi, although for the latter a reverse effect was also reported.[290]

An important role is played by the carbon substrate[394,534] which may even change the direction of the effect of the mineral on the respiration rate.

### *9. Mineral Fertilization*

The relevant papers[149,242,594,632,633,637,638] show the significant influence of mineral fertilization on soil respiration, but there is no agreement, even with respect to basic NPK nutrients.

Schinner et al.[638] fertilizing silty alluvial soil with NPK found, after 6 weeks, a 50% lowering in carbon dioxide evolution as compared to the control. On the other hand, Schaffer[637] observed a stimulation of respiratory activity at NPK doses up to 60 kg ha$^{-1}$ while a higher fertilization level caused a reverse effect.

Most of the papers, however, report an increase in soil respiratory activity under the influence of fertilization. Thus Repnevskaya[594] found that NPK application and liming in a forest soil enhanced carbon dioxide evolution by 50 to 250%. Similar effects of NPK[149] and NPKCa[242] were also observed immediately after the application.

Liming of acid forest soils increased carbon dioxide production both in the field and in laboratory conditions.[408]

Urea application, in laboratory conditions,[85,633] may greatly increase oxygen uptake by the soil (Figure 15). In this case, however, the stimulatory effect is connected not only with an increased N supply, but also with a rise in soil pH.

### *10. Heavy Metals*

Some heavy metals are known to be bacteriocides and have been applied in medicine for centuries. It is obvious that they influence, at certain concentrations, microbial life in the soil.

Concerning the influence of heavy metals on soil respiration, the most attention has been paid to lead.[160,161,175,216,352,480,688,745] It was stated that lead decreases the respiration rate in soils with low cation exchange capacity, the effect lasted even for up to several years.[161] However, in soils with high cation exchange capacity, respiratory activity is not modified by lead in concentrations up to several thousand milligrams per kilogram.

Among other heavy metals, an inhibitory effect on soil and litter respiration was reported for mercury (at concentrations above 100 mg kg$^{-1}$ of soil), zinc and copper (above 1000 mg kg$^{-1}$),[175,374,688] cadmium (above 1000 mg kg$^{-1}$), nickel (above 1000 mg kg$^{-1}$),[688] and manganese (above 1000 mg kg$^{-1}$).[374]

A negative correlation between the carbon dioxide evolution rate from a humus horizon of a podzolized forest soil in the vicinity of a brass plant and the sum of Zn + Cu in the soil was found by Tyler,[745] shown in Figure 16.

*11. Pesticides*

The use of chemicals in agriculture and forestry introduces a further factor modifying microbial life in soil, namely, pesticides.[656] Most of the numerous papers on the effect of pesticides on soil respiration concern herbicides. Those compounds are added to soil at rates of several kilograms per hectare which corresponds to several milligrams per kilogram assuming their distribution within the upper 5 cm of the soil.

The effects of the herbicides on soil respiration are not always the same, and depend on the type of the compound, application rate, soil type, organic matter content, time, etc.[42,109,200,206,286,387,392,491,575,666,677,766,780,786]

A stimulating effect of simazine was observed at a concentration of 10 mg $kg^{-1}$,[677] but opposite effects, at field doses, were also reported.[109,392]

Atrazine applied to chernozem and podzolic soils at rates of 5 to 25 mg $kg^{-1}$ decreased carbon dioxide evolution by 20 to 30%. At higher doses of the herbicide up to 625 mg $kg^{-1}$ respiratory activity was not further reduced.[491]

Another herbicide — barbane and its commercial form Carbyne® — increased soil respiration,[42,575] however, in the presence of sugars an opposite effect was observed.[575]

Moreover a decrease in soil respiration activity was observed in soils treated with field and higher doses of diurone, EPTC, and 2,3,6-TBA,[109] ioxynil, dalapan, macrocrop, amitrole-T, and dichlorprop.[766] No effect was noted for dichlorprop[786] while an inhibitory effect of MCPA and dichlorprop applied together was observed at herbicide rates ten times higher than the normal one used in agricultural practice.

There are also many reports in the related literature where no influence on respiratory activity was found with herbicides applied to soil at field rates. This concerns 2,4-D,[387,392,786] 2,4-DB,[206,392,786] 2,5-D,[392] 2,4,5-T,[392,786] amitrole,[786] atrazine, TCA, monurone, DNPP and MCPA,[392] and DCPA.[200]

The influence of insecticides on soil respiration was studied by Tu,[741,742] who found an increase in oxygen uptake due to phorate, and no significant effect of leptophos, terbufos, WL 24073, dieldrin, chlorpyrifos, and Vorlex®.

*12. Other Factors*

There are also some other factors occasionally or locally influencing soil microbial respiration.

One of them is salinity influencing osmotic potential of the soil. However, this effect has been little studied. Wilson and Griffin[804] found that the respiration intensity of different soil fungus colonies was many times lower at osmotic potentials from −4 to −8 MPa, in comparison to 0.1 MPa.

Another factor that affects soil microflora is methane from underground leakages of natural gas which can significantly increase oxygen demand in soil, resulting in the formation of anoxic zones around the pipes.[9,316,317]

Soil respiratory activity can also be increased by soil conditioners,[530,823] contamination with crude and mineral oils,[159] and by heat and plant ash.[579]

## B. Root Respiration

Plant roots respire much more intensively than does the soil. The values of root respiration of different plants are shown in Table 4. By comparing the data from this table with those from Table 3, one can see that the root respiration rate is higher than that of mineral soils by even three orders of magnitude. Thus, plant roots form in soil "threads" with a high oxygen demand. Due to this, although the plant roots constitute only a small proportion of the soil mass, their respiration rate can greatly exceed that of the soil in which the plant is grown (Figure 17).

The energy of root respiration is used for maintenance of life processes and the functions of the roots (maintenance respiration) and for root growth (growth respiration).[297,298,410-412]

Root growth respiration is the amount of oxygen used to produce 1 g of dry root mass. The values of root growth respiration measured for several plant species by Lambers and Steingröver[411,412] were found to be from 390 to 1430 mg $O_2$ per gram dry weight. It was also found that for the same plant (*Senecio aquaticus*) grown in anoxic conditions, root growth respiration was three times lower than in the aerated culture.[411]

Most of the investigations concerning root respiration were performed in the laboratory on excised root segments[51,52,104,320,421,445,455,518,527,646] or on intact root systems grown in nutrient solutions,[410-412] and only a few on intact roots of living plants grown in soil.[595,644,646]

Respiratory activity of roots depends on some external factors — oxygen, temperature, carbon dioxide, soil mechanical impedance — and also on internal factors specific for a given plant, connected with its genetic background, physiological age, metabolic reaction rate, etc.

### *1. External Factors Influencing Root Respiration*

#### ***a. Oxygen***

The influence of oxygen concentration at different temperatures on the respiration rate of excised onion roots is shown in Figure 18. The curves in this figure have two basic parts: a range of respiration limitation by oxygen concentration and a flat part where respiration rate is not dependent on oxygen concentration. Respiration rate in this range of oxygen concentrations achieves its maximum value for a given temperature and will be called maximum respiration ($q_{max}$). The curves in Figure 18 are similarly shaped to that for the respiration of aerobic microorganisms (cf. Figure 10), but the critical oxygen concentration (COC), below which limitation occurs, is much higher than for microorganisms. The COC values increase with temperature and are, for excised onion roots, equal to 0.10 to 0.15 $m^3m^{-3}$ at 15°C.[51,52]

Equally high COC values for the respiration of excised roots were found for *Eriophorum angustifolium* (0.15 $m^3m^{-3}$), pea (about 0.20 $m^3m^{-3}$),[15] and *S. aquaticus* (0.15 $m^3m^{-3}$, at 20°C).[409]

Because the activity of cytochrome oxidases decreases abruptly at very low oxygen concentrations (below 0.01 $m^3m^{-3}$) one can assume that with a COC at the root surface oxygen concentration at the root axis approaches zero.

COC depends not only on the maximum root respiration ($q_{max}$), but also on other factors, which is shown by the equation:

$$\mathrm{COC} = \frac{q_{max}\, R^2}{4D_i} \qquad (15)$$

where R = root radius and $D_i$ = coefficient of radial diffusion of oxygen in root tissue.

The thicker the root and the lower the oxygen diffusion coefficient within it, the higher the COC value (for details see Chapter 2, Section II.E.1).

In addition to COC, a half respiration concentration of oxygen (HRC), analogous to Michaelis constant, is used. The HRC values for excised root segments are within 0.07 to 0.12 $m^3m^{-3}$ for maize, 0.05 to 0.25 $m^3m^{-3}$ for rice at 25°C,[445] and 0.025 $m^3m^{-3}$ for *S. aquaticus.*[409] The HRC values estimated from data of Berry and Norris[52] for excised onion roots are within 0.04 to 0.11 $m^3m^{-3}$.

There is an opinion[25,26] that those high COC and HRC values obtained in vitro for excised roots are an artifact of the measurement conditions due to filling of the intercellular air spaces by liquid. For an intact root system, the oxygen diffusion coefficient should be much higher, implying lower COC values (cf. Equation [15]).

This point of view is supported by the results obtained by Armstrong and Gaynard[26] who found, for roots of intact rice and *E. angustifolium* plants, COC values within 0.016 to 0.027 $m^3m^{-3}$ and HRC values below 0.001 $m^3m^{-3}$.

It can be seen that these values are almost one order of magnitude lower than those quoted above for excised roots. The results of Armstrong and Gaynard[26] are more convincing, because the same authors found, for excised roots of *E. angustifolium,* COC equal to 0.15 $m^3m^{-3}$ and HRC equal to 0.03 $m^3m^{-3}$.

In addition Lambers and Smakman[410] reported, for intact root systems of six *Senecio* species grown in nutrient solution (but cut off from the tops), relatively low HRC values amounting to about 0.02 $m^3m^{-3}$, although their COC values are rather high (about 0.20 $m^3m^{-3}$).

Similar HRC values (0.005 to 0.02 $m^3m^{-3}$) for intact roots of mustard seedlings were reported by Greenwood,[276] and Greenwood and Goodman.[281]

The COC values obtained with roots grown in soil (Table 5) resemble those found for both excised and intact roots in solution.

However, HRC values for intact roots in soil are apparently lower than for excised roots. This is in agreement with the view of Armstrong and Gaynard[26] although the HRC values for intact roots are not so low as these authors suggest.

Because it is difficult to assess the thickness of the moisture films usually present on the root surface in soil conditions and their role in impeding the diffusion of oxygen, it is not easy to assign a value to the real oxygen concentration on root epidermas. This is the reason for the higher COC and HRC values observed in soil conditions as compared with those predicted by biochemistry or found with the use of special experimental techniques.[26] The involvement of the water films on root surfaces makes it difficult to directly relate root respiration and other plant responses to oxygen content in soil air. This will be treated more extensively in Chapter 2, Section II.E.1).

It should, however, be emphasized that a partial limiting of the root respiration rate does not necessarily mean a reduction in the root elongation rate.[281] The effects of oxygen availability on plant response are discussed further (Chapter 5).

#### *b. Carbon Dioxide*

It is known that carbon dioxide has a detrimental effect on the respiration of plant tissues, but this is only pronounced at high concentrations. This effect on plant root respiration, however, has been little studied. Harris and van Bavel[301] reported that 0.05 $m^3m^{-3}$ of carbon dioxide in soil air entirely inhibited root respiration of tobacco, maize, and cotton. However, Norris et al.[515] found that onion root respiration is less sensitive to carbon dioxide than is microbial respiration.

#### *c. NPK Fertilization*

The influence of mineral fertilization on root respiration has been little studied. Figliolia et al.[201] recorded that the respiration of oat roots in a nutrient solution was lowered by a deficiency on N and P, but was not influenced by K availability.

#### *d. Ethylene*

There are some papers[591,685] indicating a stimulatory effect of ethylene on respiration of potato tubers, at concentrations which may occur in soil. A suppressing effect of 5 $cm^3m^{-3}$ of ethylene in soil air on respiration of grapevine roots was reported by Ishii et al.[331]

#### *e. Temperature*

Respiration of roots, like that of soil microorganisms, increases exponentially with temperature (Figure 19). Initially, this increase is rapid ($Q_{10}$ about two or more) when the respiration rate is limited by biochemical reactions. At higher temperatures the transport of substrates and products of the metabolism (sugars, oxygen, carbon dioxide), based mainly on the diffusion process, becomes the limiting factor ($Q_{10}$ about 1.2). At temperatures above 35°C the protoplasm system is starting to breakdown.

Limitation of respiration through the physical processes of diffusive transport may also occur at lower temperatures if the oxygen concentration is low, which is shown in Figure 19 by the curves corresponding to 0.05 and 0.10 $m^3m^{-3}$ of $O_2$. These curves have been plotted on the basis of the data from Figure 18. The curves 1 and 2 corresponding to oxygen concentration $\leq 0.1$ $m^3m^{-3}$ illustrate the range of respiration limitation by oxygen, and the curve 3 concerning pure oxygen illustrates the limitation of respiration by internal factors in the tissue.

In order to describe the influence of temperature on root respiration, the Arrhenius equation was applied and in most cases[52,104,133,421] a change in the slope of the plot of lnq vs. 1/T was observed (cf. Figure 20). This change of the slope, which indicates a change of activation energy, denotes a change of the limiting factor with temperature. The values of activation energy for respiration of roots reported in the literature are very differentiated. Thus for onion roots, Berry and Norris[52] found values within 7 to 59 kJ $mol^{-1}$, while Lemon and Wiegand[421] reported 45 to 75 kJ $mol^{-1}$. In turn Carey and Berry[104] found 49 to 116 kJ $mol^{-1}$ for maize and 54 kJ $mol^{-1}$ for barley roots.

### *f. Soil Mechanical Impedance*

The plant roots need more energy while penetrating a compact soil layer. Thus their oxygen requirements in that situation should be higher. This has been proven in experiments on plants grown in nutrient solution with and without mechanical obstacles of glass beads, where oxygen uptake by bean roots has been studied,[644] as well as in compacted soil in which carbon dioxide production by cotton roots was followed.[724] It should be emphasized, in the latter case, that although a fivefold increase in carbon dioxide production per root mass unit, resulting from soil compaction, was observed, the mass of the entire root system and its total carbon dioxide production were negatively correlated with soil bulk density (Figure 21).

### *g. Soil Salinity*

One of the external factors influencing root respiration is soil salinity, though its effect has not been studied extensively. It was found[357] that the respiration rate of excised root segments of *Tamarix tetragyna* decreased linearly with the increase of NaCl concentration in the solution.

### 2. *Internal Factors Affecting Root Respiration*

Limitation of root respiration by internal factors results from the properties of the respiring tissue (these being related to the genetic background and physiological age of the plant) as well as from the effect of the shoots.

It was found that the respiration intensity is highest in the apical part of the root and decreases with the distance from the root tip,[51,52,445,455] as shown in Figure 22.

Root respiration during the vegetation season undergoes specific changes independently of the effect of temperature or other external factors. Thus the root respiration of the perennial desert shrub *Atriplex confertifolia,* measured at constant temperature, showed spring and autumn maxima.[320] According to the authors of that paper[320] this phenomenon may be connected with the reflection of environmental preconditioning, as well as with an overall genetically based program of the plant development.

In the case of maize grown in controlled conditions,[19] maximum respiration of the entire root system was 800 $cm^3$ $CO_2$ per plant per day during the appearance of panicles. Later the respiration rate decreased to about 400 $cm^3$ $CO_2$ per day per plant, and oscillated around this value until plant senescence.

Besides seasonal dynamics, plant roots also exhibit a diurnal variation in respiration rate. Huck et al.[325] found that the root respiration of plants (maize, soybean, *Derris elliptica*) grown at constant temperature was 25 to 50% higher during the day than at night. On the other hand, root segments excised from the plants at different times of the day and night showed no differentiation in respiratory activity. Also no correlation was found between respiration rate and sugar content in the root, which eliminates substrate level as a cause of the variation observed. The mechanism of the variation has not been fully explained, but it is undoubtedly connected[297,298] with the action of light and the presence of shoots.

Studies by Hansen[297] revealed that the root respiration of *Lollium multiflorum* L. during 2 to 3 day periods of high illumination was 3 to 4 times higher that at low illumination. After defoliation, root respiration decreased and was not affected by the illumination level until the regrowth of new leaves.

Some differentiation in root respiration rate was also found among the plants of different climatic zones.[133] Roots of plants of cool regions showed higher respiration rates, at the same temperatures, as compared with plants of warmer zones. According to Crawford and Palin[133] high respiration rates at low temperatures can be an advantage for an early resumption of growth but at high temperatures they can lead to an excessive consumption of carbohydrate reserves.

Lambers[409] reported that the root respiration rate of *S. aquaticus* (flood-tolerant species) was, at 20°C, lower than that of *S. jacobaea* (flood-intolerant). This would suggest that the low respiratory activity of roots may be an adoptive feature increasing plant tolerance to flooding but, of course, it requires confirmation.

### C. Respiration of Germinating Seeds in Soil

There is only sparse information concerning the respiration of germinating seeds in soil conditions. Germinating seeds, during inhibition greatly increase their respiration rate as compared to the dormancy period (Figure 23).

The seed respiration intensity and the rate of its increase depend on soil moisture and temperature.

It was found that during germination at least two different respiratory pathways exist,[817] of which the first "alternative" is insensitive to cyanide but sensitive to hydroxamic acids and to decreased oxygen concentration while the other is a "normal" respiration through the cytochrome system being sensitive to cyanide and only slightly sensitive to decreased oxygen concentration. The transition from one to another, in the case of soya, occurs within the first few hours after imbibition.

Unlike plant roots, the respiration rate of seeds increases with oxygen within the concentration range existing in the soil.[150,730,772,783,817] An example of this relationship is shown in Figure 24. This linear relation indicates that oxygen diffusion is a limiting factor throughout the entire concentration range.

## III. SOIL RESPIRATION IN THE FIELD OR IN NEAR FIELD CONDITIONS

In field conditions the yearly soil oxygen demand is of the order of several thousand kilograms per hectare (Table 6).

Oxygen is utilized during microbial oxidation of organic matter and in respiration of plant roots; the proportion of these two factors varying within a wide range (see further).

The rate of soil respiration in field conditions varies with space and time depending on the factors previously discussed such as moisture, organic and inorganic amendments, etc. as well as on the kind of soil, the kind and growth rate of plant cover, and agricultural measures.

The daily values of respiratory activity in different soils are compiled in Table 7.

### A. Respiration Within the Soil Profile

Although it seems obvious that respiration rate should decrease sharply with depth as a result of decrease of organic matter content, root mass, and, frequently, temperature drop, this hypothesis has not been experimentally confirmed in field conditions because of methodic difficulties. Most of the available data was obtained in the laboratory, usually at a relatively high temperature,[268,269,768] and some other confirmations of the above view come from calculations of respiration distribution with depth on the basis of measured carbon dioxide content and its diffusion coefficient distribution within the soil profile.[154,598] This data indicates that over 90% of soil respiratory activity is concentrated in the humus horizon of the soil.

### B. Diurnal and Seasonal Dynamics of Soil Respiration in the Field

Respiration in the field undergoes both daily and seasonal dynamics.

An example of diurnal variation in field respiration is presented in Figure 25. The course of diurnal field respiration variation is almost identical with that of soil temperature, with a maximum in the early afternoon and a minimum before sunrise.[80,145,507,598]

Field respiration also varies during the year according to soil temperature, moisture content, and to the status of plant cover and also tillage practices, application of organic amendments, etc. Most of the published data concerns fields in the temperate climatic zone and shows a distinct respiration maximum in summer.[145,152,153,507,733,793,819] A typical example of such a yearly course of soil respiration is shown in Figure 26.

### C. Effect of Cropping on Soil Respiration

The influence of crop cover on oxygen and carbon dioxide balance in soil varies during the vegetation period.[47,79,144,145,152,491,509] In the period of highest vital activity of the plants (the most intensive growth of plants, flowering) their contribution to carbon dioxide production in the soil may exceed that of the soil itself (Figure 26).

Usually, however, the increase in respiration rate caused by the presence of plants amounts to 40 to 100% of the respiration of the soil alone.

It should be pointed out that not all the increase produced by plants can be ascribed to the respiration of the living roots, as an important role is also played by the decomposition of root exudates and roots which died during vegetation.

Sauerbeck and Johnen[634] found that, for wheat and mustard grown in an atmosphere containing $^{14}CO_2$, the amount of carbon mineralized in the soil during the vegetation period was three times greater than that remaining in the soil at harvest time. Only about 1/5 of the $CO_2$ evolved from the soil originated from the respiration of living roots, the remaining part being the result of the decomposition of root exudates and dead roots (3/5) and of the native organic matter present in the soil taken for the experiment (about 1/5).

### D. Other Factors

Among other factors affecting respiration in the field one should mention agrotechnical measures, land utilization, and relief.

The influence of tillage on soil respiration is not always unambiguous for it is connected with many factors. Loosening the soil should increase respiration due to better accessibility of oxygen, but other effects of tillage practices like changes in moisture and temperature are not easy to predict. Grabert[259] observed an increase in total respiration rate with increase in plowing depth. On the other hand Richter[598] found higher carbon dioxide evolution in three soils at zero-tillage in comparison to rototillage.

De Jong[152] compared respiratory activity of a loamy soil under cereals, grassland, and fallow and found the highest activity under grassland and the lowest under fallow.

This author[152] also studied the effect of relief and observed an increase of respiration rate down the slope; carbon dioxide evolution being doubled at the foot of the hill as compared to its top.

Hall and Grossbard[291] studying the effect of grazing and cutting on meadow soil respiration found higher respiration in the grazed soil.

## IV. SUMMARY

The aim of soil respiration studies, from the point of view of soil aeration, is the quantitative description of respiration (in space and time) at any status of the dynamically changing conditions in the soil environment.

At the present time, numerous factors determining the soil respiration rate have been recognized and for those which are of highest importance such as temperature, moisture content, and oxygen concentration, quantitative descriptions have been attempted. The results of these attempts are not altogether satisfactory and do not include extreme conditions such as anoxic periods, overheating, overdrying, or freezing the soil.

In addition there are numerous reports concerning the effect of organic matter on soil respiration, but the variety of organic materials does not permit the quantitative description of these relations.

Studies concerning the other factors, both direct (e.g., nutrient content, pH, pollutants) and indirect (e.g., cultivation measures), were of a fragmentary or preliminary character.

It should be emphasized that the description of the effect of all these factors on respiration is complicated by the phenomenon of hysteresis (common in biological systems), which has been practically ignored in the studies to date.

Respiration of the soil environment comprises two basic components: the soil microbial respiration and the respiration of plant roots. The contribution of the root component varies widely and may in some cases exceed that of the soil microbes. It should be pointed out that these two components of the respiration of the soil environment are interrelated. Thus the presence of roots in soil contributes not only by their own respiration but also stimulates microbial respiration of the soil itself due to the incorporation of root exudates and the decaying residues of dead roots.

Moreover, both the components of the total respiratory activity of soil environment, i.e., the root and the microbial components depend essentially on the same factors although not necessarily in the same manner. Thus, for example, microorganisms are more drought resistant than the roots of higher plants. Some of the factors influence only one of the two components, e.g., soil resistance to penetration affects only the roots, while in turn organic matter affects only the microbial respiration.

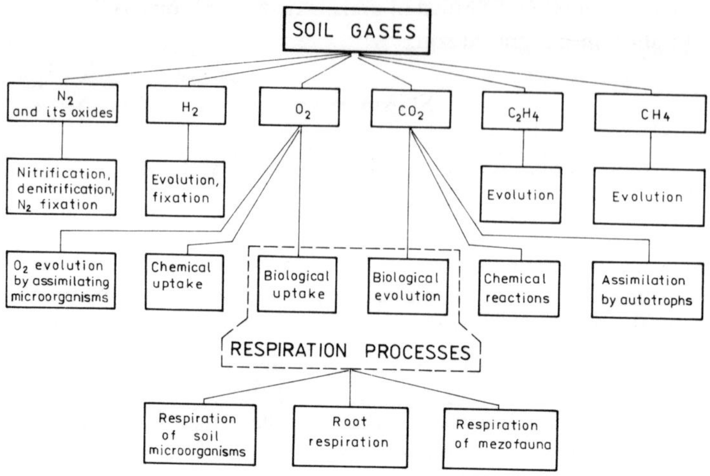


FIGURE 1. A scheme of soil processes affecting the composition of soil air.

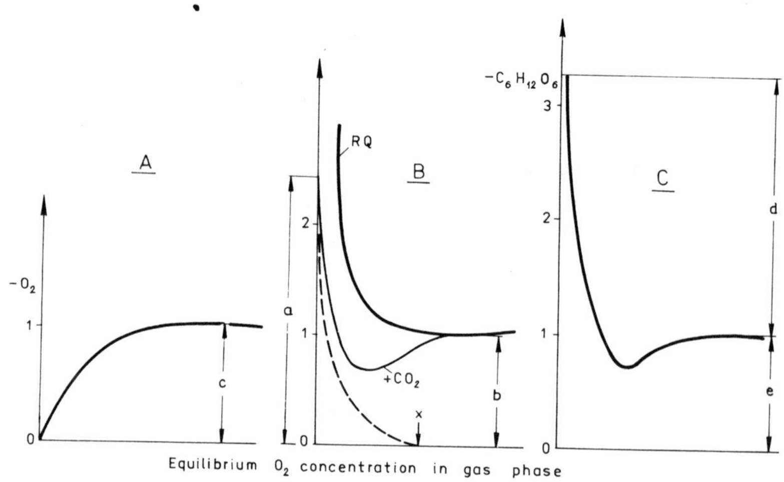


FIGURE 2. Influence of $O_2$ concentration on $O_2$ uptake (A), $CO_2$ production and respiratory quotient (B), and glucose use (C) by higher plant tissues and facultative anaerobe cells. X marks the point showing the limiting $O_2$ concentration below which anaerobic respiration begins (RQ > 1). Dashed line determines amount of $CO_2$ evolved due to anoxic processes. MQ = (a − b)/c, PQ = d/e. (Data from Zurzycki, J. and Michniewicz, M., Eds., *Plant Physiology*, PWRiL, Warsaw, 1977, 263.)

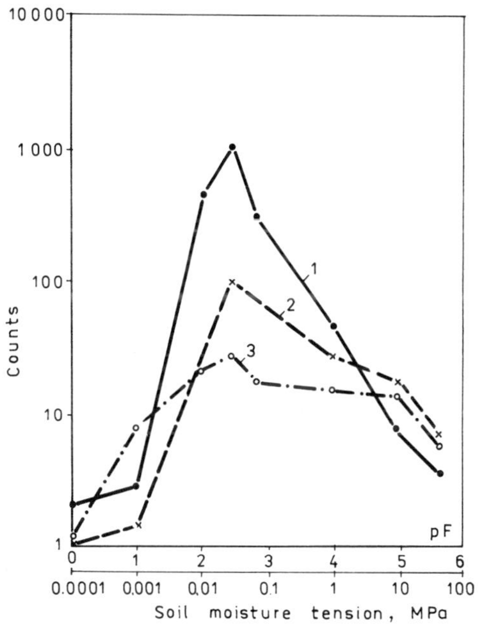


FIGURE 3. Dilution plate counts of bacteria (1), actinomyces (2), and fungi (3) in Freshfield $A_1$ soil horizon held at different moisture tension. (Data from Williams, S. T., Shameemullah, M., Watson, E. T., and Mayfield, C. I., *Soil Biol. Biochem.*, 4, 215, 1972.)

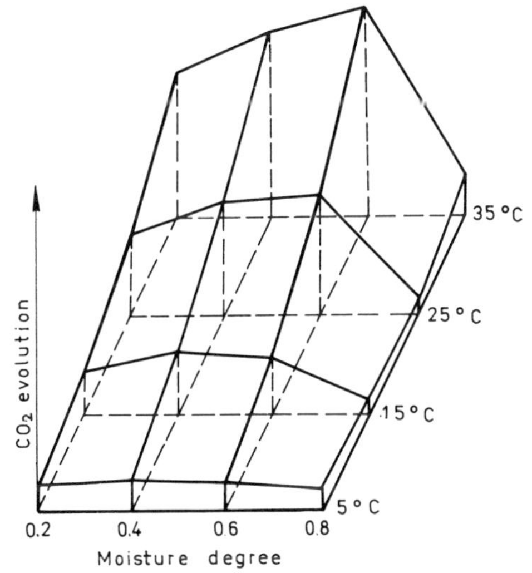


FIGURE 4. Dependence of $CO_2$ evolution from sandy loam soil on its moisture and temperature. (Modified from Tamm, E. and Krzysch, G., *Z. Acker Pflanzenbau.*, 117, 359, 1963.)

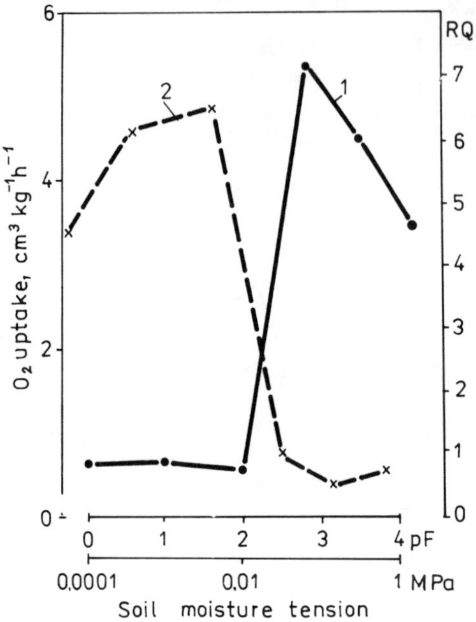


FIGURE 5. Relationship between moisture tension and $O_2$ consumption (1) and RQ (2) for clay soil. (Data from Rixon, A. J., *J. Soil Sci.*, 19, 56, 1968.)

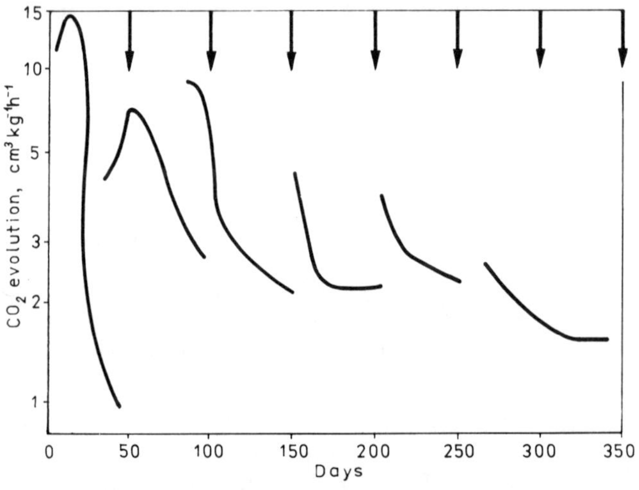


FIGURE 6. Dynamics of $CO_2$ evolution from peat incubated in a dessicator with exchange of air (arrows) when $O_2$ concentration dropped to 10%. (Data from Trybis, W. P. and Aviedeev, L. B., *Pochvovedenie*, 7, 101, 1978.)

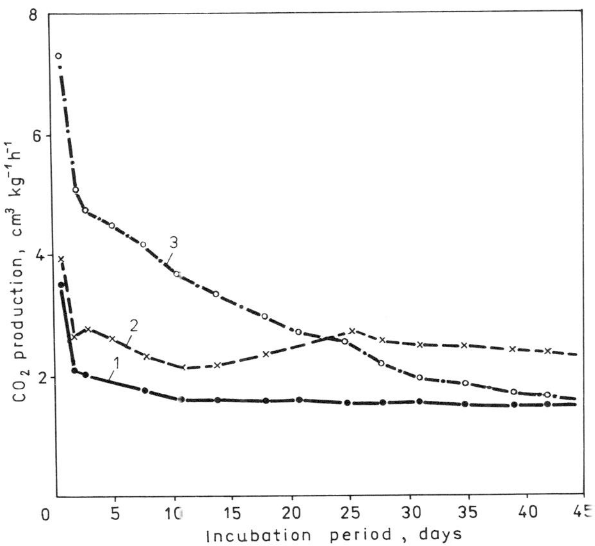


FIGURE 7. Rate of $CO_2$ production by different mineral soils of New Zealand, at 20°C. 1 — Tawhiti silt loam. 2 — Tima silt loam. 3 — Harihari loamy sand. (Data from Ross, D. J. and Cairns, A., *N. Z. J. Sci.*, 21, 581, 1978.)

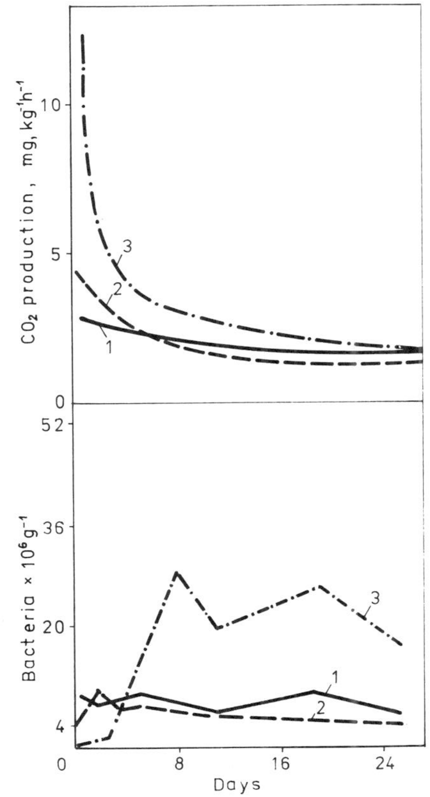


FIGURE 8. Carbon dioxide production and numbers of bacteria grown on the plates after a 3-day incubation of a sandy soil remoistened after drying at different temperatures. 1 — fresh soil; 2 — soil dried at 30°C, 3 days; 3 — soil dried at 85°C, 1 day. (Data from Jager, G., in *Proc. Coll. Dynam. Soil Commun., Progr. Biol.*, North-Holland, Amsterdam, 1967, 178.)

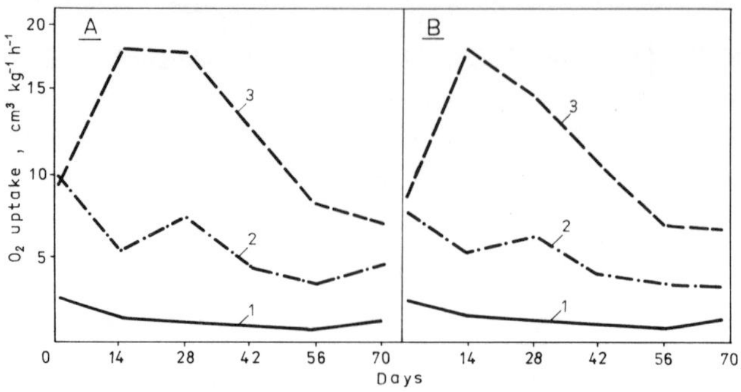


FIGURE 9. Influence of crop residues of oats (A) and cotton (B) on total respiratory activity of a fine sandy loam soil. 1 — check, 2 — 1% of crop residues, 3 — 4% of crop residues. (Data from Lyda, S. D. and Robinson, G. D., *Soil Sci. Soc. Am. Proc.*, 33, 92, 1969.)

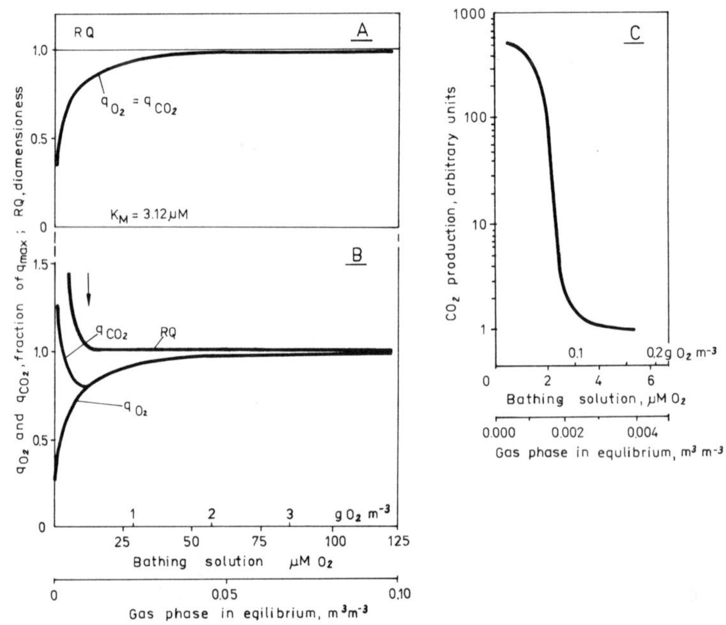


FIGURE 10. Idealized patterns of resipratory quotient (RQ), $CO_2$ production ($q_{CO_2}$) and $O_2$ uptake ($q_{O_2}$) by microbes as a function of $O_2$ content, at 20°C. A — for obligatory aerobes ($q_{O_2}$ calculated from Michaelis-Menten equation assuming the highest $K_M$ value given for bacteria by Longsmuir[437]), B — for facultative anaerobes at the same $K_M$ value, C — for obligate anaerobes. The arrow denotes $O_2$ concentration at which RQ exceeds 1.

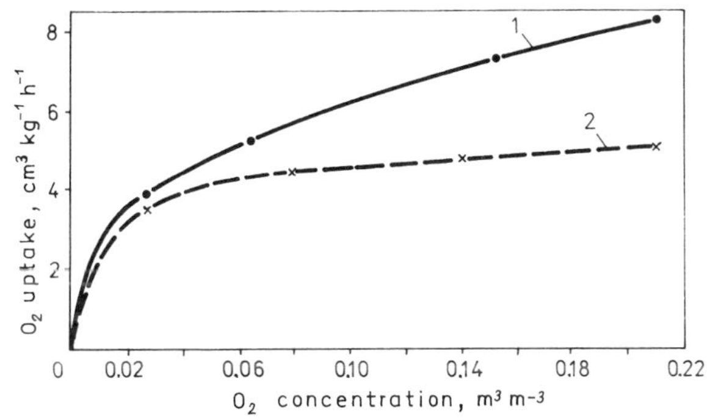


FIGURE 11. Influence of $O_2$ concentration in soil air on the microbial activity of meadow loamy soil (3 to 7 cm clods from a depth of 5 to 15 cm) at a soil moisture tension of 11 kPa. (1) First day after sampling; (2) second day after sampling. (Data from *Report 1975 ICW*, Wageningen, The Netherlands, Mededeling 153, 1975, 54.)

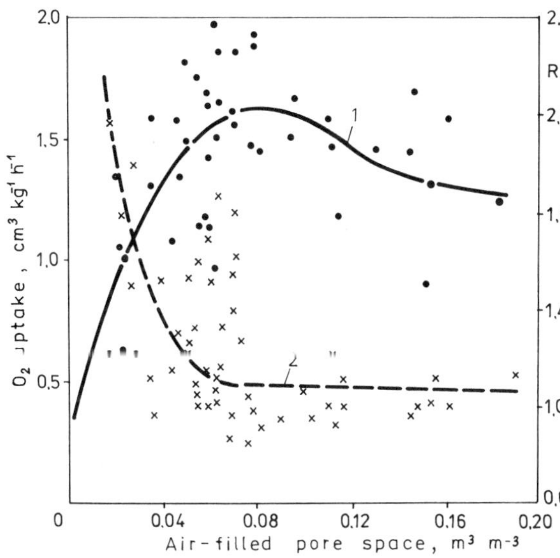


FIGURE 12. Dependence of $O_2$ uptake (1) and RQ (2) on the air-filled pore space for loamy soil. The range of porosity corresponds to moisture tensions 0 to 30 kPa. (Data from Bridge, B. J. and Rixon, A. J., *J. Soil Sci.*, 27, 279, 1976.)

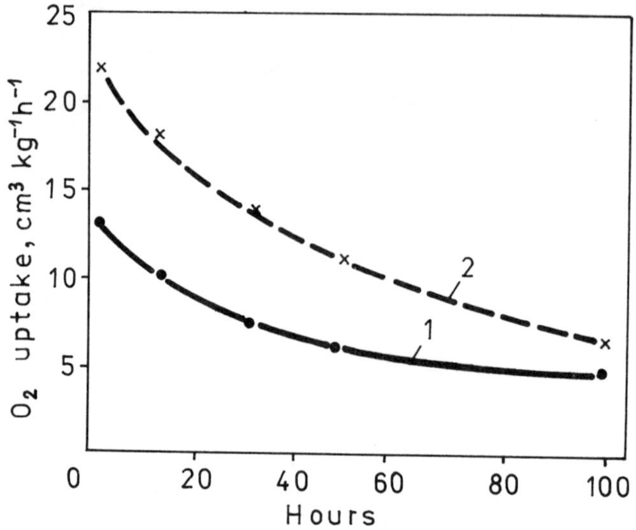


FIGURE 13. Influence of aggregate size on the respiratory activity of soil. 1 — aggregates < 2 mm, 2 — aggregates < 0.18 mm. (Data from Croswell, E. T. and Waring, S. A., *Soil Biol. Biochem.*, 4, 436, 1972.)

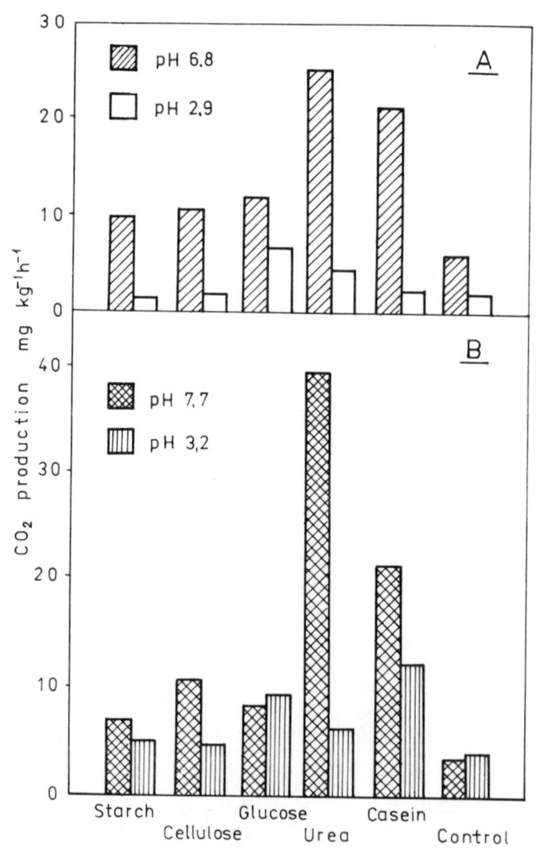


FIGURE 14. Effect of acidification on respiration of two soils with different organic amendments. A — soil acidified in field conditions due to long-term effect of the adjacent sulfur block, B — soil acidified in laboratory with $H_2SO_4$. (Data from Bryant, R. D., Gordy, E. A., and Laishley, E. J., in *Water, Air and Soil Pollution*, D. Reidel, Dodrecht, 1979, 437.)

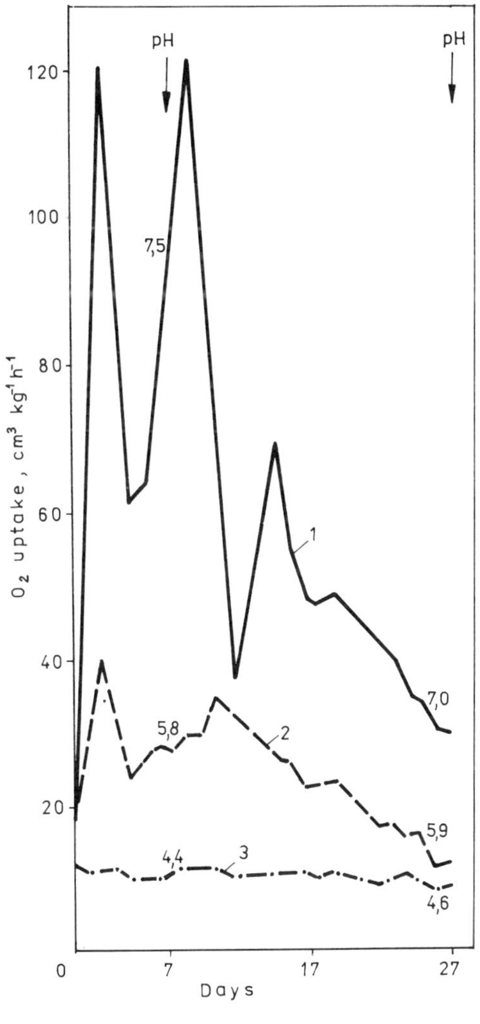


FIGURE 15. Respiration (at 20°C) of a soil treated with urea. 1 — soil + 16 mg N $g^{-1}$, 2 — soil + 5 mg N $g^{-1}$, 3 — nonfertilized soil (control). (Data from Salonius, P. O. and Mahendrappa, M. K., *Soil Sci.*, 127, 358, 1979.)

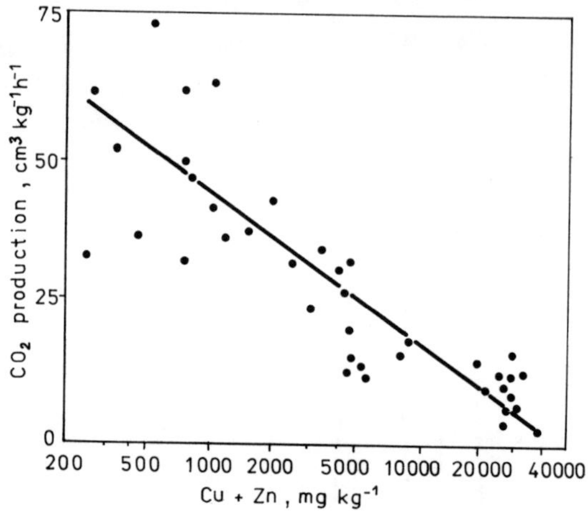


FIGURE 16. Relationship between respiration of forest soil in the vicinity of a brass plant and Cu + Zn concentration in this soil. (Modified from Tyler, G., *Plant Soil*, 41, 303, 1974.)

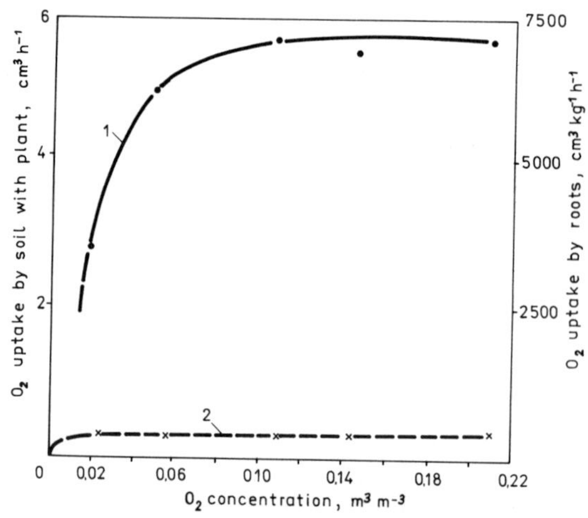


FIGURE 17. Oxygen uptake by intact roots of maize in soil (1) and by soil itself (2) at a moisture tension of 13 kPa as a function of $O_2$ concentration. (Data from *Report 1975 ICW*, Wageningen, The Netherlands, Mededeling 153, 1975, 54.)

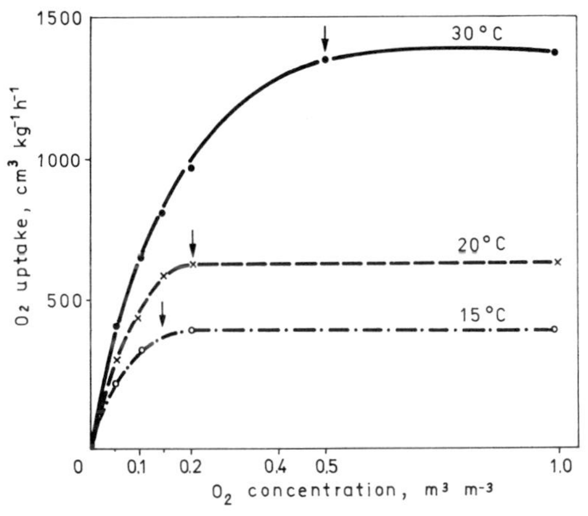


FIGURE 18. Influence of oxygen concentration, at different temperatures, on the respiration of onion roots — excised segments 0 to 5 mm from the tip. Arrows mark critical oxygen concentration, below which limitation of respiration by oxygen availability appears. (Data from Berry, L. J. and Norris, W. E., *Biochim. Biophys. Acta*, 3, 593, 1949.)

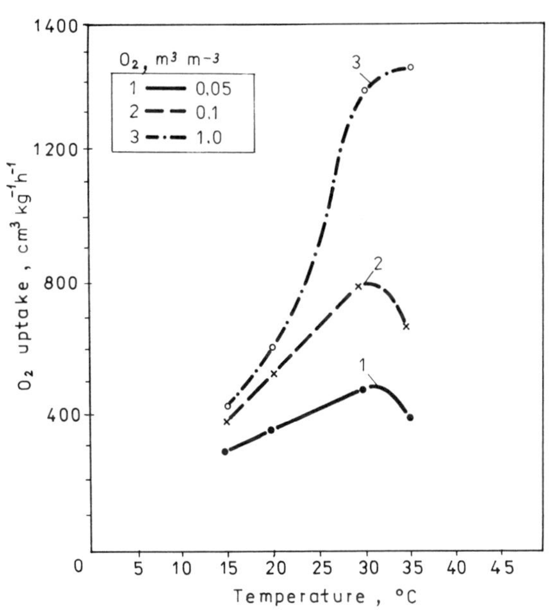


FIGURE 19. Respiration of excised segments of onion roots (0 to 5 mm from tip), at different $O_2$ contents, as related to temperature. (Data from Berry, L. J., *J. Cell. Comp. Physiol.*, 33, 41, 1949; Berry, L. J. and Norris, W. E., *Biochim. Biophys. Acta*, 3, 593, 1949.)

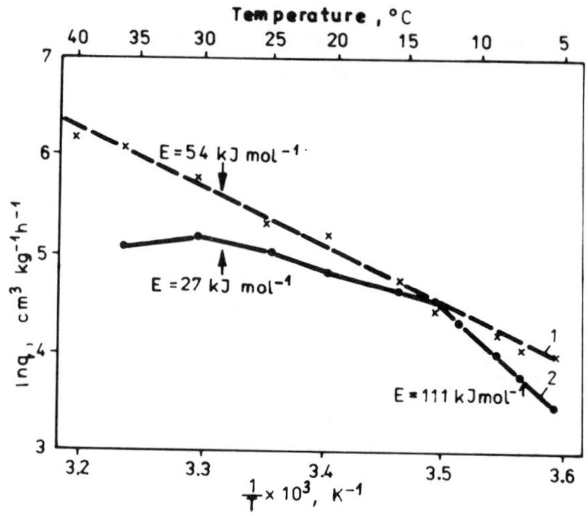


FIGURE 20. Arrhenius plot for respiration of excised roots of barley (1) and maize (2). (Data from Carey, R. W. and Berry, L. J., *Plant Physiol.*, 61, 858, 1978.)

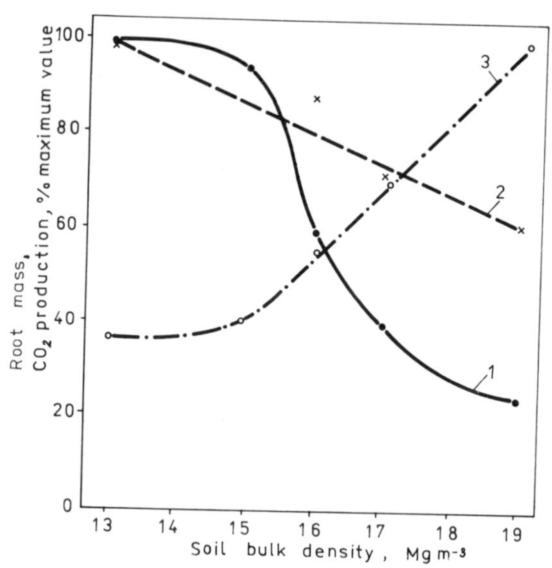


FIGURE 21. Influence of soil compaction on the respiration rate and weight of cotton roots in the percent of the maximum values. 1 — dry root mass, 2 — total $CO_2$ production by root system, 3 — $CO_2$ evolution per root dry mass unit. (Data from Tackett, J. L. and Pearson, R. W., *Soil Sci. Soc. Am. Proc.*, 28, 741, 1964.)

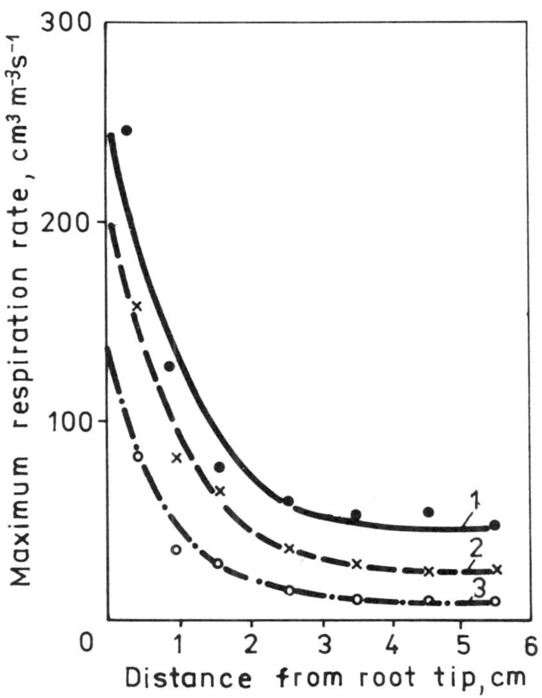


FIGURE 22. Respiration rate of rice roots in the atmosphere containing the following concentrations of oxygen: 1 — 0.8 $m^3m^{-3}$, 2 — 0.208 $m^3m^{-3}$, 3 — 0.044 $m^3m^{-3}$. (Data from Luxmoore, R. J., Stolzy, L. H., and Letey, J., *Agron. J.*, 62, 317, 1970.)

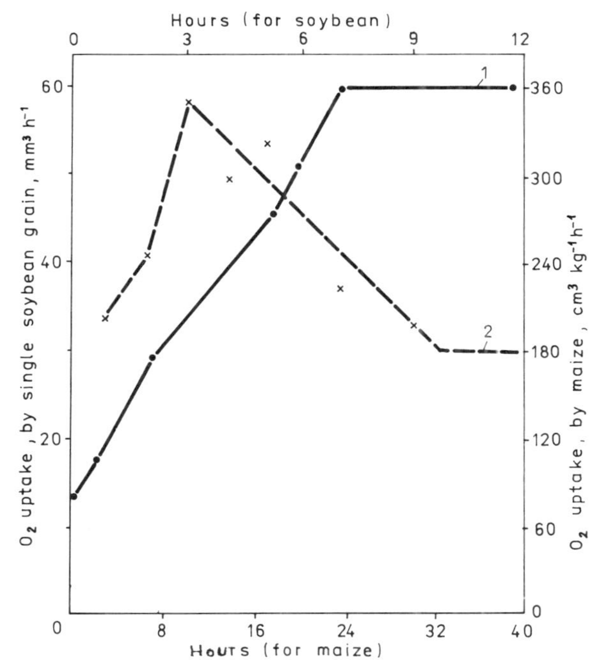


FIGURE 23. Respiration of germinating seeds of maize (1) and soybean (2) after previous soaking in water. (Data from Wengel, R. W., *Agron. J.*, 58, 69, 1966; Yentur, S. and Leopold, A. C., *Plant Physiol.*, 57, 274, 1976.)

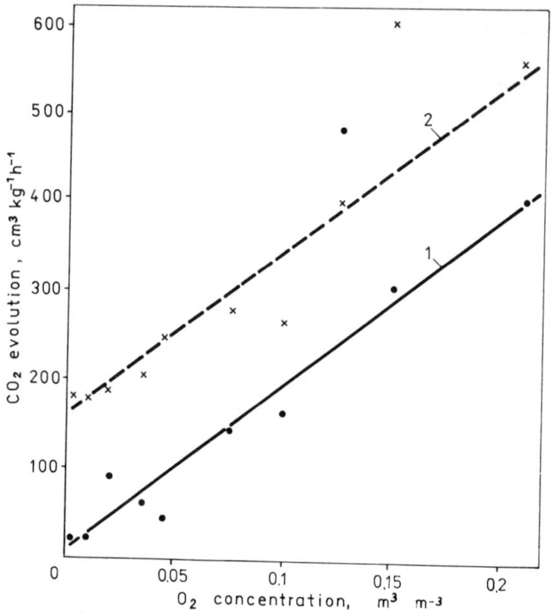


FIGURE 24. Influence of $O_2$ concentration on the respiration of germinating seeds of wheat (1) and crested wheatgrass (2). (Data from Dasberg, S., Enoch, H., and Hillel, D., *Agron. J.*, 58, 206, 1966.)

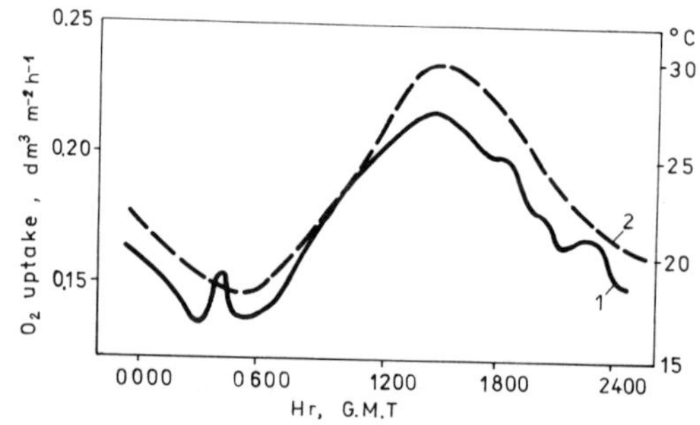


FIGURE 25. Diurnal variation of soil respiration (1) and temperature (2) in Rothamsted Exp. Station (U.K.). (Data from Currie, J. A., *Tech. Bull. Min. Agric., Fish. Food*, 29, 461, 1975.)

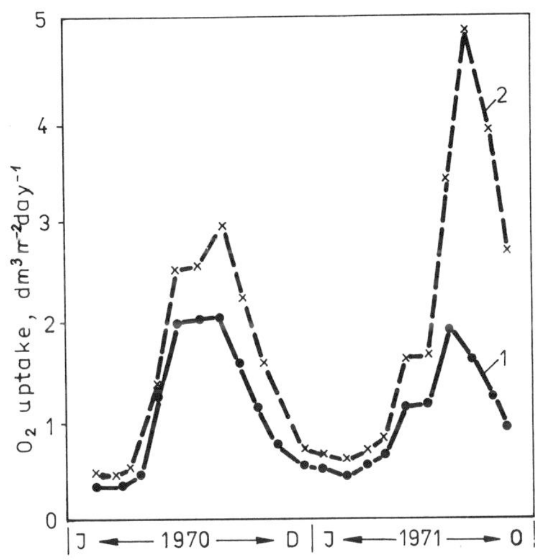


FIGURE 26. Yearly variation of respiration in uncropped (1) and cropped (2) soil in England. (Data from Currie, J. A., *Tech. Bull. Min. Agric., Fish. Food*, 29, 461, 1975.)

## Table 1
## GROUPING OF MICROBES ACCORDING TO THEIR OXYGEN REQUIREMENTS

| Aerobes | Facultative anaerobes | Obligate anaerobes |
|---|---|---|
| Most of fungi (*Micromycetes*), *Actinomycetes* and *Protozoa*<br>Most of chemoautotrophs: nitrifying (*Nitrosomonas* sp., *Nitrobacter* sp.), sulfuric (*Thiobacillus thioparus, T. cillus thiooxidans*), hydrogen (*Hydrogenomonas facilis, H. agilis*), iron (*Thiobacillus ferrooxidans, Gallionella ferruginea*) and carboxyoxide (*Carboxydomonas*) bacteria<br>*Myxobacteria*<br>Heterotrophs fixing nitrogen asymbiotically (*Azotobacter, Arthrobacter, Beijerinckia, Derxia, Azotomonas*) and symbiotically (*Rhizobia*)<br>*Pseudomonas fluorescens* | Chemoautotrophic *Thiobacillus denitrificans*<br>Hydrogen oxidizing *Micrococcus denitrificans*<br>Heterotrophs reducing manganese, iron, and nitrates<br>Most yeasts | Chemoautotrophic *Desulfovibrio desulfuricans* oxidizing hydrogen at the cost of sulfates, sulfites, or thiosulfates<br>Photoautotrophic bacteria (*Rhodospirillaceae, Chromatiaceae, Chlorobiaceae*)<br>Heterotrophs fixing nitrogen asymbiotically (e.g., *Clostridium*) |

**Table 2**
**COMPARISON OF RESPIRATION RATE OF DIFFERENT ORGANISMS**

| Organism | $O_2$ uptake ($cm^3$ $hr^{-1}$ kg body$^{-1}$) |
|---|---|
| Hibernating mammals | 8 — 1,500 |
| Hibernating birds | 60 — 1,240 |
| Nonhibernating big mammals (body weight > 100 kg) | 110 — 395 |
| Nonhibernating small mammals (body weight < 1 kg) | 530 — 5,300 |
| Reptiles | 5 — 1,100 |
| Amphibia | 14 — 130 |
| Fish | 15 — 540 |
| Crustacea | 10 — 360 |
| Insects | 37 — 15,000 |
| Bacteria | up to 1,200,000[a] |
| Fungi | up to 10,000[a] |
| Algae | up to 40,000[a] |
| Lichens | up to 900[a] |
| Plant roots (cf. Table 4) | up to 27,100[a] |
| Plant roots (cf. Table 4) | up to 1,600 |
| Normal soils (cf. Table 3) | up to 20 |

[a] On dry weight basis.

Data from References 189, 305, and 464.

## Table 3
## RESPIRATORY ACTIVITY OF SOILS

| Soils | $O_2$ uptake ($cm^3\ kg^{-1}\ hr^{-1}$) | $CO_2$ evolution ($cm^3\ kg^{-1}\ hr^{-1}$) | Ref. |
|---|---|---|---|
| Dark-grey soil | | 0.26—2.00 | 651 |
| Clay soils | 2.2—28.0 | | 137 |
| Mineral soils | 0.5—1.0 | | 573 |
| | 0.8—6.0 | | 617 |
| Loamy sand | 0.9 | 0.9 | 242 |
| Heavy loam | 2.0—10.4 | | 532 |
| Soddy-podzolic soil | 0.4—6.0 | 0.4—4.0 | 268 |
| Sandy loam soil | | 0.14—1.63 | 727 |
| Sandy loams and clays | 0.15—0.52 | | 79 |
| Mineral soils | | 0.6—90.0 | 614 |
| | 1.45—17.0 | | 615 |
| Silty loam | 1.0—4.0 | | 449 |
| Silty loam plus plant residues | 4.0—20.0 | | 449 |
| Grassland gley and pseudogley soils | | 2.4—5.0 | 768 |
| Peat | 1.04—1.18 | | 79 |
| Peats | | 1.0—15.0 | 740 |
| | 2.4—117.0 | | 616 |
| Forest litter | 219—480 | | 268 |
| | | 10—250 | 818 |
| Litter of coniferous forest | 3.0—85.0 | | 632 |
| | | 30—250 | 688 |
| Forest litter at different temperatures | | 40—400 | 755 |
| Humoferric podzol | 1.0—12.0 | | 633 |

## Table 4
## ROOT RESPIRATORY ACTIVITY

| Plant | $O_2$ or $CO_2$ | $mm^3\ g^{-1}\ hr^{-1}$ or $cm^3\ kg^{-1}\ hr^{-1}$ | | $mm^3\ cm^{-3}\ hr^{-1}$ | $mm^2\ cm^{-2}\ hr^{-1}$ | |
|---|---|---|---|---|---|---|
| | | Root dry weight | Root fresh weight | Root volume | Root surface area | Ref. |
| *Phaseolus vulgaris* | $O_2$ | 5,050—5,400 | 250—325 | 225—435 | 4—5.5 | 644 |
| Mung bean | $O_2$ | | | 257—373 | | 421 |
| Soybeans | $O_2$ | | 330—470 | | | 518 |
| Lupine | $O_2$ | | | 200—813 | | 421 |
| Maize | $O_2$ | 2,850—6,226 | | | | 595 |
| | $O_2$ | | 360—580 | | | 518 |
| | $O_2$ | | | 240—630 | | 421 |
| | $O_2$ | | | 47—475 | | 445 |
| | $O_2$ | | 30—110 | | | 325 |
| Barley | $O_2$ | | 100—650 | | | 104 |
| | $O_2$ | | | 205—443 | | 455 |
| Rice | $O_2$ | | | 25—900 | | 445 |
| *Atriplex confertifolia* | $CO_2$ | 20—380 | | | | 320 |
| Oats | $CO_2$ | | 0—1,500 | | | 201 |
| *Derris elliptica* | $O_2$ | 145—165 | | | | 325 |
| | $CO_2$ | 125—160 | | | | 325 |
| *Senecio aquaticus* | $O_2$ | 190—380 | | | | 410 |
| Alfalfa | $O_2$ | | | 135—590 | | 421 |
| Carrot | $O_2$ | | 115—190 | | | 518 |
| Onion | $CO_2$ | 5,100—27,100 | 300—1,600 | | | 52 |
| | $O_2$ | 3,700—24,000 | 100—1,350 | | | 52 |
| Potato (excised tubers) | $O_2$ | | 41—51 | | | 518 |
| | $O_2$ | | 240—800 | | | 527 |
| Potato tubers | $O_2$ | | 2.3—3.2 | | | 685 |
| | $CO_2$ | | 2.8—3.0 | | | 685 |
| Birch (cut roots $\Phi < 1$ cm) | $CO_2$ | 50—300 | | | | 485 |

## Table 5
## COMPARISON OF THE CRITICAL OXYGEN CONCENTRATION (COC) AND HALF RESPIRATION CONCENTRATION (HRC) VALUES FOR PLANT ROOTS

| Roots | Plants | COC $m^3m^{-3}$ | HRC $m^3m^{-3}$ | Ref. |
|---|---|---|---|---|
| Excised | Onion | 0.10—0.50 | 0.04—0.11 | 51, 52 |
| | *Eriophorum angustifolium* | 0.15 | 0.03 | 25 |
| | Peas | 0.20 | | 15 |
| | Maize | | 0.07—0.12 | 445 |
| | Rice | | 0.05—0.25 | 445 |
| Whole in solution | *Senecio* | 0.15—0.20 | 0.02—0.025 | 410 |
| | *E. angustifolium* and rice | 0.016—0.027 | <0.001 | 25 |
| | Mustard | ~0.20 | 0.005—0.10 | 276, 281 |
| Intact | Maize | 0.10 | ~0.02 | 595 |
| | Cotton | 0.21 | < 0.01 | 724 |
| | Dwarf peas | 0.15 | 0.02—0.04 | 646 |

## Table 6
## SEASONAL SOIL RESPIRATION

| Soils | $CO_2$ evolved (kg ha$^{-1}$ year$^{-1}$) | $O_2$ uptake computed, at the assumption RQ = 1 (kg ha$^{-1}$ year$^{-1}$) | Ref. |
|---|---|---|---|
| Soils formed from silty loams, Canada | 1,800—23,800 | 1,300—17,300 | 152 |
| Soils of mixed mesophytic forest, Tennessee | 38,100 | 27,700 | 177, 589 |
| Soils of tallgrass prairie, Missouri | 16,750 | 12,180 | 391 |
| Soils of cool-temperate forests, Minnesota | 26,000—29,000 | 18,900—21,200 | 592, 593 |
| Soils of tropical rain forest, Malaysia | 52,100 | 38,000 | 517 |
| Soils of evergreen oak forests of warm-temperate zone, Japan | 30,900—47,200 | 22,500—34,300 | 368, 502, 819 |
| Submountain and mountain forest soils, Japan | 20,300—30,100 | 14,077—21,900 | 358 |
| Sandy, loess and loamy soils, Germany | 9,400—12,800 | 6,800—9,300 | 598 |
| Soils of England | 8,000—12,200 | 5,800—8,900 | 145 |

**Table 7**
**DIURNAL SOIL RESPIRATION IN THE FIELD**

| Soil | $CO_2$ evolution $dm^3m^{-2}$ $day^{-1}$ | Ref. |
|---|---|---|
| Peat soils | | 47 |
| Without plants | 0.6—1.1 | |
| Under different plants | 1.2—2.2 | |
| Soddy-podzolic soils under forest | 0.1—1.5 | 62 |
| Different soils in lysimeters | 1.8—11.0[a] | 79 |
| Very fine sandy loam supporting a sugar beet crop | 2.3—3.9 | 80 |
| Mineral soils | 0.1—5.1 | 145 |
| Heavy clay soil | 0.3—2.8 | 153 |
| Fine sandy loams | 1.1—7.9 | 153 |
| Sandy soil, cropped | 0.2—1.0 | 231 |
| Different soils | 1.6—28.4 | 439 |
| Soils | | 491 |
| Bare | 1.8—6.9 | |
| Cropped | 4.0—20.2 | |
| Soil under maize | 4.2 | 495 |
| Soddy-podzolic soils | | 507 |
| Under forest | 3.9—12.7 | |
| Under grass | 4.2—14.3 | |
| Under rye | 3.5—10.4 | |
| Soils of England | 0.17—0.38 | 535 |
| Soils | | 598 |
| Sandy | 12.1—15.8 | |
| Loess | 4.8—6.1 | |
| Loamy | 6.1—7.3 | |
| Tropical soils | 3.7—31.0 | 643 |
| Soils of tropical forests, Java | 2.0—2.4 | 776 |
| Grassland soil of the arid shrub-steppe | 0.3—2.2 | 793 |
| Forest soil | 1.2—13.3 | 819 |
| Heavy-clay grassland soils | 0.0—2.4 | 588 |
| Brown forest soil, stony loam differently used | 4.5—6.3 | 184 |

[a] $O_2$ uptake.

Chapter 2

# GAS TRANSPORT IN THE SOIL ENVIRONMENT

The exchange of considerable amounts of gases, particularly oxygen and carbon dioxide, between the soil and the atmosphere (in the range of several to several tens milligrams per hectare per year [cf. Chapter 1]) takes place under the influence of both the pressure gradient (mass flow) and the concentration gradient (diffusion flow). Both these kinds of flow may take place in the soil pores and also in the tissue of plants (internal exchange). The relative contribution of the two fundamental mechanisms of gas exchange, i.e., mass and diffusion flow, as well as of the two flow pathways, i.e., through the soil and through the plant, may vary within relatively broad limits; however, the diffusive flow through the soil is of primary significance for mesophytes, and through the plant for aquatic plants.

## I. MASS FLOW

The flow of gas through a porous medium under the influence of the pressure gradient may be of a laminar character (free of turbulence) or of a turbulent character. In a soil medium we deal almost entirely with laminar flow for which the corresponding Reynold's number (Re) is less than 1.[144] The Reynold's number is given by:

$$Re = \frac{2\bar{v}\, R\, \rho}{\eta} \tag{1}$$

where $\bar{v}$ = mean linear flow rate, m $s^{-1}$; R = capillary radius, m; $\rho$ = density of liquid or gas, kg $m^{-3}$; $\eta$ = dynamic viscosity of liquid or gas, kg $m^{-1}s^{-1}$.

In the laminar range the flow is described by Darcy's equation:

$$\frac{dV}{dt} = -A\frac{k}{\eta}\frac{dp}{dx} \tag{2}$$

where dV/dt = volumetric rate of gas flow, $m^3s^{-1}$; A = cross-section area of the porous medium, $m^2$; $\eta$ = dynamic viscosity of the gas, kg $m^{-1}s^{-1}$; dp/dx = pressure gradient, Pa $m^{-1}$ = kg $m^{-2}s^{-2}$; k = gas permeability, $m^2$.

In the case of a single capillary of a radius R, the volumetric gas flow is described by Poiseuille's equation:

$$\frac{dV}{dt} = -\frac{\pi R^4}{8\eta}\frac{dp}{dx} \tag{3}$$

and the formula for the mean linear flow rate $\bar{v}$ has the form:

$$\bar{v} = \frac{R^2}{8\eta}\frac{dp}{dx} \tag{4}$$

The lower limit of applicability of Poiseuille's law is constituted by capillaries of diameters close to the mean free path (MFP) of the gas molecules at a given pressure. The viscosity $\eta$ does not normally depend on gas pressure and only when the MFP is comparable with the diameter of the capillaries (which can be achieved either by decreasing the pressure or by decreasing the capillary diameter) is a decrease in the value of $\eta$ observed.[569] In such a case we deal with what is known as visco-molecular flow, constituting a transition from normal flow, i.e., viscous, to the Knudsen flow, i.e., the flow of rarefied gas.[569] The MFP of soil gases under normal conditions is up to 0.1 μm (cf. Table 2, Chapter 3). This means that Equation (3) is applicable only to soil pores of effective diameters above 0.1 μm, i.e., to pores which are emptied at soil moisture tension values of up to 3 MPa. Thus the range of applicability of the laws of both Poiseuille and Darcy in relation to the flow of air in the soil covers the full range of moisture available to plants. By combining Equations (2) and (3) and introducing the surface area of the capillary for its cross-section $A_c = \pi R^2$, we obtain the gas permeability of the capillary ($k_c$) equal $R^2/8$. The effect of the pore size on the gas permeability of the soil is shown in Table 1.

Let us consider a single capillary of radius R = 1 mm, for which the product of its cross-section area ($\pi\ 10^{-6} m^2$) and gas permeability $k_c$ ($1.25 \cdot 10^{-7}$ $m^2$) equals $3.93 \cdot 10^{-13}$ $m^4$. The flow through this capillary is the same as through 1 $m^2$ of soil of a permeability k = $3.93 \cdot 10^{-13}$ $m^2$. If, in turn, we consider 100 identical capillaries of radius R = 0.1 mm, despite the fact that their joint cross-section area is the same, their gas permeability, and therefore also the product of permeability, and the joint cross-section area ($100\ \pi R^4/8$) is 100 times less ($3.93 \cdot 10^{-15}$ $m^4$) than those for a single capillary of a radius R = 1 mm. Thus the flow of gas through a single capillary of a radius R = 1 mm is equivalent to the gas flow through 10,000 capillaries of a radius 10 times less, the joint cross-section area of which is, however, 100 times greater.

In a porous medium composed of parallel capillaries of the same radius, permeability (along the axis of the capillary) is proportional to the product of the fraction of the cross-section area of the medium available for gas flow and the square of the capillary radius. In the case of an anisotropic medium of differentiated pores, such as soil, the gas permeability can be calculated from the pore size distribution[35,462] or determined through direct measurements (cf. Chapter 6, Section II).

### A. Air Permeability of Soil

The values of the air permeability of soils fall usually within a range from 0.01 to $500 \cdot 10^{-12}$ $m^2$ (Table 2) and depend directly on the gas porosity of the soil and on the square of the soil pore radii, and indirectly on all factors affecting these two properties, such as moisture, compaction, or the specifics of a given soil system.

#### *1. Effect of Soil Moisture Tension*

Figure 1 presents the effect of soil moisture tension on the air permeability of a sandy loam soil at various levels of soil compaction achieved under laboratory conditions. A rapid increase in the air permeability can be seen with increasing soil moisture tension, but the values and the rate of the increase are greater for the less compacted soil.

#### *2. Effect of Soil Bulk Density*

Turski et al.[744] found on various soils that the effect of soil bulk density on the air permeability, at a constant soil moisture tension, follows a curvilinear course with a clear maximum (Figure 2). This relationship is rather surprising, as the air permeability decreases at the lowest level of soil compaction.

#### *3. Effect of Air-Filled Porosity of Soil*

The effect of the air-filled porosity of a soil on its air permeability, at different soil compaction levels, also follows a course eluding explanation (Figure 3).

As can be seen, the same value of the constant k, e.g., $5 \cdot 10^{-12}$ $m^2$, is reached at a much higher value of the air-filled porosity in a loose soil (0.34 $m^3m^{-3}$) than in a compacted soil (0.12 $m^3m^{-3}$). This seems to indicate a greater effectiveness of pores in a compacted soil than in a loose soil. A possible explanation of this phenomenon lies in the discontinuity of pores under conditions of low moisture tension of the soil, as the air-filled porosity of 0.34 $m^3m^{-3}$ corresponds to a lower value of soil moisture tension in a loose soil than 0.12 $m^3m^{-3}$ in a compacted soil.

## Table 1
## COMPARISON OF AIR PERMEABILITIES OF CAPILLARIES OF DIFFERENT RADII

| Capillary radius R mm | Capillary cross-section $A_c = \pi R^2$ $m^2$ | Air permeability of the capillary $k_c = R^2/8$ $m^2$ | $k_c \cdot A_c = \pi R^4/8$ $m^4$ | k of soil ($m^2$) equiv to the presence per 1 $m^2$ of | | | |
|---|---|---|---|---|---|---|---|
| | | | | 1 Capillary | $10^2$ Capillaries | $10^4$ Capillaries | n Capillaries |
| 0.001 | $\pi \cdot 10^{-12}$ | $1.25 \cdot 10^{-13}$ | $3.93 \cdot 10^{-25}$ | $3.93 \cdot 10^{-25}$ | $3.93 \cdot 10^{-23}$ | $3.93 \cdot 10^{-21}$ | $n \cdot 3.93 \cdot 10^{-25}$ |
| 0.01 | $\pi \cdot 10^{-10}$ | $1.25 \cdot 10^{-11}$ | $3.93 \cdot 10^{-21}$ | $3.93 \cdot 10^{-21}$ | $3.93 \cdot 10^{-19}$ | $3.93 \cdot 10^{-17}$ | $n \cdot 3.93 \cdot 10^{-21}$ |
| 0.1 | $\pi \cdot 10^{-8}$ | $1.25 \cdot 10^{-9}$ | $3.93 \cdot 10^{-17}$ | $3.93 \cdot 10^{-17}$ | $3.93 \cdot 10^{-15}$ | $3.93 \cdot 10^{-13}$ | $n \cdot 3.93 \cdot 10^{-17}$ |
| 1.0 | $\pi \cdot 10^{-6}$ | $1.25 \cdot 10^{-7}$ | $3.93 \cdot 10^{-13}$ | $3.93 \cdot 10^{-13}$ | $3.93 \cdot 10^{-11}$ | $3.93 \cdot 10^{-9}$ | $n \cdot 3.93 \cdot 10^{-13}$ |

**Table 2**
**AIR PERMEABILITIES (k) REPORTED BY DIFFERENT AUTHORS FOR SOILS**

| | Unit | | |
|---|---|---|---|
| **Soil** | **Original** | **Unified in $\mu m^2$** | **Ref.** |
| Dune sand | $7 \cdot 10^{-9}$—$8 \cdot 10^{-7}$ $cm^2$ | 0.7—80 | 35 |
| Sandy loam soil | $10^{-10}$—$10^{-6}$ $cm^2$ | 0.01—100 | 35 |
| Pasture silty soil | 2—500 $\mu m^2$ | 2—500 | 272 |
| Gray-wooded soil | 1—30 D | 1—30 | 759 |
| Five different soils | 2—400 $m^4 g^{-1}$ $min^{-1}$ ($k/\eta$ values) | 0.62—124 | 162, 744 |
| Several soils | 10—140 $\mu m^2$ | 10—140 | 213 |

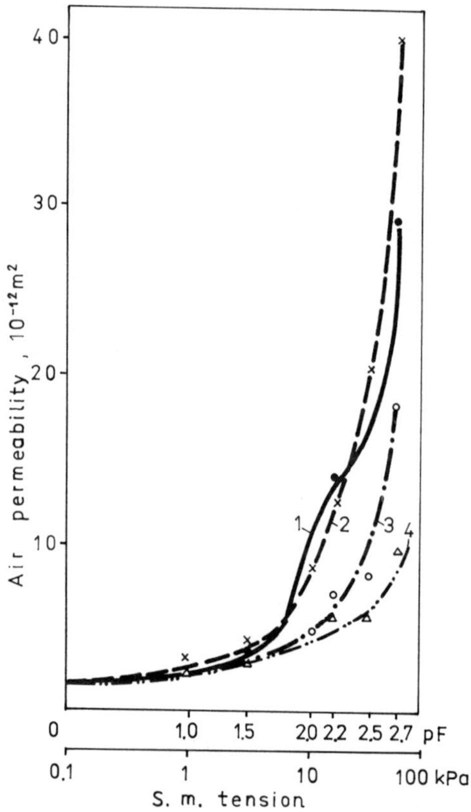


FIGURE 1. Air permeability of a brown soil formed from sandy loam as a function of soil moisture tension at different bulk densities; (1) 1.09 Mg $m^{-3}$, (2) 1.16 Mg $m^{-3}$, (3) 1.21 Mg $m^{-3}$, (4) 1.41 Mg $m^{-3}$. (Data from Turski, R., Domżał, H., and Słowińska-Jurkiewicz, A., *Roczn. Glebozn.*, 29, 3, 1978.)

## B. Factors Inducing Mass Flow in Soil

The factors that cause mass flow in the soil are variations in the soil temperature, atmospheric pressure, soil moisture (rain, irrigation, evaporation, movements of the ground water table), and wind.

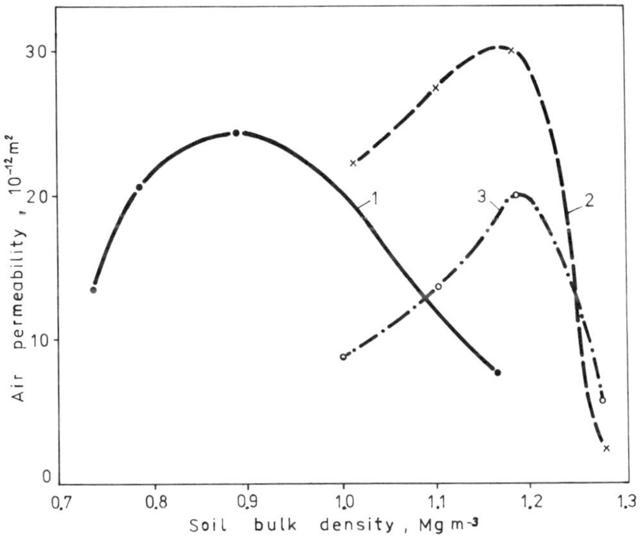


FIGURE 2. Air permeability as a function of bulk density of (1) rendzina soil formed from chalkstone (at moisture tension 150 hPa, pF 2.2), (2) chernozem formed from loess (at moisture tension 150 hPa, pF 2.2), and (3) podzolic soil formed from sand (at moisture tension 500 hPa, pF 2.7). (Data from Turski, R., Domżał, H., and Słowińska-Jurkiewicz, A., *Roczn. Glebozn.*, 29, 3, 1978.)

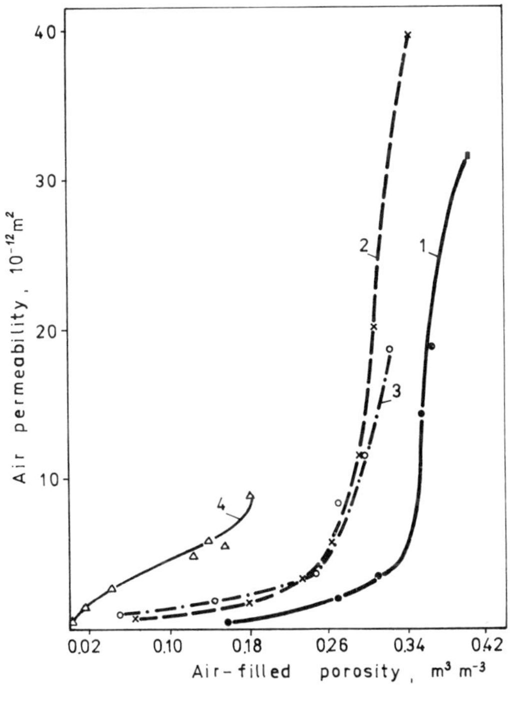


FIGURE 3. Air permeability of a brown soil formed from sandy loam as a function of air-filled porosity at different bulk densities; (1) 1.09 Mg $m^{-3}$, (2) 1.16 Mg $m^{-3}$, (3) 1.21 Mg $m^{-3}$, (4) 1.41 Mg $m^{-3}$. (Data from Turski, R., Domżał, H., and Słowińska-Jurkiewicz, A., *Roczn. Glebozn.*, 29, 3, 1978.)

*1. Variations in Soil Temperature*

Soil temperature variations, particularly in the day-night cycle, result in a cyclic expansion and contraction of the soil air and thus its partial exchange with the atmosphere.

If we assume that the day-night variation in the the temperature of the 0 to 20 cm soil top layer reaches, at an extreme, 20°C, then the cooling of the soil, e.g., from 27 to 7°C (i.e., from 300 to 280K) will result, according to the gas laws, in the contraction of the volume of air in the soil layer cooled by approximately 7%. This means that the soil air in a surface layer 1.4 cm thick (7% of 20 cm) will be replaced by the atmospheric air. Practically, the temperature factor does not have any significance for the whole soil profile, which is confirmed by other authors.[44,561]

*2. Fluctuation of Atmospheric Pressure*

Another factor causing the mass flow of gases in the soil is variation in the atmospheric pressure resulting in similar variation in the pressure of the soil air. McCool and Bouyoucos (quoted after Currie[144]), as early as 1924, noted, that by placing a barometer in the soil at a depth of 3 m, pressure changes of the soil air follow those in the atmosphere without any significant delay. Therefore Boyle's law can be applied in the calculation of the volume of air forced into the soil as a result of changes in the atmospheric pressure.

Since the day-night changes in the atmospheric pressure reach no more than 4%, the amount of air exchanged in the soil profile does not exceed 4% of the volume of air contained in the soil. This means that in a profile of uniform porosity, underlayed by an impervious horizon (soil water, compacted layer, bedrock) at a depth of, e.g., 1 m, only a thin 4 cm surface layer is involved in the exchange with fresh atmospheric air. The depth of this layer is frequently less due to the decrease in the air-filled porosity with the depth of the soil profile.

In the case of soils in which the limitation of air-filled porosity occurs at considerable depths (e.g., soils derived from loess), the aeration of the profile due to variations in the atmospheric pressure may reach much deeper layers, even down to several tens of centimeters.

*3. Changes of Soil Moisture*

During irrigation, or precipitations, water percolates into the soil forcing out an equivalent volume of soil air. Water brings a small amount of oxygen into the soil with it (approximately 30 times less than in an equivalent volume of air at a temperature of 20°C, cf. Table 4). In turn, due to evaporation and percolation the volume of water in the soil profile decreases and is replaced by atmospheric air. A water loss of 1 cm results in the exchange of air in the surface horizon of the soil 3 cm thick at an air-filled porosity of 0.33 $m^3m^{-3}$, and 10 cm thick at an air-filled porosity of 0.1 $m^3m^{-3}$.

Water losses of 24 hr reach up to 0.4 cm under the conditions of a moderate climate, e.g., France,[555] up to 0.7 cm in the Southern states of the U.S.,[138] and up to 0.9 cm in a hot climate like that of Egypt.[520] Therefore the maximum depth of air exchange in a soil profile due to evaporation may be down to 10 cm in a hot climate and about half of that value in a moderate climate (assuming that the air-filled porosity of the soil is 0.1 $m^3m^{-3}$). At higher values of the air-filled porosity the depths will be correspondingly less.

Changes in the ground water table affect the soil air like a piston, forcing it out of the soil or sucking it in. Under natural conditions these changes are slow and only sporadically play a significant role in intensifying the movement of soil air.

In special cases where the rate of infiltration is very high, for example, during the periodic infiltration of sewage through very permeable sands, the participation of mass flow may cover 30 to 40% of the soil requirement for oxygen.[413]

*4. Wind Action*

In some cases the wind may have a significant effect on the exchange of gases between the soil and the atmosphere, particularly at shallow depths and in rather porous materials.[366] Usually, however, the effect of wind is rather insignificant, the wind affects the gas exchange indirectly, through increasing water evaporation, its direct effect being limited to a shallow surface horizon. Romell (quoted after Kimball and Lemon[366]) reports that the contribution of wind to the gas exchange of a sandy soil is less than 0.1% of the total exchange. His findings are supported by contemporary authors.[144,366,561]

## II. DIFFUSION

Another mechanism of the gas exchange in the soil, of much greater significance than the mass flow, is the concentration diffusion. This is a continuous process of gas transport in the direction of decreasing concentration.

In contrast to mass flow, the occurrence of diffusion flow is not induced by the presence of external factors, but only by the gradient of concentration which is a result of respiration processes.

In considering the process of gas diffusion in the soil it is necessary to distinguish macrodiffusion (i.e., the diffusion in the whole soil profile disregarding the heterogeneity of soil in the microscale), the diffusion in particular crumbs, microdiffusion through water films surrounding plant roots, and diffusion through the plants themselves.

### A. Principles of Gas Diffusion in a Porous Body

The diffusion flow ($f_x$) of any diffusing agent through a unit of surface area of the medium in which the diffusion takes place, in a unit of time, in a uniaxial system, is proportional to the gradient of concentration (dC/dx) constituting the driving force of the flow and to the diffusion coefficient D (the dimensions of which are the square of length per unit of time) characterizing the mobility of the agent diffusing in a given medium:

$$f_x = -D\frac{dC}{dx} \qquad (5)$$

Applying Equation (5) to the description of oxygen diffusion in soil, oxygen concentration C usually refers to the unit of soil air volume. It is in this sense that the symbol C will be used in the present text. The coefficient D means the effective diffusion coefficient is related to the soil as a whole (and not to the pores filled with air), and the flow $f_x$ is related to a unit surface area of the soil. This is emphasized in order to avoid confusion when comparing publications by various authors. If, in Equation (5), the oxygen concentration in soil air C is replaced by the content of oxygen in the soil G,[207,707] we get:

$$f_x = -D_G\frac{dG}{dx} \qquad (6)$$

where:

$$G = C\,(E_g + \alpha_B\theta) \qquad (7)$$

and, as has already been mentioned, C is the concentration of oxygen in the gaseous phase of the soil, and $E_g$ and $\theta$ are the contents of air and water, respectively, expressed as fractions of the soil volume, and $\alpha_B$ is Bunsen's coefficient of solubility.

The values of the solubility coefficients, together with the diffusion coefficients and diffusion constants (see further in the text) for $O_2$ and $CO_2$, are presented in Tables 3 and 4, and for other gases in Table 2, Chapter 3.

## Table 3
## SOLUBILITY COEFFICIENTS, DIFFUSION COEFFICIENTS, AND DIFFUSION CONSTANTS OF OXYGEN IN AIR AND WATER, AT DIFFERENT TEMPERATURES

| | Solubility coeff | | | | | Diffusion coeff | | Diffusion constants | |
|---|---|---|---|---|---|---|---|---|---|
| | In air at NP | | In water | | | In air | In water | In air | In water |
| | $\alpha_a$ | | $\alpha_w$ | | $\alpha_B$[a] | $D_o$[b] | $D_w$[b] | $K_o$ | $K_w$ |
| Temp. °C | $\times 10^{-2}$ g $m^{-3}Pa^{-1}$ | $\times 10^{-4}$ mol $m^{-3}Pa^{-1}$ | $\times 10^{-4}$ g $m^{-3}Pa^{-1}$ | $\times 10^{-5}$ mol $m^{-3}Pa^{-1}$ | $m^3m^{-3}$ | $\times 10^{-5}$ $m^2s^{-1}$ | $\times 10^{-9}$ $m^2s^{-1}$ | $\times 10^{-9}$ mol $m^{-1}s^{-1}Pa^{-1}$ | $\times 10^{-14}$ mol $m^{-1}s^{-1}Pa^{-1}$ |
| 0 | 1.410 | 4.41 | 6.9 | 2.16 | 0.0489 | 1.78 | 0.99 | 7.85 | 2.14 |
| 5 | 1.385 | 4.43 | 6.0 | 1.89 | 0.0436 | 1.84 | 1.27 | 7.97 | 2.40 |
| 10 | 1.360 | 4.25 | 5.4 | 1.68 | 0.0394 | 1.89 | 1.54 | 8.04 | 2.59 |
| 15 | 1.337 | 4.18 | 4.8 | 1.51 | 0.0360 | 1.95 | 1.82 | 8.15 | 2.75 |
| 20 | 1.314 | 4.11 | 4.4 | 1.37 | 0.0333 | 2.01 | 2.10 | 8.26 | 2.87 |
| 25 | 1.292 | 4.04 | 4.0 | 1.25 | 0.0309 | 2.07 | 2.38 | 8.36 | 2.97 |
| 30 | 1.270 | 3.97 | 3.7 | 1.15 | 0.0290 | 2.14 | 2.67 | 8.50 | 3.07 |

*Note:* $\alpha_B = \alpha_w/\alpha_a$ (Bunsen's absorption coeff = gas solubility per unit vol of water at given temp and NP (normal pressure) = 101.3 kPa).

[a] Data from *Handbook of Physicochemistry* (Polish), Wyd. Naukowo-Techniczne, Warsaw, 1974.

[b] Data from Armstrong, W., *Adv. Bot. Res.*, 7, 325, 1979.

## Table 4
## SOLUBILITY COEFFICIENTS, DIFFUSION COEFFICIENTS, AND DIFFUSION CONSTANTS OF CARBON DIOXIDE IN AIR AND WATER, AT DIFFERENT TEMPERATURES

| | Solubility coeff | | | | | Diffusion coeff | | Diffusion constants | |
|---|---|---|---|---|---|---|---|---|---|
| | In air at NP | | In water | | | In air | In water | In air | In water |
| | $\alpha_a$ | | $\alpha_w$ | | $\alpha_B$[a] | $D_o$[b] | $D_w$[b] | $K_o$ | $K_w$ |
| Temp. °C | $\times 10^{-2}$ g $m^{-3}Pa^{-1}$ | $\times 10^{-4}$ mol $m^{-3}Pa^{-1}$ | $\times 10^{-2}$ g $m^{-3}Pa^{-1}$ | $\times 10^{-4}$ mol $m^{-3}Pa^{-1}$ | $m^3m^{-3}$ | $\times 10^{-5}$ $m^2s^{-1}$ | $\times 10^{-9}$ $m^2s^{-1}$ | $\times 10^{-9}$ mol $m^{-1}s^{-1}Pa^{-1}$ | $\times 10^{-13}$ mol $m^{-1}s^{-1}Pa^{-1}$ |
| 0 | 1.95 | 4.43 | 3.34 | 7.59 | 1.713 | 1.38 | 1.15 | 6.11 | 8.73 |
| 5 | 1.92 | 4.35 | 2.78 | 6.30 | 1.450 | 1.43 | 1.30 | 6.22 | 8.20 |
| 10 | 1.88 | 4.28 | 2.33 | 5.29 | 1.238 | 1.48 | 1.46 | 6.33 | 7.72 |
| 15 | 1.85 | 4.20 | 1.99 | 4.51 | 1.075 | 1.53 | 1.63 | 6.43 | 7.35 |
| 20 | 1.82 | 4.13 | 1.71 | 3.89 | 0.942 | 1.59 | 1.77 | 6.57 | 6.89 |
| 25 | 1.79 | 4.06 | 1.48 | 3.36 | 0.829 | 1.64 | 1.92 | 6.66 | 6.45 |
| 30 | 1.76 | 4.00 | 1.30 | 2.95 | 0.738 | 1.70 | 2.08 | 6.80 | 6.14 |

*Note:* For explanations see Table 3.

[a] Data from *Handbook of Physicochemistry* (Polish), Wyd. Naukowo-Techniczne, Warsaw, 1974.
[b] Data from Armstrong, W., *Adv. Bot. Res.*, 7, 325, 1979.

Coefficient $D_G$ (defined by Equation [6]) is related to coefficient D (defined by Equation [5]) as follows:

$$D_G = \frac{D}{E_g + \alpha_B\theta} \tag{8}$$

In the case of a dry soil or of a soil fully saturated with water, we have

$$D_G = \frac{D}{E_o} \tag{9}$$

where $E_o$ is total porosity of the soil. In this case $D_G$ is the coefficient of diffusion in soil pores only.

In a soil partly saturated with water, for gases of low solubility in water (e.g., $O_2$, see Table 3 and Figure 3, Chapter 3), an approximated relationship can be applied with the exception of very high moisture contents.

$$D_G \approx \frac{D}{E_g} = D^* \tag{10}$$

where: D* (frequently called the apparent diffusivity) has the physical sense of the diffusion coefficient in air filled pores alone.

The concentration of a gas in water $C_\theta$ is related to its partial pressure P by the following relationship:

$$C_\theta = \alpha_w P \tag{11}$$

known as Henry's law, where: $\alpha_w$ is solubility coefficient of the gas in water (see Tables 3 and 4).

The relationship between the gas concentration in air (C) and its partial pressure is similar in form:

$$C = \alpha_a P \tag{12}$$

where $\alpha_a$ means "solubility" in the gaseous phase, which is known to be to 1/RT, where R = gas constant and T = absolute temperature (see Tables 3 and 4).

Combining Equation (5) with Henry's law, we obtain the formula for diffusion flow, expressed by means of the gradient of partial pressure:

$$f_x = -K\frac{dP}{dx} \tag{13}$$

where K is known as the Krogh's diffusion constant[207,209,354,482,506,706,707,749] related to diffusion coefficient as follows:

$$K = \alpha_a D \frac{E_g + \alpha_B\theta}{E_o} \tag{14}$$

Equations (5) and (13) are two forms of Fick's first law, analogous to Darcy's law describing the flow of liquids and gases under the effect of the gradient of pressure, or to Fourier's law describing the transfer of heat under the effect of the gradient of temperature.

The analogy of Fick's law, after substitution of the diffusion coefficient with diffusion resistance (see Equation [148]) to Ohm's law describing the flow of electric current, is also significant for the construction of electrical analogues of the processes of diffusion flow.

The diffusion constant of gas in air, $K_o$, is therefore:

$$K_o = \alpha_a D_o \tag{15}$$

and the diffusion constant in water, $K_w$, is:

$$K_w = \alpha_w D_w \tag{16}$$

where $D_o$ and $D_w$, and $\alpha_a$ and $\alpha_w$ are the coefficients of diffusion and solubility of the considered gas, in air and water, respectively.

By combining Equations (15) and (16) we get:

$$K_w = \frac{D_w \alpha_w}{D_o \alpha_a} K_o \tag{17}$$

and, having considered that

$$\frac{\alpha_w}{\alpha_a} = \alpha_B \tag{18}$$

we have

$$K_w = \alpha_B \frac{D_w}{D_o} K_o \tag{19}$$

As can be seen from Equation (13), the gas diffusion in water, at the same gradient of partial pressure, takes place $\alpha_B D_w/D_o$ times slower than in air.

In the case of oxygen it is approximately 300,000 times slower, and in the case of carbon dioxide it is about 10,000 times slower (see Tables 3 and 4, data for a temperature of 20°C).

Throughout the remainder of the present text, the diffusion equations for oxygen will include its concentration in the gaseous phase C, diffusion coefficients in soil as defined by Equation (5), and the transition to a form containing partial pressure P can always be made by means of Equations (12) and (14).

Similarly, the transition to formulae including the oxygen content per unit of soil volume G and the diffusion coefficient $D_G$ as defined by Equation (6) can be made by means of Equations (7) to (10).

Equation (5) is a simplified form of a much more complex equation which, for the mutual counterdiffusion of components A and B in a two-component mixture, has the form:[706]

$$f_A = - D_A (C_A + C_B) \left( X_A \frac{d \ln \gamma A}{dX_A} + 1 \right) \frac{dX_A}{dx} \tag{20}$$

where $D_A$ = diffusion coefficient of component A, $C_A$ and $C_B$ = concentrations of components A and B (mol $m^{-3}$), $X_A$ = molar fraction of component A in the mixture equal to $C_A/(C_A + C_B)$, $\gamma_A$ = activity coefficient of component A, and x = diffusion path.

In the case when the diffusion coefficients of components A and B are not equal (which is the situation most frequently met in the soil), mass flow equal to $X_A(f_A + f_B)$ occurs as well, and then the equation describing the total flow f* (mass and diffusion) of component A has the form:

$$f_A^* = - D_A (C_A + C_B) \left( X_A \frac{d \ln \gamma A}{dX_A} + 1 \right) \frac{dX_A}{dx} + X_A (f_A + f_B) \tag{21}$$

In the case of equimolar diffusion, i.e., when the diffusion flows of components A and B are identical in value but opposite in direction, the mass flow term disappears and we get Equation (20).

If the concentrations of the diffusing gases are low, the activity coefficient equals one and Equation (20) is simplified to:

$$f_A = - D_A (C_A + C_B) \frac{dX_A}{dx} \tag{22}$$

Considering that $X_A = C_A/(C_A + C_B)$, we get the following equation:

$$f_A = - D_A \frac{dC_A}{dx} \tag{23}$$

which is identical to the Fick's first law (5). When using Fick's law one should keep in mind the above assumptions concerning equimolarity and low concentrations of the diffusing gases. The applicability of Fick's law is better when the concentrations of diffusing gases are low and the difference in their molecular weights is small (e.g., $N_2$ and $O_2$). Such conditions are met in well-aerated soil, while in the case of soil with insufficient oxygenation the assumption of equimolar diffusion is not met due to the formation of such gases as methane, hydrogen sulfide, or excessive amounts of carbon dioxide.

The value of the diffusion coefficient can be determined by means of direct measurements, or by means of a formula proposed by Reid and Sherwood[590] for the diffusion coefficient of component 1 diffusing in a homogeneous mixture of n components:

$$D_1 = \frac{(1 - X_1)}{\sum_{j=2}^{n} \left(\frac{X_j}{D_{1j}}\right)} \tag{24}$$

where $X_1$ = molar fraction of component 1, $X_j$ = molar fraction of component j, $D_{1j}$ = diffusion coefficient for a two-component mixture of components 1 and j calculated according to the formula proposed by Bird et al.[54] (modified for SI units):

$$D_{1j} = 1.883 \cdot 10^{-2} \frac{T^{3/2} \sqrt{\frac{1}{M_1} + \frac{1}{M_j}}}{p\, \sigma_{1j}^2 \Omega} \ m^2 s^{-1} \tag{25}$$

where $D_{1j}$ = diffusion coefficient of a two-component mixture of components 1 and j of molecular masses $M_1$ and $M_j$, $\sigma_{1j}$ = coefficient determined by the size of molecules of components 1 and j, $\Omega$ = coefficient proportional to the collision probability of gas molecules, and p = total pressure of gas in Pa.

The values of coefficients $\sigma$ and $\Omega$ have been tabulated by Hirschfelder et al.[314,315] It follows from Equation (25) that effect of temperature and pressure on the value of the diffusion coefficient can be assessed by means of the following equation:

$$D_{T_2 p_2} = D_{T_1 p_1} \frac{p_1}{p_2} \left(\frac{T_2}{T_1}\right)^{1.5} \tag{26}$$

where $D_{T_1 p_1}$ and $D_{T_1 p_1}$ = diffusion coefficients at pressure and temperature values indicated by the subscripts.

According to the conservation law, the change in the quantity of gas dG diffusing in a unit of soil volume during time dt is equal to the change taking place in flow $df_x$ on the diffusion path dx minus the consumption (e.g., $O_2$) or plus the production (e.g., $CO_2$, $C_2H_4$) of the gas q in the soil, according to the formula:

$$\frac{\partial G}{\partial t} = - \frac{\partial f_x}{\partial x} \pm q \tag{27}$$

Considering that the amount of gas G in a unit of soil volume is equal to the sum of its contents in the liquid and gaseous phases of the soil, i.e., $G = C (E_g + \alpha_B \theta)$, and combining Equations (27) and (5), we obtain:

$$\frac{\partial[C\,(E_g + \alpha_B\theta)]}{\partial t} = \frac{\partial\left(D\dfrac{\partial C}{\partial x}\right)}{\partial x} \pm q \tag{28}$$

where C = concentration of the diffusing component in the soil air, $\alpha_B$ = Bunsen's solubility coefficient of the gas, $\theta$ = soil moisture content, x = diffusion path, q = respiratory activity related to the soil volume unit, and $E_g$ = air-filled porosity of the soil.

In the case of gases which are not soluble in the soil solution ($\alpha_B = 0$), Equation (28) simplifies to:

$$\frac{\partial(C \cdot E_g)}{\partial t} = \frac{\partial\left(D\dfrac{\partial C}{\partial x}\right)}{\partial x} \pm q \tag{29}$$

In a soil profile all the parameters considered may be affected both by the depth (x) and time (t), which results from the heterogeneity of the profile and from the variability of the properties of the soil with time. If we express this effect mathematically, we obtain:

$$\frac{\partial}{\partial t}\{C[E_g(\theta,t,x) + \alpha_B\,\theta(t,x)]\} = \frac{\partial}{\partial x}\left[D(x,t,\theta)\frac{\partial C}{\partial x}\right] \pm q(x,t,\theta,C) \tag{30}$$

The above general form of the uniaxial diffusion equation is the starting point for the consideration of specific situations.

In many cases, when D, $E_g$, q, and $\theta$ are constant, e.g., in a particular horizon of the soil profile, the application of a simplified equation is possible. If we consider that D (within the range of applicability of Fick's law) does not depend on the concentration of gas and if we neglect the solubility of the gas in the liquid phase of the soil (which is possible, e.g., for oxygen, for which at a temperature of 20°C $\alpha_B = 0.033$), this equation will have the form.

$$\frac{\partial C}{\partial t} = \frac{D}{E_g}\frac{\partial^2 C}{\partial x^2} \pm \frac{q}{E_g} \tag{31}$$

or

$$\frac{\partial C}{\partial t} = D^*\frac{\partial^2 C}{\partial x^2} \pm q^* \tag{32}$$

where $D^* = D/E_g$ and $q^* = q/E_g$ are apparent diffusivity and respiration, respectively.

For spatial diffusion of oxygen this equation, in any system of coordinates, has the form:

$$\frac{\partial C}{\partial t} = \frac{D}{E_g}\nabla^2 C \pm q^* \tag{33}$$

where $\nabla^2$ = div grad.

In the perpendicular system of coordinates x, y, z:

$$\nabla^2 = \frac{\partial^2}{\partial x^2} + \frac{\partial^2}{\partial y^2} + \frac{\partial^2}{\partial z^2} \tag{34}$$

while in a spherical system of coordinates r, $\Theta$, $\varphi$:

$$\nabla^2 = \frac{1}{r^2}\frac{\partial}{\partial r}\left(r^2\frac{\partial}{\partial r}\right) + \frac{1}{r^2\sin\Theta}\frac{\partial}{\partial\Theta}\left(\sin\Theta\frac{\partial}{\partial\Theta}\right) + \frac{1}{r^2\sin^2\Theta}\frac{\partial^2}{\partial\varphi^2} \tag{35}$$

where the relationship between the perpendicular and spherical coordinates is given as:

$$\begin{aligned} x &= r \sin \Theta \cos \varphi \\ y &= r \sin \Theta \sin \varphi \\ z &= r \cos \Theta \end{aligned} \tag{36}$$

For diffusion in a cylindrical system of coordinates ρ, φ, z, expression (34) has the form:

$$\nabla^2 = \frac{1}{\rho}\frac{\partial}{\partial \rho}\left(\rho \frac{\partial}{\partial \rho}\right) + \frac{1}{\rho^2}\frac{\partial^2}{\partial \varphi^2} + \frac{\partial^2}{\partial z^2} \tag{37}$$

where the relationship between the coordinates is given as:

$$\begin{aligned} x &= \rho \cos \varphi \\ y &= \rho \sin \varphi \\ z &= z \end{aligned} \tag{38}$$

Under steady state conditions, Equations (31) and (32) assume the form of Poisson's equation for uniaxial diffusion:

$$\frac{\partial^2 C}{\partial x^2} = \pm \frac{q}{D} = \pm \frac{q^*}{D^*} \tag{39}$$

In the absence of sources or sinks of the diffusing gas (e.g., no respiratory activity, i.e., $q = 0$), Equation (31) is reduced to the form:

$$\frac{\partial C}{\partial t} = \frac{D}{E_g}\frac{\partial^2 C}{\partial x^2} \tag{40}$$

which is known as the Fick's second law and in the case of a steady state diffusion ($\partial C/\partial t = 0$), to an equation of the Laplace type:

$$\frac{\partial^2 C}{\partial x^2} = 0 \tag{41}$$

## B. Macrodiffusion (Diffusion Within the Soil Profile)

### *1. Gas Diffusion Coefficient in Soil*

The gas diffusion coefficient in soil is a fraction of the gas diffusion coefficient in free atmosphere, as it depends not only on the kind of gas, pressure, and temperature, but on the amount of air-filled pores, and also on their continuity and shape, which in turn are determined by the characteristics of the given soil as a porous system of particles and by its moisture content.

The diffusive properties of a soil medium are usually characterized by means of the relative diffusion coefficient $D/D_o$, where $D$ = gas diffusion coefficient in soil and $D_o$ = diffusion coefficient of the same gas in atmospheric air, under the same conditions of pressure and temperature. An advantage of this way of expressing the gas diffusion coefficient in soil is that it does not depend on the temperature, pressure, or the kind of diffusing gas.

The values of $D_o$ for oxygen can be found in Table 3, and those for carbon dioxide can be found in Table 4. Values for other gases that may be included in the composition of soil air can be found in Table 2, Chapter 3. The value of $D/D_o$, for any value of pressure or temperature, can be calculated by means of Equation (26).

The relationship of $D/D_o$ to the soil moisture tension, at different levels of compaction, is presented in Figure 4. As can be seen, this coefficient increases, in a curvilinear manner, with increase in the soil moisture tension, and decreases rapidly with increase in the soil compaction level.

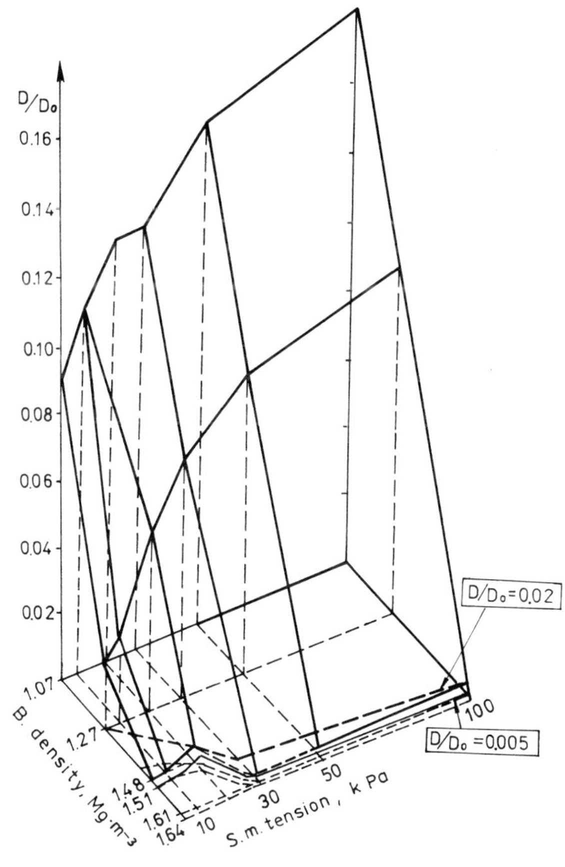


FIGURE 4. Relationship of $D/D_0$ in a loamy textured black earth to soil moisture tension and bulk density. (Modified from Stępniewski, W., *Pol. J. Soil Sci.*, 14, 3, 1981.)

Soil compaction has also a strong effect on the relationship of $D/D_0$ to soil moisture content $\theta$, which can be seen in Figure 5. The lowest moisture content values correspond to the state of air-dry soil, and the corresponding $D/D_0$ values fall within a range from 0.03 to 0.2. This range covers the values for air-dry soils most frequently quoted in the literature; only in exceptional cases, for very loose soils, these values may be higher and even reach 0.5

Generally speaking, the values of $D/D_0$ for air-dry soils, not much different from those for absolutely dry soils, are within 0.02 to 0.5 of the value of $D_0$. These correspond (see Tables 3 and 4) to values of D for oxygen from $4 \cdot 10^{-7}$ to $10^{-5}$ $m^2s^{-1}$, and for carbon dioxide from $3.2 \cdot 10^{-7}$ to $8 \cdot 10^{-6}$ $m^2s^{-1}$ (at 20°C).

The Figure 5 does not permit an accurate assessment of the values of $D/D_0$ at full saturation of the soil with water. There is also a lack of accurate measurement data for this range, as the measurement methods for the diffusion coefficient are suited to medium moisture contents and at low values of D tend to include a greater determination error. We may suppose, however, that the degree of retardation of diffusion by the solid phase of the soil is the same both in the case when the diffusion takes place in the gaseous phase (dry soil) and when the diffusion occurs in the liquid phase (soil saturated with water). Thus the relative coefficient of gas diffusion in a soil fully saturated with water related to the coefficient of diffusion in pure water ($D_w$) should be the same as the relative coefficient of diffusion with respect to the atmospheric air in the case of a dry soil. Therefore, we can assume that in a

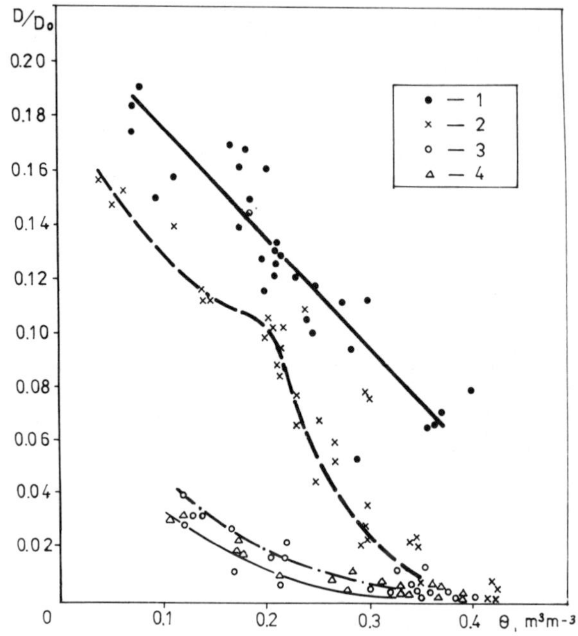


FIGURE 5. Relationship of $D/D_o$ to volumetric moisture content θ of a loamy textured black earth (the same as in Figure 4) at different bulk densities; (1) 1.07 Mg $m^{-3}$, (2) 1.27 Mg $m^{-3}$, (3) 1.48 Mg $m^{-3}$, (4) 1.51 Mg $m^{-3}$. (From Stępniewski, W., unpublished).

soil saturated with water the diffusion coefficient falls within the range of 0.02 to 0.5 $D_w$, which corresponds (see Tables 3 and 4) to a range of D values from $4.2 \cdot 10^{-11}$ to $10^{-9}$ $m^2s^{-1}$ for oxygen, and from $3.5 \cdot 10^{-11}$ to $8.8 \cdot 10^{-10}$ $m^2s^{-1}$ for carbon dioxide (at 20°C).

In contrast to the relationships of $D/D_o$ (soil moisture tension) and $D/D_o$ (soil moisture content) discussed above, it has been found[264,596,698] that the relationship of $D/D_o$ to the air-filled porosity of soil is not clearly dependent on soil compaction.

A typical example of this relationship can be found in Figure 6. With the increase in $E_g$ a nonlinear increase in $D/D_o$ is observed, but after exceeding $E_g$ values of 0.3 to 0.4 $m^3m^{-3}$ [141,144,264,484,698] further increase in $D/D_o$ assumes a straightlinear character and is slower than in the range of low $E_g$ values. This is explained by the fact that the increase in the $E_g$ values occurs at a higher range of the air-filled porosity as a result of dewatering of the intra-aggregate micropores whose contribution in gas diffusion is lower (see further in the text) than that of the interaggregate macropores.

The phenomenon of hysteresis in the relationship of $D/D_o$ to $E_g$ has not been extensively investigated in relation to soils, and the results obtained,[650] concerning sand only, indicate an absence of this effect.

Various mathematical models have been applied to the description of the relationship of $D/D_o$ to $E_g$, the simplest of them being:

$$\frac{D}{D_o} = \tau E_g \qquad (42)$$

where τ is the coefficient of pore effectiveness for gas diffusion.

This coefficient indicates what fraction is constituted by the actual gas diffusion coefficient in the soil in relation to a situation in which all the soil pores formed a system of straight capillaries parallel to the direction of diffusion.

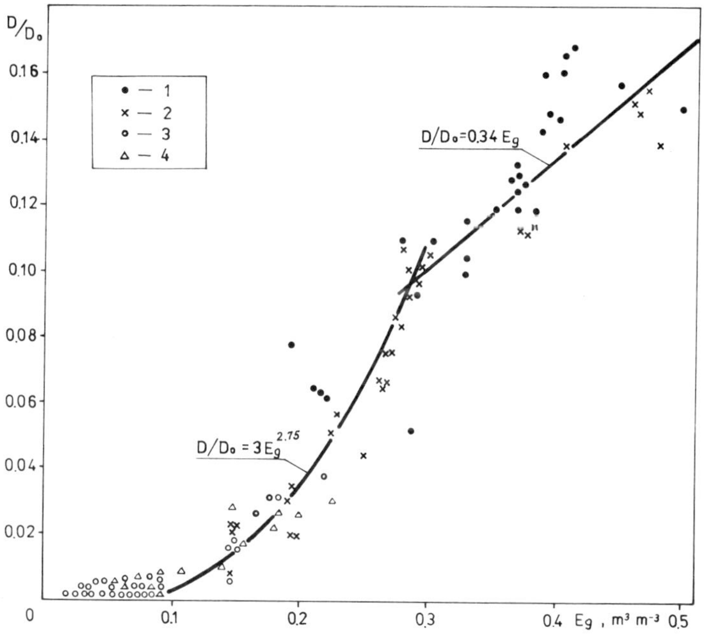


FIGURE 6. Relationship of $D/D_o$ to air-filled porosity $E_g$ of a loamy textured black earth (the same as in Figure 4) at different bulk densities; (1) 1.07 Mg $m^{-3}$, (2) 1.27 Mg $m^{-3}$, (3) 1.48 Mg $m^{-3}$, (4) 1.51 Mg $m^{-3}$. (Modified from Stępniewski, W., *Pol. J. Soil Sci.*, 14, 3, 1981.)

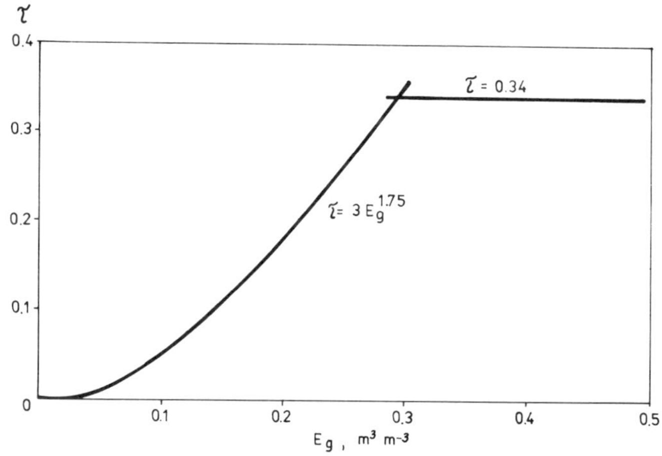


FIGURE 7. Dependence of the coefficient of pore effectiveness to gas diffusion $\tau$ on air-filled porosity $E_g$ drawn on the basis of Figure 6.

In any given soil, the value of the coefficient of pore effectiveness for diffusion is not constant, which reduces its applicability, in practice limited to interpretation problems. The relationship of $\tau$ to $E_g$, presented in Figure 7, indicates that the effectiveness of pores for diffusion first increases with soil dewatering, reaching a maximum in a range corresponding to dewatering the large interaggregate pores, and then decreases when the smaller pores are subsequently emptied.

The cause of this drop in the effectiveness of air-filled pores for diffusion seems to lie in the diameters of the pores available for diffusion. The diffusion coefficient does not depend on the size of the capillaries or pores (unlike to the air permeability which is related to the square of the pore radius), provided that the pore diameters are greater than the MFP of the molecules as the effects of the collision of gas molecules with the capillary walls have no real significance, as long as this condition is satisfied.

It can be assumed that the limit is reached at pore diameters of 0.10 μm (cf. Table 2, Chapter 3) which, in the case of soil, means pores which are emptied of water at soil moisture tension levels over 3 MPa (pF > 4.5), i.e., at moisture contents below the wilting point.

Therefore, it is the gradual transition from normal diffusion to diffusion of rarefied gas that seems to be the main reason for the decrease in the diffusive effectiveness of small pores which become emptied of water at high values of the soil moisture tension.

More useful, in practice, than the model described by Equation (42) is the linear model:[30,60,95,265,604,619,731,784,785]

$$\frac{D}{D_o} = 1\,(E_g - m) \tag{43}$$

where m has the physical sense of dead porosity, i.e., the volume of pores unavailable for diffusion, and l denotes the effectiveness, for diffusion, of the remaining pores filled with air. However, because of the generally curvilinear character of the relationship of $D/D_o$ to $E_g$, a better description is provided by the power model[36,87,140,141,264,265,365,407,463,483] in the form:

$$\frac{D}{D_o} = \gamma\, E_g^{\mu} \tag{44}$$

where $\gamma$ and $\mu$ are empirical coefficients characterizing the material under investigation.

This model has been used more frequently than the others. It also permits the determination of the relationship of the effectiveness coefficient to the air-filled porosity:

$$\tau = \gamma\, E_g^{\mu-1} \tag{45}$$

Recently, Troeh et al.[737] proposed a third model, also containing two empirical parameters (u and h), which according to the authors combines the advantages of both the previous models and permits a more reliable extrapolation of experimental data. This model has the form:

$$\frac{D}{D_o} = \left(\frac{E_g - u}{1 - u}\right)^h \tag{46}$$

A review of the values of the coefficients for all the three models formulated in Equations (43), (44), and (46), as obtained by various authors, is compiled in Table 5.

Stępniewski[696,698] has found that the empirical parameters $\gamma$ and $\mu$ in Equation (44) are interrelated, presented in Figure 8. Although these parameters were obtained for various soils and by various researchers, they display an interrelation which can, to a good approximation, be exposed by the following equation:

$$\gamma = 0.1\,\mu^3 \tag{47}$$

The values obtained by Lai et al.,[407] Marshall,[463] and Millington and Quirk[483] who assumed that $\gamma = 1$, i.e., adopted the model $D/D_o = E_g^{\mu}$, are not included in Figure 8 as they are too far off the curve.

As the parameters $\gamma$ and $\mu$ from the empirical Equation (44) are interrelated by the formula (47), the relationship of $D/D_o$ to $E_g$ may be described by means of only one parameter characterizing a given porous medium. Combining Equations (44) and (47) we obtain:

## Table 5
## VALUES OF EMPIRICAL PARAMETERS OF THREE MODELS DESCRIBING RELATIONSHIP OF $D/D_o$ TO AIR FILLED POROSITY OF SOIL

| Material tested | Range of $E_g$ $m^3m^{-3}$ | Equations and coefficients | | | | Ref. |
|---|---|---|---|---|---|---|
| | | $D/D_o = l(E_g - m)$ | | $D/D_o = [(E_g - u)/(1-u)]^h$ | | |
| | | l | m | u | h | |
| Sand, glass spheres | 0.15—0.4 | 0.66 | 0 | 0 | 1.3 | 552 |
| Rothamsted subsoil | 0.43—0.63 | 0.81 | 0.11 | 0.1 | 1.5 | 552 |
| Brookston clay | 0.10—0.30 | 1.3 | 0.1 | 0.05 | 1.1 | 60 |
| Paulding clay | 0.05—0.33 | 0.71 | −0.01 | 0 | 1.2 | 60 |
| Loam | 0.10—0.45 | 0.67 | 0 | 0 | 1.4 | 731 |
| Sand | 0.10—0.45 | 0.97 | 0.09 | 0 | 1.3 | 731 |
| Soils at different depths | 0.20—0.40 | 0.37 | 0.17 | 0.15 | 2.0 | 619 |
| Several soils | 0.14—0.39 | 0.66 | 0.1 | 0.07 | 1.6 | 95 |
| Soils | 0.15—0.60 | 0.90 | 0.14 | — | — | 785 |
| Kaolin (wet) | 0.14—0.54 | 0.83 | 0.13 | — | — | 265 |
| Kaolin (dry) | 0.30—0.57 | 0.94 | 0.21 | — | — | 265 |
| Soils | >0.10 | 0.9 | 0.11 | — | — | 784 |
| | 0.25—0.52 | 0.88 | 0.16 | — | — | 604 |
| | 0.07—0.30 | 0.53 | 0.05 | 0.03 | 1.5 | 30 |
| | | $D/D_o = \gamma E_g^{\mu}$ | | | | |
| | | γ | μ | | | |
| Sand, soils | 0.16—0.61 | 1.0 | 2.0 | 0.15 | 1.7 | 87 |
| Porous solids | 0.25—0.5 | 1.0 | 3/2 | — | — | 463 |
| Sand, glass spheres | 0.17—0.43 | 1.9 | 1.4 | 0 | 1.4 | 140 |
| Soils | 0.1—0.5 | 1.0 | 4/3 | 0.05 | 1.4 | 483 |
| Sand, glass spheres | 0.3—0.4 | 1.75 | 2.1 | 0.15 | 1.1 | 141 |
| Silt loam topsoils | 0.02—0.4 | 0.49 | 1.5 | 0 | 2.0 | 265 |
| Webster loam | 0.00—0.18 | 40.6 | 4.0 | 0.05 | 1.5 | 365 |
| Silty clay loam | 0.15—0.43 | 5.25 | 3.36 | 0.15 | 1.1 | 264 |
| Puddled topsoils | 0.04—0.4 | 2.0 | 3.0 | 0.10 | 2.0 | 36 |
| Nonpuddled topsoils | 0.04—0.4 | 0.85 | 2.0 | 0.03 | 2.0 | 36 |
| Soils | 0.12—0.37 | 1.0 | 5/3 | 0.05 | 1.4 | 407 |
| Several soils | 0.06—0.54 | 0.15—10 | 1.6—4 | — | — | 35 |
| Brown loess soils | 0.0—0.45 | 0.21—2.7 | 0.84—3.53 | — | — | 696 |
| Soils | 0.0—0.35 | 2.5—5 | 2.75—4 | — | — | 698 |

$$\frac{D}{D_o} = 0.1\,\mu^3 E_g^{\mu} \tag{48}$$

or

$$\frac{D}{D_o} = \gamma E_g^{\sqrt[3]{10\,\gamma}} \tag{49}$$

A graphic interpretation of Equation (48) is presented in Figure 9. A general tendency appears here that in soils of a higher value of coefficient μ the values of $D/D_o$ are lower in the low $E_g$ range and greater at higher $E_g$ values.

Attempts have also been made at the application, for the description of the relationship of $D/D_o$ to $E_g$, of more complex empirical models,[140,143,146] and also at a theoretical derivation of this relationship.[484]

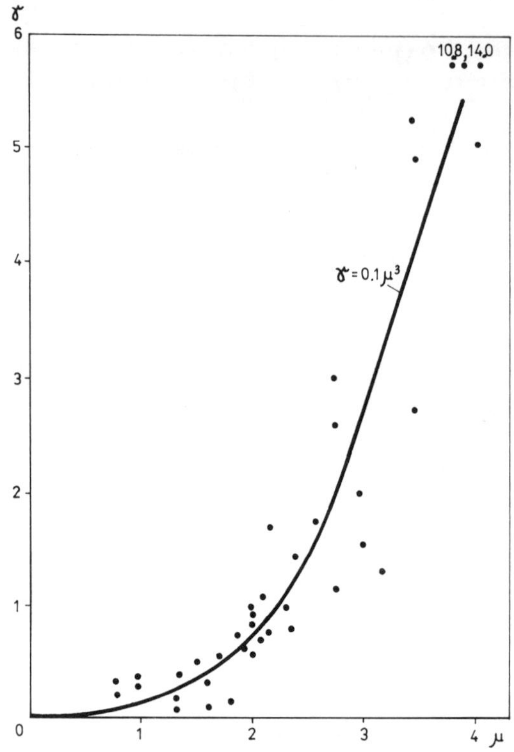


FIGURE 8. Relationship between parameters γ and μ from Equation (44). Values γ and μ are taken from the literature.[35,36,87,141,264,265,365,696,698]

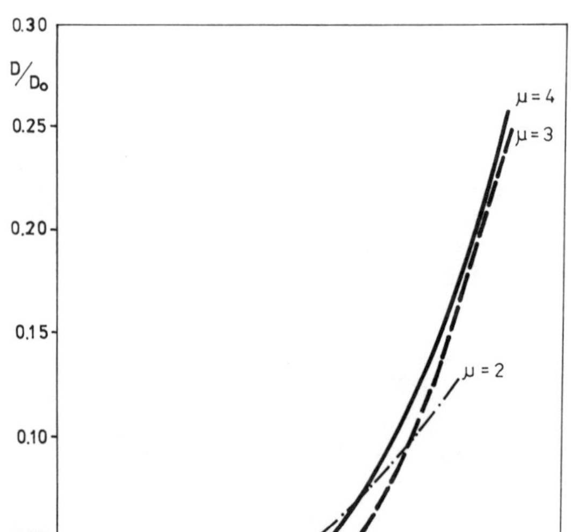


FIGURE 9. Graphical interpretation of the relationship of $D/D_o$ on air-filled porosity described by the equation $D/D_o = 0.1\mu^3E_g^\mu$ at different values of μ.

*2. Gas Diffusion Constant in Soil*

In a soil system, depending on the moisture content, diffusion may take place only in pores filled with air (dry soil), jointly in pores filled with water, and in pores filled with air, or else (in the case of total saturation of the soil with water) only in pores filled with water. Hence for the description of the diffusive properties of soil in the full range of its moisture contents it is convenient to employ a diffusion constant instead of diffusion coefficient. Diffusion equations are then related to partial pressures, which permits the comparisons of diffusion rates at all levels of soil moisture contents without the risk of confusion as to the manner of expressing the concentration of the diffusing agent (per a unit of soil air volume or of soil water volume).

In a moisture content range when the pores filled with air form a continuous system, which is decisive for the overall diffusion rate, the relative diffusion coefficient $D/D_o$ is equal to the relative diffusion constant $K/K_o$. Hence Figures 4 to 6 presenting $D/D_o$ in soil at the same time present the same relationships for $K/K_o$.

By following considerations analogous to those concerning the diffusion coefficient, it should be stated that the range of variability of $K/K_o$ for dry soils is 0.02 to 0.5, e.g., for oxygen at 20°C K it is from $1.65 \cdot 10^{-10}$ to $4.13 \cdot 10^{-9}$ mol $m^{-1}s^{-1}Pa^{-1}$. In the case of total saturation of soil with water, the value of K should be from 0.02 to 0.5 of the value of oxygen diffusion constant in water $K_w$ ($2.87 \cdot 10^{-14}$ mol $m^{-1}s^{-1}Pa^{-1}$ at 20°C), i.e., from $5.7 \cdot 10^{-16}$ to $1.44 \cdot 10^{-16}$ mol $m^{-1}s^{-1}Pa^{-1}$. Therefore the full range of variability of K in soil extends from $5.7 \cdot 10^{-16}$ to $4.13 \cdot 10^{-9}$ mol $m^{-1}s^{-1}Pa^{-1}$, i.e., covers over six orders of magnitude.

Although the diffusion constant has a broader applicability than the diffusion coefficient, a majority of researchers have applied the diffusion coefficient in relation to soils, and only a few publications[207,209,706,707] quote values of K obtained from measurements in soils. All of these papers quote only the K values measured by Flühler[207] for eight soils, and the range of values obtained is from $2 \cdot 10^{-15}$ (at full saturation with water) to $2 \cdot 10^{-9}$ mol $m^{-1}s^{-1}Pa^{-1}$ (dry soil).

*3. Solution of Equations for Oxygen Diffusion in Soil*

The solutions of diffusion equations to be found in the literature have been obtained in a form containing the diffusion coefficient D as defined in Equation (5) or the diffusivity coefficient $D^* = D/E_g$ and concentration C in soil air. This original form has been retained in the present text. It should be kept in mind, however, that these equations are applicable only for a range of soil moisture contents at which diffusion takes place in the gaseous phase and no other. In the case of discontinuity of pores filled with air (which occurs in the range of $E_g$ values below 0.1 to 0.2 $m^3m^{-3}$) it should be remembered that oxygen concentration must be related not to the gaseous phase but to the liquid phase:

$$C_\theta = \alpha_B C \tag{50}$$

but it is most convenient to apply the diffusion constant and partial pressure, using Equations (11) to (16).

In the range of analytical solutions of diffusion equations, in relation to soil under steady-state conditions (constant in time), there are solutions for a multilayer profile, while for a soil under nonsteady-state conditions, solutions have been obtained only for a two-layer profile.

### *a. Steady State*

#### (1) Multilayer Profile

Oxygenation in a multilayer soil profile of a depth $x = L$ composed of n layers each of which has a determined depth $a_i$, and a respiratory activity $q_i$ and oxygen diffusion coefficient

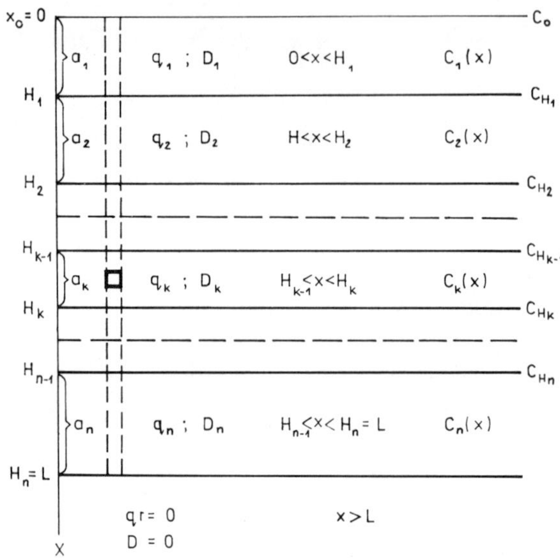

FIGURE 10. Scheme of the multilayered soil profile. (From Stępniewski, W., Influence of Mineral Fertilization and Watering of Arable Soil on its Oxygenation, Ph.D. dissertation, Agricultural University, Lublin, Poland, 1975.)

$D_i$ has been considered by Stępniewski.[693] His approach is a generalization of that of Kowalik[377] for a two-layer profile. A multilayer profile is presented in Figure 10. It should be stressed that within every layer the values of $q_i$ and $D_i$ do not change with time or profile depth. For $x > L$ the profile does not show any respiratory activity and the oxygen diffusion coefficient equals zero, which means that the bedrock is nonpermeable to gas.

Let us consider a unit element of soil volume at depth x in layer k, marked in Figure 10 with a square. The oxygen requirement at this depth, $Q_x$, is equal to the oxygen uptake by deeper layers, and therefore:

$$Q_x = \sum_{i=k}^{n} a_i q_i - q_k (x - H_{k-1}) \tag{51}$$

On the other hand, the oxygen supply will be, according to Fick's first law:

$$f_x = - D_k \frac{dC_k}{dx} \tag{52}$$

In the case of the steady-state conditions, there is an equilibrium between oxygen consumption and supply at any depth, and therefore:

$$\sum_{i=k}^{n} a_i q_i + q_k H_{k-1} - q_k x = - D_k \frac{dC_k}{dx} \tag{53}$$

As $D_k$ and $q_k$ are constant, this equation can be integrated within limits from $H_{k-1}$ to x and from $C_{H_{k-1}}$ to $C_k(x)$:

$$\left[\sum_{i=k}^{n} q_i a_i + q_k H_{k-1}\right] \int_{H_{k-1}}^{x} dz - q_k \int_{H_{k-1}}^{x} z\, dz = - D_k \int_{C_{H_{k-1}}}^{C_k(x)} du \tag{54}$$

As a result we get:

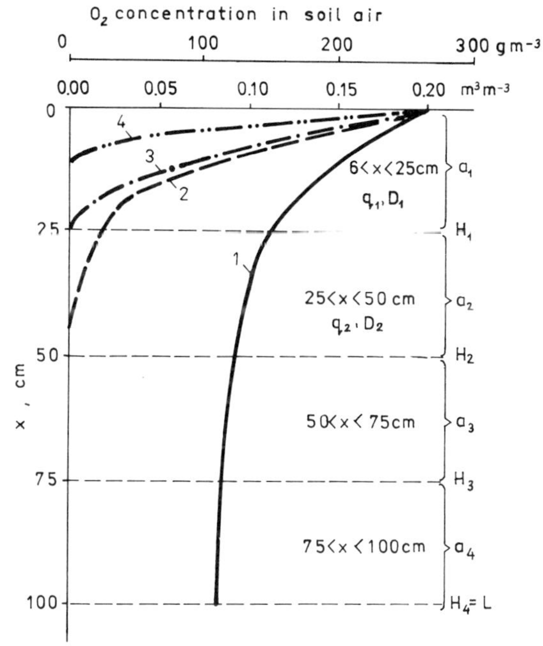


FIGURE 11. Steady-state distribution of oxygen concentration in soil air calculated from Equations (55) and (56) for a four-layered soil profile of equal thickness of the layers: $a_1 = a_2 = a_3 = a_4 = 25$ cm (temp. 20°C) and of the following values of the respiration rate and diffusion coefficient in the particular layers: Curve 1: $q_1 = 0.5$ mg $m^{-3}s^{-1}$, $q_2 = 0.1$ mg $m^{-3}s^{-1}$, $q_3 = 0.05$ mg $m^{-3}s^{-1}$, $q_4 = 0.01$ mg $m^{-3}s^{-1}$, $D_1 = D_2 = D_3 = D_4 = 2 \cdot 10^{-7}$ $m^2s^{-1}$; Curve 2: $q_1 = 1.2$ mg $m^{-3}s^{-1}$, $D_1 = 2 \cdot 10^{-7}$ $m^2s^{-1}$, $q_2 = 0.3$ mg $m^{-3}s^{-1}$, $D_2 = 2.1 \cdot 10^{-8}$ $m^2s^{-1}$; Curve 3: $q_1 = 1.92$ mg $m^{-3}s^{-1}$, $D_1 = 2 \cdot 10^{-7}$ $m^2s^{-1}$; Curve 4: $q_1 = 10$ mg $m^{-3}s^{-1}$, $D_1 = 2 \cdot 10^{-7}$ $m^2s^{-1}$.

$$C_k(x) = C_{H_{k-1}} - \frac{2\sum_{i=k}^{n} q_i a_i (x - H_{k-1}) - q_k (x - H_{k-1})^2}{2D_k} \tag{55}$$

For $x = H_k$, i.e., for the lower boundary of layer k, we have:

$$C_{H_k} = C_{H_{k-1}} - \frac{2\sum_{i=k}^{n} q_i a_i a_k - q_k a_k^2}{2D_k} \tag{56}$$

And at the lower boundary of the profile ($x = L = H_n$) we obtain:

$$C_L = C_{H_n} - C_{H_{n-1}} - \frac{q_n a_n^2}{2D_n} \tag{57}$$

For an example, the distribution of oxygen concentration in a four layer soil profile is presented in Figure 11. As can be seen from this figure, the highest gradients of oxygen concentration occur in the surface layer of the soil, i.e., in the layer of the greatest respiratory activity. In deeper layers, of lower respiratory activity, the differentiation in oxygen concentration, despite the differences in the values of the diffusion coefficient, is only slight.

Therefore, in the case when the last (bottom) layer is biologically inactive $Q_n = 0$, there will be no decrease in oxygen concentration in this layer, i.e., $C_L = C_{H_{n-1}}$.

On the basis of Equations (55) and (56) we can find the distribution of oxygen concentration under steady-state conditions in a miltilayer soil profile with n number of layers.

**(2) Monolayer Profile**

Case 1: D and q are constant in time and do not change with depth.

In the simplest case, if we deal with a monolayer profile in which the respiratory activity q is constant in time and does not vary with depth, i.e., $k = n = 1$ and $a_1 = H_1 = L$, Equation (55) is reduced to:

$$C = C_o - \frac{q\,(2Lx - x^2)}{2D} \tag{58}$$

This result is identical to the solution obtained by other authors.[377,378,532,610,807]

At the boundary of the biologically active zone for $x = L = H_n$ (irrespective of whether the underlying layer is permeable or impermeable for gases) we have:

$$C_L = C_o - \frac{qL^2}{2D} \tag{59}$$

If we denote that level of oxygen concentration which is critical for the activity of microorganisms as $C'$, then from Equation (59) we can determine the critical value of respiratory activity $q'$ at which the oxygen concentration at the lower boundary of the biologically active zone ($x = L$) reaches the value of $C'$:

$$q' = \frac{2D\,(C_o - C')}{L^2} \tag{60}$$

or alternatively the critical value of the diffusion coefficient $D'$ which ensures oxygen concentrations above $C'$ in the whole profile down to depth L:

$$D' = \frac{qL^2}{2\,(C_o - C')} \tag{61}$$

Many authors maintain that $C'$ is practically equal to zero which, if it is introduced into Equations (60) and (61), will give:

$$q' = \frac{2C_oD}{L^2} \tag{62}$$

and

$$D' = \frac{qL^2}{2C_o} \tag{63}$$

In the case when not all the biologically active profile is oxygenated ($q < q'$), the depth of the oxic zone $L_{an}$ is

$$L_{an} = \sqrt{\frac{2C_oD}{q}} \tag{64}$$

The replacement of oxygen concentration in atmospheric air $C_o$ in this equation with its partial pressure $P_o$, and the diffusion coefficient D with the constant K, gives $L_{an}$ (see Equations [12] and [14]):

$$L_{an} = \sqrt{\frac{2P_oK}{q}} \tag{65}$$

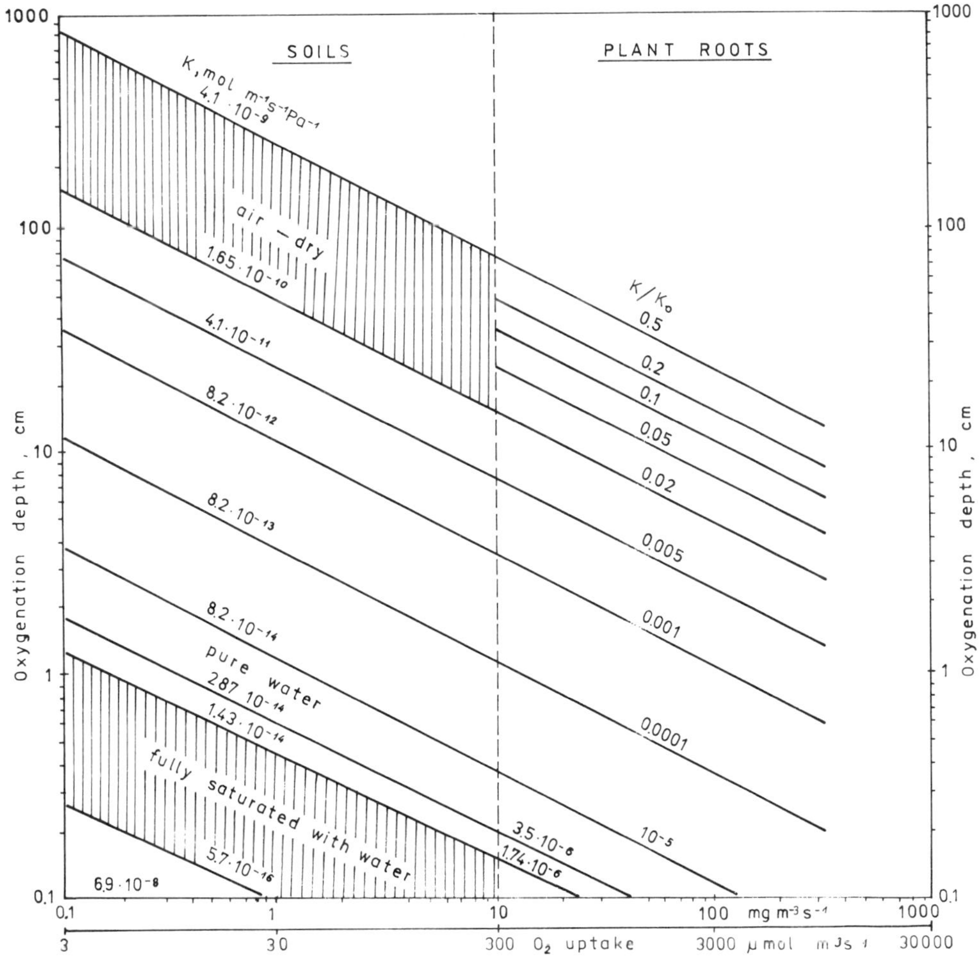


FIGURE 12. Soil oxygenation depth under steady-state conditions vs. respiration rate q, at different values of oxygen diffusion constant K calculated from Equation (65) (q and D do not vary with the depth and with oxygen concentration). The diagram can be related to oxygenation of the root via the internal diffusion, assuming q and D constant along the root and a lack of radial oxygen exchange between the root and the environment. *Note:* for the diffusion in the gas phase $K/K_o = D/D_o$.

The relationship of the depth of the soil anoxic zone, $L_{an}$, to the oxygen consumption q, at different values of the diffusion coefficient D, calculated on the basis of Equation (65), is presented in Figure 12. This figure includes the range of q values normally found in soils (0.1 to 10 mg $m^{-3}s^{-1}$), and also higher values (up to 300 mg $m^{-3}s^{-1}$) that may occur in the roots of living plants and in litters and soils after the addition of organic substrates that easily decompose (cf. Chapter 1). It also includes the full range of values of the oxygen diffusion constant K that may occur in the soil and we see that K values from $4.1 \cdot 10^{-9}$ to $1.65 \cdot 10^{-10}$ mol $m^{-1}s^{-1}Pa^{-1}$, equivalent values of $K/K_o$ (and therefore also $D/D_o$) being from 0.5 to 0.02, can be found in air dry soils, the lower values occurring in more compacted soils and the higher values occurring in looser soil. The depth of oxygenation in this case is from 0.08 to 0.8 m at the highest values of q normally occurring in soils, and from 0.8 to 8 m at low q values.

Values of K approximately 300,000 times lower, i.e., from $1.43 \cdot 10^{-14}$ to $5.7 \cdot 10^{-16}$ mol $m^{-1}s^{-1}Pa^{-1}$ characterize soils fully saturated with water. The depth of the oxic zone in this case is only down to 1.5 mm at the higher q values and down to 15 mm at the lower q

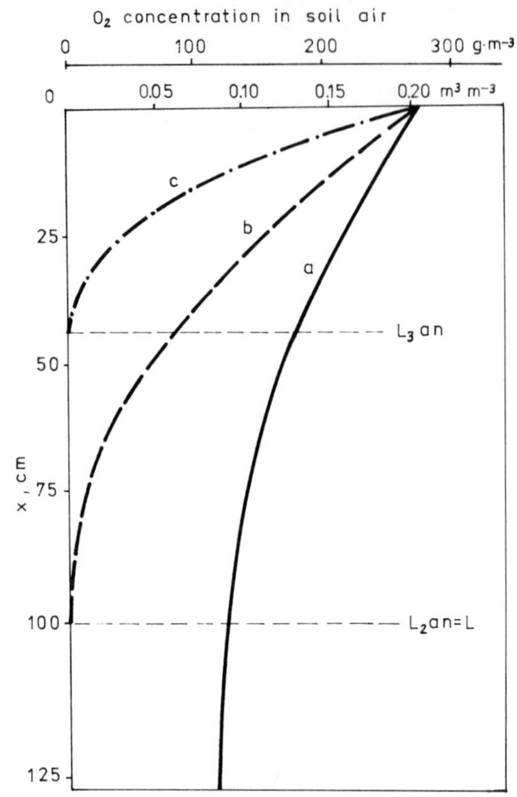


FIGURE 13. Oxygen distribution in a homogenous soil profile at $q = 0.594$ mg $m^{-3}s^{-1}$ and differentiated values of oxygen diffusion coefficient D. (a) $D = 2 \cdot 10^{-6}$ $m^2s^{-1}$ $> D'$, (b) $D = 10^{-6}$ $m^2s^{-1} = D'$, (c) $D = 2 \cdot 10^{-6}$ $m^2s^{-1}$ $< D'$. (From Stępniewski, W., Influence of Mineral Fertilization and Watering of Arable Soil on its Oxygenation, Ph.D. dissertation, Agricultural University, Lublin, Poland, 1975.)

values found in the soil (the soil, obviously, being in contact with atmospheric air). Morphological properties of the soil profile and experimental data[277] confirm these calculations.

As an example, the distribution of oxygen concentration in a homogenous soil profile, at various values of diffusion coefficient, calculated from Equation (58), can be found in Figure 13. This figure presents three cases:

1. Curve "a" illustrates a situation in which the soil diffusion coefficient is higher than the critical value, i.e., the whole profile is oxygenated and the oxygen concentration at the bottom of the biologically active zone is above zero.
2. Curve "b" presents a situation in which the oxygen concentration decreases to zero at the bottom of the biologically active zone.
3. Curve "c" shows a situation in which the depth of the oxic zone is less than the extent of the biologically active zone.

Case 2: D is constant in time and depth, q is constant in time but decreases with increasing depth.

In a monolayer profile, it is also possible to describe the distribution of oxygen concentration for a case in which the respiratory activity q(x) is constant in time, but changes with depth according to the equation:

$$q(x) = q_o \left[1 - \left(\frac{x}{L}\right)^k\right] \tag{66}$$

where $q_o$ = respiratory activity on the soil surface and k = constant.

Following a line of reasoning analogous to that for the multilayer profile, i.e., assuming that at depth x the supply of oxygen f(x) = −D dC/dx is equal to its consumption Q(x) within the deeper soil layers (with the assumption of oxygenation of the whole profile down to a depth L), that is

$$Q(x) = q_o \int_x^L \left[1 - \left(\frac{z}{L}\right)^k\right] dz \tag{67}$$

we get:

$$C(x) = C_o - \frac{q_o}{D(k + 1)L^k}\left(kL^{k+1}x + \frac{x^{k+2}}{k + 2}\right) - \frac{qx^2}{2D} \tag{68}$$

The above general equation, if k = 1, gives us the solution obtained by Romell,[610] and if k = 0.25 the equation gives us the solution reported by Wesseling[784] for the diffusion of carbon dioxide.

Case 3: D and q vary with depth but are constant in time.

When in a soil profile both the respiratory activity q(x) and the diffusion coefficient D(x) are variable, additional assumptions are necessary and we shall, after Wesseling,[784] adopt the following:

1. Oxygen consumption varies according to formula (66) with k = 0.25, which gives a total consumption for a unit of soil surface area B = 0.2 $q_oL$
2. Gas porosity in the soil profile decreases gradually according to the equation:

$$E_g(x) = E_{go}(1 - ax^k) \tag{69}$$

   where $E_{go}$ = air-filled porosity at the soil surface, a and k = constants; Wesseling[784] considered cases for k = 1 (linear decrease of air-filled porosity with depth) and for k = 2 (parabolic decrease in porosity)
3. For x = L:

$$E_g(x) = E_{go}(L) = 0.1 \tag{70}$$

   that is

$$E_{go}(1 - aL^k) = 0.1 \tag{71}$$

4. $$D(x) = D_o[E_g(x) - 0.1] \tag{72}$$

5. $$C = C_o \quad \text{for } x = 0$$

   and

$$\frac{dC}{dx} = 0 \quad \text{for } x = L \tag{73}$$

**Table 6**
**VALUES OF $J_1$ AND $J_2$ FROM EQUATIONS (53) AND (54)**

| | x/L | | | | | |
|---|---|---|---|---|---|---|
| | **0.0** | **0.2** | **0.4** | **0.6** | **0.8** | **1.0** |
| $J_1$ | 0 | 0.163 | 0.274 | 0.347 | 0.388 | 0.401 |
| $J_2$ | 0 | 0.149 | 0.236 | 0.284 | 0.309 | 0.316 |

Data from Wesseling, J., *Neth. J. Agric. Sci.*, 10, 109, 1962.

With the above assumptions, oxygen concentration in the soil profile is described by the equation:

$$C(x) = C_o - \frac{q_o L J_1}{4.5 D_o E_{go} a} \tag{74}$$

for k = 1,
and by the equation:

$$C(x) = C_o \frac{q_o L J_2}{4.5 D_o E_{go} a} \tag{75}$$

for k = 2, where values of $J_1$ and $J_2$ have been calculated and tabulated by Wesseling (Table 6).

### *b. Transient State*

#### (1) Two-Layer Profile with Respiring Top Layer

For the transient state, the most general solution of the diffusion Equation (32) has been found for a two-layer profile in which a biologically active surface layer $O \leq x \leq H$ having an apparent respiratory activity $q^*(t)$ and a constant diffusivity $D_1^*$ is underlayed by a biologically inactive layer $H \leq x \leq L$ of a diffusivity $D_2^*$, under which there is an impermeable layer at a depth $x \geq L$. Six cases can be distinguished here.

Case 1: $D_1^* \neq D_2^*$; $q^* = a + b \exp(k+)$.

The values of diffusivity are different in the two layers, and $q^*$ in the biologically active layer varies in time (but not with depth), according to the equation: $q^* = a + b \exp(kt)$ where a, b, k = parameters characterizing the soil.

Oxygen diffusion in this case is described by two equations:

$$\frac{\partial C_1}{\partial t} = D_1^* \frac{\partial^2 C_1}{\partial x^2} - [a + b \exp(kt)] \tag{76}$$

for the top layer, biologically active, and

$$\frac{\partial C_2}{\partial t} = D_2^* \frac{\partial^2 C_2}{\partial x^2} \tag{77}$$

for the underlying layer, biologically inactive. For x = H

$$C_1(H,t) = C_2(H,t) \tag{78}$$

and

$$D_1^* \left(\frac{\partial C_1}{\partial x}\right) = D_2^* \left(\frac{\partial C_2}{\partial x}\right) \tag{79}$$

For the following initial and boundary conditions:

$$\begin{array}{llll} C_1 = C_o & 0 < x \leq H & t = 0 \\ C_2 = C_o & H < x \leq L & t = 0 \\ C_1 = C_o & x = 0 & t > 0 \\ \dfrac{\partial C_2}{\partial x} = 0 & x = L & t > 0 \end{array} \tag{80}$$

the solution of Equations (76) and (77).[379] obtained by means of Laplace transforms, has the form:

$$C_1(x,t) =$$

$$C_o - a\left\{\frac{x(2H-x)}{2D_1^*} + \frac{2\sqrt{D_2^*}}{L}\sum_{n=1}^{\infty}\frac{\exp(-\omega_n t)}{\omega_n^{3/2}}\frac{[1-\cos(\beta_{1n}H)]\sin(\beta_{1n}x)}{\left[\delta-\left(\delta-\frac{1}{\delta}\right)\left(1-\frac{H}{L}\right)\cos^2\left(\beta_{1n}H\right)\right]}\right\} +$$
$$\frac{b\exp(kt)}{|k|}\left\{1-\frac{\cos[\lambda_1(H-x)]\cos[\lambda_2(L-H)] - \delta\{\sin[\lambda_1(H-x)] + \sin(\lambda_1 x)\}\sin[\lambda_2(L-H)]}{U(|k|)}\right\} -$$
$$\frac{2b\sqrt{D_2^*}}{L}\sum_{n=1}^{\infty}\frac{\exp(-\omega_n t)[1 - \cos(\beta_{1n}H)]\sin(\beta_{1n}x)}{\omega_n^{1/2}(\omega_n+k)\left[\delta-\left(\delta-\frac{1}{\delta}\right)\left(1-\frac{H}{L}\right)\cos^2\left(\beta_{1n}H\right)\right]} \tag{81}$$

and

$$C_2(x,t) = C_o - a\left\{\frac{H^2}{2D_1^*} + \right.$$
$$\left.\frac{2\sqrt{D_2^*}}{L}\sum_{n=1}^{\infty}\frac{\exp(-\omega_n t)}{\omega_n^{3/2}}\frac{[1-\cos(\beta_{1n}H)]\sin(\beta_{1n}H)[\cos\beta_{2n}(L-x)]}{\left[\delta-\left(\delta-\frac{1}{\delta}\right)\left(1-\frac{H}{L}\right)\cos^2(\beta_{1n}H)\right]\cos[\beta_{2n}(L-H)]}\right\} -$$
$$\frac{b\exp(kt)}{|k|}\frac{[1-\cos(\lambda_1 H)]\cos[\lambda_2(L-x)]}{U(|k|)} -$$
$$\frac{2b\sqrt{D_2^*}}{L}\sum_{n=1}^{\infty}\frac{\exp(-\omega_n t)[1-\cos(\beta_{1n}H)]\sin(\beta_{1n}H)\cos[\beta_{2n}(L-x)]}{\omega_n^{1/2}(\omega_n + k)\left[\delta-\left(\delta-\frac{1}{\delta}\right)\left(1-\frac{H}{L}\right)\cos^2(\beta_{1n}H)\right]\cos[\beta_{2n}(L-H)]} \tag{82}$$

where:

$$\begin{aligned} \lambda_1 &= \sqrt{|k|\ D_1^*} \\ \lambda_2 &= \sqrt{|k|\ D_2^*} \\ \beta_{1n} &= \sqrt{\omega_n/D_1^*} \\ \beta_{2n} &= \sqrt{\omega_n/D_2^*} \\ \delta &= \sqrt{D_2^*/D_1^*} \end{aligned} \tag{83}$$

and moreover:

$$U(\omega) = \cos\left(H\sqrt{\frac{\omega}{D_1^*}}\right)\cos\left[\sqrt{\frac{\omega}{D_2^*}}(L-H)\right] - \delta\sin\left(H\sqrt{\frac{\omega}{D_1^*}}\right)\sin\left[(L-H)\sqrt{\frac{\omega}{D_2^*}}\right] \tag{84}$$

$\omega = \omega_n$ — value of $\omega$ where the function $U(\omega)$ reaches its nth positive zero, which can be determined by means of standard procedure.

Case 2: $D_1^* \neq D_2^*$; $q^* = b\exp(kt)$

Two-layer profile having differing diffusion coefficients in the two layers and an apparent respiratory activity variable in time in the surface layer. Initial and boundary conditions as in Equation (80).

In this case a = 0 and Equations (81) and (82) are simplified[379] to the form:

$$C_1(x,t) = C_o + \frac{b\exp(kt)}{|k|}\left\{1 - \frac{\cos[\lambda_1(H - x)]\cos[\lambda_2(L-H)] - \delta\{\sin[\lambda_1(H-x)] + \sin(\lambda_1 x)\}\sin[\lambda_2(L-H)]}{U(|k|)}\right\} - \frac{2b\sqrt{D_2^*}}{L}\sum_{n=1}^{\infty}\frac{\exp(-\omega_n t)[1-\cos(\beta_{1n}H)]\sin(\beta_{1n}x)}{\omega_n^{1/2}(\omega_n + k)\left[\delta - \left(\delta - \frac{1}{\delta}\right)\left(1 - \frac{H}{L}\right)\cos^2(\beta_{1n}H)\right]} \tag{85}$$

and

$$C_2(x,t) = C_o - \frac{b\exp(kt)}{|k|}\,\frac{[1-\cos(\lambda_1 H)]\cos[\lambda_2(L-x)]}{U(|k|)} - \frac{2b\sqrt{D_2^*}}{L}\sum_{n=1}^{\infty}\frac{\exp(-\omega_n t)[1-\cos(\beta_{1n}H)]\sin(\beta_{1n}H)\cos[\beta_{2n}(L-x)]}{\omega_n^{1/2}(\omega_n + k)\left[\delta - \left(\delta - \frac{1}{\delta}\right)\left(1 - \frac{H}{L}\right)\cos^2(\beta_{1n}H)\right]\cos[\beta_{2n}(L-H)]} \tag{86}$$

Case 3: $D_1^* \neq D_2^*$, $q^* = a$, $b = 0$

Respiratory activity term $q^*$ in the surface layers of the profile does not vary in time. Initial and boundary conditions are as in Equation (80). Equations (81) and (82) are simplified[379] to the form:

$$C_1(x,t) = C_o - a\left\{\frac{x(2H - x)}{2D_1^*} + \frac{2\sqrt{D_2^*}}{L}\sum_{n=1}^{\infty}\frac{\exp(-\omega_n t)[1-\cos(\beta_{1n}H)]\sin(\beta_{1n}x)}{\omega_n^{3/2}\left[\delta - \left(\delta - \frac{1}{\delta}\right)\left(1 - \frac{H}{L}\right)\cos^2(\beta_{1n}H)\right]}\right\} \tag{87}$$

and

$$C_2(x,t) = C_o - a\left\{\frac{H^2}{2D_1^*} + \frac{2\sqrt{D_2^*}}{L}\sum_{n=1}^{\infty}\frac{\exp(-\omega_n t)[1-\cos(\beta_{1n}H)]\sin(\beta_{1n}H)\cos[\beta_{2n}(L-x)]}{\omega_n^{3/2}\left[\delta - \left(\delta - \frac{1}{\delta}\right)\left(1 - \frac{H}{L}\right)\cos^2(\beta_{1n}H)\right]\cos[_{2n}(L-H)]}\right\} \tag{88}$$

Case 4. $D_1^* = D_2^* = D^*$; $q^* = a + b\exp(kt)$

Diffusivity in both layers of the soil profile is the same, and the apparent respiratory activity in the surface layer varies in time (but not with depth). Equations (81) and (82), with the initial and boundary conditions as in Equation (80) and allowing for the fact that:

$$\lambda_1 = \lambda_2 = \lambda = \sqrt{\frac{|k|}{D^*}}$$

$$\beta_{1n} = \beta_{2n} = \beta_n = \frac{(2n-1)\pi}{2L} = \sqrt{\frac{\omega_n}{D^*}} \tag{89}$$

$$\delta = \sqrt{\frac{D_2^*}{D_1^*}} = 1$$

assume the following forms:[379]

$$C_1(x,t) = C_o - \frac{ax(2H-x)}{2D^*} + \frac{b\exp(kt)}{|k|}\left\{1 - \frac{\cos[\lambda(L-x)] - \sin[\lambda(L-H)]\sin(\lambda x)}{\cos(\lambda L)}\right\} - \frac{4}{\pi}\sum_{n=1}^{\infty}\frac{[1-\cos(\beta_n H)]\sin(\beta_n x)}{2n-1}\exp(-\beta_n^2 D^* t)\left[\frac{a}{\beta_n^2 D^*} + \frac{b}{\beta_n^2 D^* + k}\right] \tag{90}$$

and

$$C_2(x,t) = C_o - \frac{aH^2}{2D^*} - \frac{b\exp(kt)}{|k|}\frac{[1 - \cos(\lambda H)\cos[\lambda(L-x)]}{\cos(\lambda L)} - \frac{4}{\pi}\sum_{n=1}^{\infty}\frac{[1-\cos(\beta_n H)]\sin(\beta_n x)\cos[\beta_n(L-x)]}{(2n-1)\cos[\beta_n(L-H)]}\exp(-\beta_n^2 D^* t)\left[\frac{a}{\beta_n^2 D^*} + \frac{b}{\beta_n^2 D^* + k}\right] \tag{91}$$

Case 5: $D_1^* = D_2^* = D^*$; $q^* = b\exp(kt)$, i.e., $a = 0$
Diffusivity in both layers is the same and does not vary in time, but the apparent respiratory activity in the surface layer varies in time. Equations (90) and (91), with the initial and boundary conditions as in Equation (80), are simplified to:

$$C_1(x,t) = C_o + \frac{b\exp(kt)}{|k|}\left[1 - \frac{\cos[\lambda(L-x)] - \sin[\lambda(L-H)]\sin(\lambda x)}{\cos(\lambda L)}\right] - \frac{4b}{\pi}\sum_{n=1}^{\infty}\frac{[1 - \cos(\beta_n H)]\sin(\beta_n x)\exp(-\beta_n^2 D^* t)}{(2n-1)(\beta_n^2 D^* + k)} \tag{92}$$

and

$$C_2(x,t) = C_o - \frac{b\exp(kt)}{|k|}\frac{[1-\cos(\lambda H)\cos[\lambda(L-x)]}{\cos(\lambda L)} - \frac{4b}{\pi}\sum_{n=1}^{\infty}\frac{[1-\cos(\beta_n H)]\sin(\beta_n x)\cos[\beta_n(L-x)]\exp(-\beta_n^2 D^* t)}{(2n-1)(\beta_n^2 D^* + k)\cos[\beta_n(L-H)]} \tag{93}$$

where $\beta_n$ and $\lambda$ are determined by Equation (89).

Case 6: Identical levels of diffusivity in both the layers, apparent respiratory activity of the top layer constant in time and the initial and boundary conditions as in Equation (80). Equations (90) and (91) are simplified to:

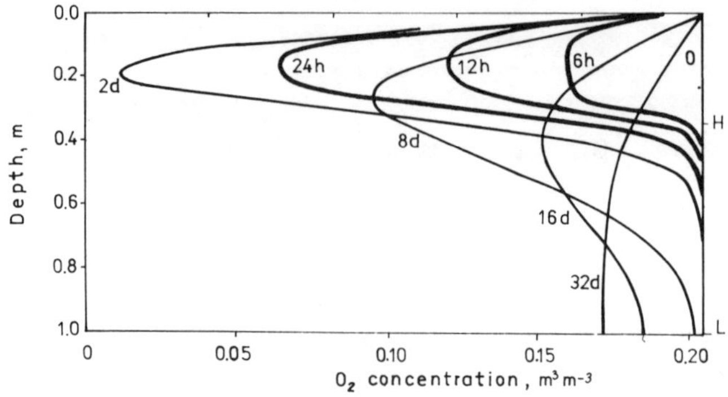


FIGURE 14. Oxygen distribution in a soil profile of a depth L = 1 m, with the thickness of the biological active surface layer H = 0.3 m, diffusivity $D^* = 6 \cdot 10^{-7}$ $m^{2-1}$ within the whole profile, apparent respiratory activity $q^* = b\exp(kt)$, where $k = -0.5$ $day^{-1}$ and $b = 0.07$ kg $m^{-3}$ $day^{-1}$, at different time (d [days], h [hours]) from the beginning of the experiment (pig slurry application). (Modified from Kowalik, P., Barnes, C. J., and Smiles, D. E., *Soil Sci. Soc. Am. J.*, 43, 255, 1979.)

$$C_1(x,t) = C_o - \frac{ax(2H - x)}{2D^*} - \frac{4a}{\pi D^*}\sum_{n=1}^{\infty}\frac{[1 - \cos(\beta_n H)]\sin(\beta_n x)\exp(-\beta_n^2 D^* t)}{\beta_n^2(2n - 1)} \tag{94}$$

and

$$C_2(x,t) = C_o - \frac{aH^2}{2D^*} - \frac{4a}{\pi D^*}\sum_{n=1}^{\infty}\frac{[1-\cos(\beta_n H)]\sin(\beta_n x)\cos[\beta_n(L-x)]\exp(-\beta_n^2 D^* t)}{\beta_n^2(2n - 1)\cos[\beta_n(L - H)]} \tag{95}$$

where $\beta_n$ and $\lambda$ are determined by Equation (89).

None of the six solutions of the diffusion equations for transient state conditions presented above have been verified by comparison with experimental data. For Case 5 only results of the calculations have been published[379] and these can be found in Figure 14.

**(2) Monolayer Profile**

In the case of a monolayer profile of depth L, overlying an impervious layer (ground water, solid rock, or the bottom of a soil column) in which the apparent respiratory activity $q^*$ does not vary with depth but may change with time, diffusion of oxygen will be described by Equation (32).

Depending on the changes in $q^*$, three possible cases may be distinguished.

<u>Case 1:</u> $q^* = a + b\exp(kt)$

In this case the oxygen concentration distribution is described by the equation:

$$\frac{\partial C}{\partial t} = D^* \frac{\partial^2 C}{\partial x^2} - [a + b\exp(kt)] \tag{96}$$

where a, b, k = constants and t = time.

The solution of this equation, at initial conditions:

$$C(x,0) = C_o$$
$$C(0,t) = C_o \tag{97}$$
$$\frac{\partial C(L,t)}{\partial x} = 0$$

where $C_o$ is the oxygen concentration in the atmosphere, can be found from Equation (90), substituting H = L and C(x,t) for C(x,t). We then get:[379]

$$C(x,t) = C_o - \frac{ax(2L-x)}{2D^*} + \frac{b\exp(kt)}{|k|}\left\{1 - \frac{\cos[\lambda(L - x]}{\cos(\lambda L)}\right\} - \frac{2\sqrt{D^*}}{L}\sum_{n=1}^{\infty}\frac{\exp(-\omega_n t)}{\sqrt{\omega_n}}\left[\frac{a}{\omega_n} + \frac{b}{\omega_n + k}\right]\frac{\cos[\beta_n(L - x)]}{\sin(\beta_n L)} \tag{98}$$

where $\beta_n$ and $\lambda$ are defined in Equation (89), and $\omega_n$ is the n-th zero of the function:

$$U(\omega) = \cos L\sqrt{\frac{\omega}{D^*}} \tag{99}$$

that is

$$\omega_n = \frac{(2n - 1)^2\,\pi^2\,D^*}{4L^2} = \beta_n^2\,D^* \tag{100}$$

Applying Fourier's expansions:

$$x(2L - x) = \frac{32\,L^2}{\pi^3}\sum_{n=1}^{\infty}\frac{\sin\left[\frac{(2n-1)\pi x}{2L}\right]}{(2n - 1)^3} \tag{101}$$

and

$$1 - \frac{\cos[\lambda(L-x)]}{\cos(\lambda L)} = -\frac{4\lambda^2}{\pi}\sum_{n=1}^{\infty}\frac{\sin\left[\frac{(2n-1)\pi x}{2L}\right]}{(2n-1)\left[\frac{(2n-1)^2\pi^2}{4L^2} - \lambda^2\right]} \tag{102}$$

we get[379] the equation:

$$C(x,t) = C_o - \frac{16\,L^2}{D^*\pi^3}\sum_{n=1}^{\infty}\frac{\sin(\beta_n x)}{(2n-1)^3}\left\{a\,[1 - \exp(-\beta_n^2 D^* t)] + \frac{b\beta_n^2 D^*}{\beta_n^2 D^* + k}[\exp(kt) - \exp(-\beta_n^2 D^* t]\right\} \tag{103}$$

which is identical with that obtained by Papendick and Runkles.[533]

Results calculated by means of this equation are in satisfactory agreement with experimental data.[379,533]

An example of oxygen concentration distribution in time at various depths in the soil profile, calculated on the basis of Equation (103), can be found in Figure 15.

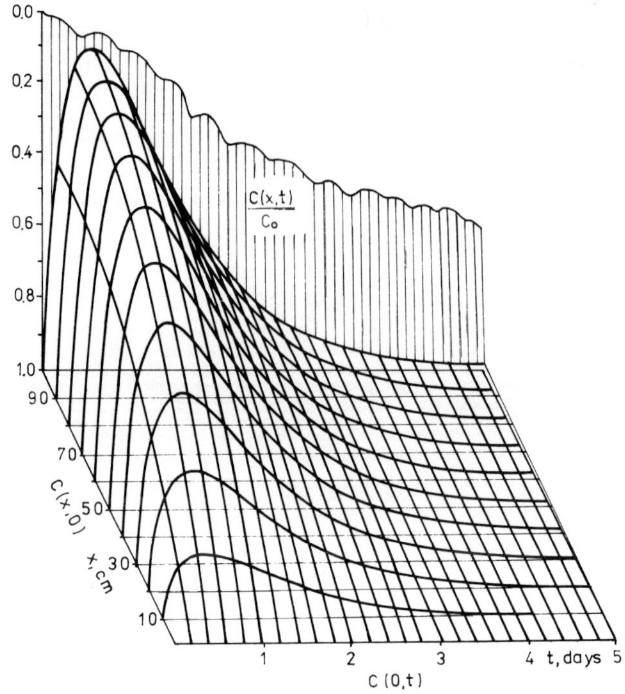


FIGURE 15. Distribution of oxygen concentration in a uniform soil profile of a depth L = 1 m with biological activity in entire depth $\alpha$ = b exp(kt), calculated from Equation (103) at the following values of empirical parameters: a = 0.0392 mg $m^{-3}s^{-1}$, b = 12 mg $m^{-3}s^{-1}$, k = $-1.37 \cdot 10^{-5} s^{-1}$ and D* = $10^{-5}$ $m^2s^{-1}$. (Modified from Kowalik, P. and Obarska-Pempkowiak, H., in Proc. 2nd Int. Conf. Phys. Prop. Agric. Mater., Gödölö, Hungary, Agricultural University, Gödölö, Hungary, 103, 1980.)

Case 2: $q^* = b \exp(kt)$

The solution of Equation (96) for this case, with initial and boundary conditions as in case 1, is obtained by putting a = 0 in Equation (103).

$$C(x,t) = C_o - \frac{16bL^2}{\pi^3} \sum_{n=1}^{\infty} \frac{\sin(\beta_n x)}{(2n-1)^3} \frac{\beta_n^2}{\beta_n^2 D^* + k} [\exp(kt) - \exp(-\beta_n^2 D^* t)] \quad (104)$$

Case 3: $q^* = a$

For this case, a solution of Equation (96)[476] has been obtained for variable initial conditions, i.e., $C(x,0) = C_1(x)$ and with the following boundary conditions:

$$C(0,t) = C_o \quad (105)$$

$$\left(\frac{\partial C}{\partial x}\right)_{L,t} = 0$$

The solution has the form:

$$C(x,t) = C_o - \frac{ax(2L - x)}{2D^*} -$$

$$[C_o - C_1(x)] \frac{4}{\pi} \sum_{n=1}^{\infty} \left[\frac{1}{2n-1} \sin(\beta_n x) \exp(-\beta_n^2 D^* t)\right] + \quad (106)$$

$$\frac{16a\,L^2}{\pi^3\,D^*} \sum_{n=1}^{\infty} \left[\frac{1}{(2n-1)^3} \sin(\beta_n x) \exp(-\beta_n^2 D^* t)\right]$$

This equation at $t \rightarrow \infty$, i.e., under steady-state conditions, is simplified to Equation (58), due to the relationship $q^*/D^* = q/D$.

When oxygen concentration in the whole soil profile is $C_o$, which means $C(x,0) = C_1(x) = C_o$, Equation (106) reduces to:

$$C(x,t) = C_o \frac{ax(2L-x)}{2D^*} + \frac{16a\,L^2}{\pi^3\,D^*} \sum_{n=1}^{\infty} \frac{\sin(\beta_n x)\exp(-\beta_n^2 D^* t)}{(2n-1)^3} \qquad (107)$$

which can also be obtained from Equation (103) by substituting b = 0, and combining with the expansion (101). The Equation (107) is identical to the solution obtained by Papendick and Runkles.[532]

Another possible case, suggested among others by Radford and Greenwood,[577] is that in which the initial concentration of oxygen is zero throughout the soil profile, i.e., $C(x,0) = C_1(x) = 0$. Then Equation (106) becomes:

$$C(x,t) = C_o - \frac{ax(2L-x)}{2D^*} - C_o \frac{4}{\pi} \sum_{n=1}^{\infty} \left[\frac{1}{2n-1} \sin(\beta_n x)\exp(-\beta_n^2 D^* t)\right] + \frac{16a\,L^2}{\pi^3\,D^*} \sum_{n=1}^{\infty} \left[\frac{1}{(2n-1)^3} \sin(\beta_n x)\exp(-\beta_n^2 D^* t)\right] \qquad (108)$$

This equation, like Equation (106), permits calculation of the apparent coefficient of gas diffusion $D^*$, at a known $q^* = a$, from measurements of the oxygen concentrations at known intervals of time.

In the simplest situation, when $q^* = a = 0$, with the initial conditions $C(x,0) = C_1(x) = 0$, this equation reduces to:

$$C(x,t) = C_o - \frac{4C_o}{\pi} \sum_{n=1}^{\infty} \left[\frac{\sin(\beta_n x)\exp(-\beta_n^2 D^* t)}{2n-1}\right] \qquad (109)$$

which is identical to the equation developed by Papendick and Runkles.[532]

### C. Diffusion Within Soil Aggregates

Diffusion coefficients for gases in soil aggregates have not been measured directly, but we may expect that they are similar to those in a nonaggregated (structureless) soil. Indirect measurements have shown that in dry aggregates of a diameter of 1 to 2 mm the intra-aggregate porosity is from 0.19 to 0.41 $m^3m^{-3}$, and the values of $D/D_o$ vary from 0.015 to 0.156.[143,146] The effectiveness coefficient of intra-aggregate pores (see Equation [42]) is from 0.06 to 0.4.

Assuming, as before, that the relative coefficient of oxygen diffusion in aggregates saturated with water $D_c/D_w$ (where $D_c$ = diffusion coefficient in the aggregate, $D_w$ = oxygen diffusion coefficient in pure water; see Table 3) in relation to water is the same as the relative coefficient of diffusion in relation to air in dry aggregates, we obtain values of the oxygen diffusion coefficient in aggregates saturated with water, at a temperature of 20°C, in the range: from $3 \cdot 10^{-11}$ to $3.3 \cdot 10^{-10}$ $m^2s^{-1}$, and from $2.6 \cdot 10^{-11}$ to $2.8 \cdot 10^{-10}$ $m^2s^{-1}$ for carbon dioxide. Equivalent values of the diffusion constant K are from $4.3 \cdot 10^{-16}$ to $4.5 \cdot 10^{-15}$ mol $m^{-1}Pa^{-1}s^{-1}$ for oxygen, and from $1.03 \cdot 10^{-14}$ to $1.07 \cdot 10^{-13}$ mol $m^{-1}Pa^{-1}s^{-1}$ for carbon dioxide.

The first to consider the problems of the oxygenation of aggregates were Currie[142] and Greenwood[273] in 1961.

The equation of diffusion (33) in the case of a dry crumb spherical in shape, assuming it is isotropic, i.e., the diffusion coefficient within the aggregate $D_c$ and its respiratory activity q are uniform in all directions within the internal space of the aggregate, has the form:

$$E_g \frac{\partial C}{\partial t} = \frac{1}{r^2} \frac{\partial}{\partial r} \left( D_c r^2 \frac{\partial C}{\partial r} \right) + q \tag{110}$$

where r = distance from the center of the aggregate and $E_g$, C, t, q, and $D_c$ = the air-filled porosity, oxygen concentration in the gaseous phase, time, respiratory activity of the aggregate, and diffusion coefficient in the aggregate, respectively.

The solution of this equation, under steady-state conditions and with q constant in time, has the form:

$$C_r = C_R - \frac{q}{6D_c} (R^2 - r^2) \tag{111}$$

where R = aggregate radius, $C_r$ = oxygen concentration at a distance r from the center of the aggregate, and $C_R$ = oxygen concentration on the surface of the aggregate.

This equation permits the calculation of a critical oxygen concentration on the surface of the aggregate $C'_R$ at which $C_r$ at the center of the aggregate drops to zero:

$$C'_R = \frac{qR^2}{6D_c} \tag{112}$$

and of a critical value of aggregate radius R′ for given values of $D_c$, $C_R$, and q:

$$R' = \sqrt{\frac{6C_R D_c}{q}} \tag{113}$$

Equations (112) and (113), after the substitution of concentrations $C'_R$ and $C_R$ with corresponding partial pressures of oxygen $P'_R$ and $P_R$, and of diffusion coefficient $D_c$ with the diffusion constant in the aggregate $K_c$ (cf. Equations [12] and [14]), assume the form:

$$P'_R = \frac{qR^2}{6K_c} \tag{114}$$

and

$$R' = \sqrt{\frac{6P_R K_c}{q}} \tag{115}$$

The value R′ determines the radius of the biggest aggregate that can still be fully oxygenated under these conditions. A graphic interpretation of Equation (115) is presented in Figure 16.

In the case when the inner part of the aggregate of a radius $r_{an}$ is anoxic, i.e., $C_r = 0$ for $0 < r < r_{an}$, then the oxygen concentration distribution in the outer, oxygenated part of the aggregate, i.e., in the range $r_{an} < r < R$, is described[142,280] by the equation:

$$C_r = C_R - \frac{q}{6D_c} \left[ R^2 - r^2 - 2r_{an}^3 \left( \frac{1}{r} - \frac{1}{R} \right) \right] \tag{116}$$

Substituting $C_r = 0$ for $r = r_{an}$, we get:[149,279,670]

$$\frac{6C_R D_c}{q} = R^2 - 3r_{an}^2 + \frac{2r_{an}^3}{R} \tag{117}$$

Equations (111) and (116) have been verified experimentally by Greenwood and Goodman[280] through direct measurements of oxygen concentration in soil aggregates. The agreement obtained was satisfactory.

Recently, there have been attempts at further modelling of the oxygenation of aggregated system.[417,668,670] In the present text, however, we shall present only the development of a model for a system of aggregates of differentiated diameters with a lognormal distribution proposed by Smith.[668—670]

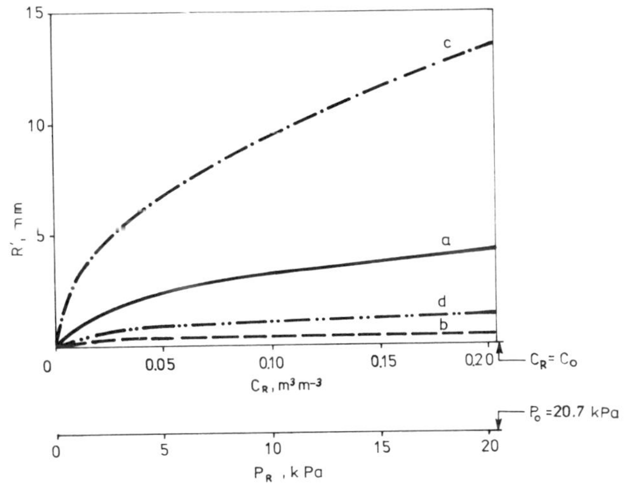


FIGURE 16. Critical aggregate radius R′ as a function of oxygen partial pressure $P_R$ or its concentration $C_R$ in the air outside the aggregate, calculated from Equation (115) for extreme values of oxygen diffusion constant $K_c$ likely to occur in aggregates fully saturated with water and extreme respiration rates q normally occurring in soil. (a) $K_{c1} = 4.32 \cdot 10^{-6}$ mol $m^{-1}s^{-1}Pa^{-1}$, $q_1 = 3 \cdot 10^{-6}$ mol $m^{-3}s^{-1}$, (b) $K_{c1} = 4.32 \cdot 10^{-6}$ mol $m^{-1}s^{-1}Pa^{-1}$, $q_2 = 3 \cdot 10^{-4}$ mol $m^{-3}s^{-1}$, (c) $K_{c2} = 4.5 \cdot 10^{-15}$ mol $m^{-1}s^{-1}Pa^{-1}$, $q_1 = 3 \cdot 10^{-6}$ mol $m^{-3}s^{-1}$, (d) $K_{c2} = 4.5 \cdot 10^{-15}$ mol $m^{-1}s^{-1}Pa^{-1}$, $q_2 = 3 \cdot 10^{-4}$ mol $m^{-3}s^{-1}$.

According to this author[670] the ratio of the radius of the anoxic zone $r_{an}$ in an aggregate to the aggregate radius R is expressed by the relationship:

$$\frac{r_{an}}{R} = \frac{-2\cos\varphi}{\cos\dfrac{\varphi}{3}} \tag{118}$$

and

$$\varphi = \frac{-1}{\sin\dfrac{R'}{R}} \tag{119}$$

where R′, as before, means the critical radius ensuring the oxygenation of the whole aggregate (see Equations [113] and [115]).

Smith[668,670] adopts the assumption of the lognormal distribution of aggregate radii, where a fraction of the total volume of all the aggregates $\delta v_R$ having a radius between $R - \delta_R/2$ and $R + \delta_R/2$ is described by the equation:

$$\delta v_R = \frac{1}{\delta_R\sqrt{2\pi}} \exp\left[-\frac{(\log R - \log \mu)^2}{2\sigma^2}\right] \delta_R \tag{120}$$

where $\mu$ = means aggregate radius and $\sigma$ = standard deviation of aggregate radii. Then the fraction of the volume of aggregate of a radius R in which anoxic conditions prevail equals the product $(r_{an}/R)^3\ \delta v_R$, and the fraction of the total volume of aggregates $\phi$, in which anoxic conditions exist, is expressed[676] by the formula:

$$\emptyset = \frac{1}{\sigma\sqrt{2\pi}} \int_{R'}^{\infty} \frac{r_{an}^3}{R^4} \exp\left[-\frac{(\log R - \log \mu)^2}{2\sigma^2}\right] dR \tag{121}$$

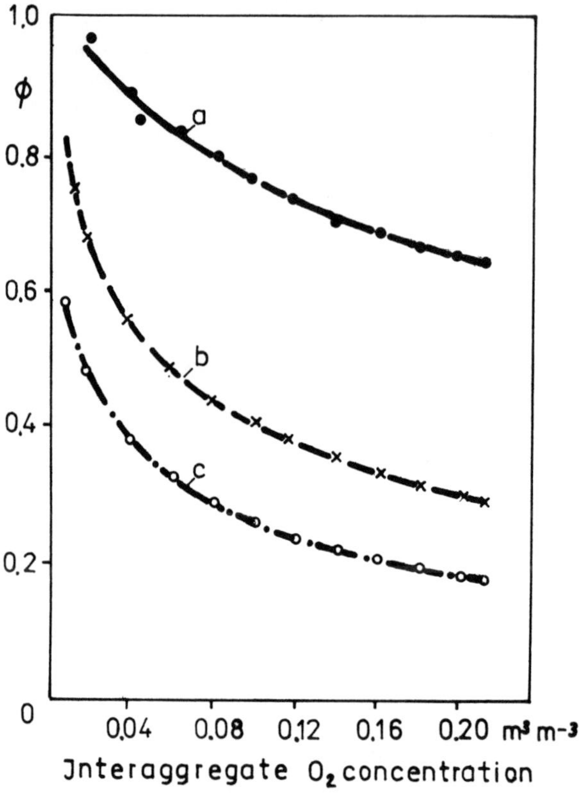


FIGURE 17. Fraction of the total aggregate volume which is anoxic (Ø) as a function of oxygen concentration outside the aggregate; $q = 2 \cdot 10^{-7}$ $m^3 m^{-3} s^{-1}$, $\sigma = 1$ throughout. (a) $D_c = 10^{-10}$ $m^2 s^{-1}$, $\log_{10}\mu = 0.7$; (b) $D_c = 10^{-9}$ $m^2 s^{-1}$, $\text{Log}_{10}\mu$ 0.7; (c) $D_c = 10^{-9}$ $m^2 s^{-1}$, $\log_{10}\mu = -0.3$ (μ in cm). (Data from Smith, K. A., *J. Soil Sci.*, 31, 263, 1980.)

This equation permits the numerical determination of the fraction of soil volume ø, which is anoxic, as a function of depth. This is presented in Figure 17. It can be seen from this figure that a tenfold change in D has a greater effect on the extent of anaerobisis (curves a and b) than an identical change in the size of aggregates (curves b and c).

Smith[670] developed this model further in relation to a soil profile composed of crumbs of a uniform lognormal size distribution and a constant respiratory activity q (per unit of crumb volume) throughout the whole depth of the profile, and a constant interaggregate porosity $E_i$, obtaining the formula:

$$x = \int_C^{0.21} \frac{1}{\sqrt{[2A(1 - Ø)\, dc]}}\, dC \qquad (122)$$

where:

$$A = \frac{q(1 - E_i)}{D_c} \qquad (123)$$

This equation permits a graphic or numerical presentation of the relationship of oxygen concentration distribution to the profile depth x, as well as the determination of the fraction of soil volume which is anoxic at a given depth of the profile. These relationships are presented in Figure 18.

This model has been applied to describe the diffusion of $N_2$, $N_2O$ and $NO_3^-$ in the denitrification process taking place in the anoxic areas of soil aggregates.[670]

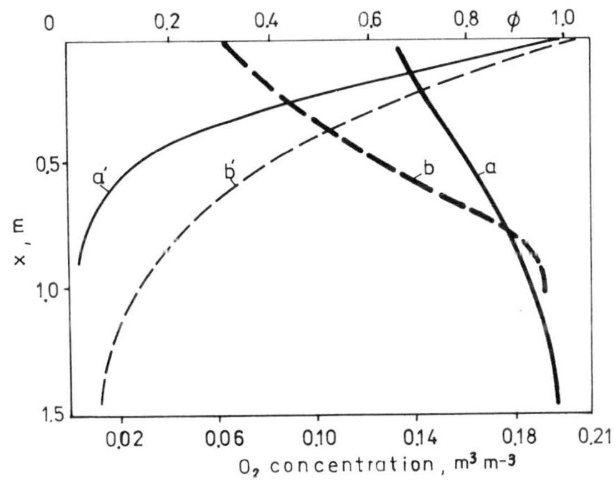


FIGURE 18. Fraction of the total aggregate volume (Ø) which is anoxic (lines a and b), interaggregate oxygen concentration (lines a′ and b′) as a function of depth in a profile of log normal aggregate size distribution; $q = 2 \cdot 10^{-7}\ m^3m^{-3}s^{-1}$, $\log_{10}\mu = 0.7$, $\sigma = 1$ throughout. (a, a′) $D_c = 10^{-9}\ m^2s^{-1}$, $D = 7.6 \cdot 10^{-8}\ m^2s^{-1}$; (b, b′) $D_c = 10^{-10}\ m^2s^{-1}$, $D = 7.6 \cdot 10^{-8}\ m^2s^{-1}$ ($\mu$ in cm). (Data from Smith, K. A., *J. Soil Sci.*, 31, 263, 1980.)

A different statistical approach to the problem of oxygenation in an aggregated soil may be found in the publications of Flühler et al.[209] and Stolzy et al.[707] These authors state that in many cases the variability of the soil properties investigated cannot be described by means of a continuous normal distribution, and in such sitations they suggest that the results obtained be classified as successes and failures in relations to a certain threshold value. As an example they quote measurements of the redox potential, where the value of 400 mV has been adopted as the boundary between oxic and anoxic metabolism. Then, instead of dealing with mean values, it is more convenient to determine the number of Eh values higher than 400 mV corresponding to oxic conditions and the number of values below that figure corresponding to anoxic conditions.

In such cases simplified statistics is applied, described by a discrete binomial distribution:

$$W(n) = \binom{W}{n} p^n (1-p)^{N-n} \tag{124}$$

where p = soil volume fraction in which anoxic conditions prevail, 1 − p = soil volume fraction with oxic conditions, N = total number of measurements, W(n) = probability that out of N measurements we get n values in the anoxic zone, n = number of values measured in the anoxic zone, and N − n = number of values measured in the oxic zone.

In this way the probability of revealing an anoxic area, in the case under discussion, increases with increase in the number of measurements taken, N, and the soil volume fraction with anoxic conditions, p.

This approach permits the use, instead of mean values, of the cumulative frequency of the result distribution for the property measured. This may be, for example, Eh, ODR, concentration of $N_2O$, $O_2$, etc.

### D. Microdiffusion

As early as 1926 Hutchins[329] expressed a conviction that is is not so much the concentration of oxygen in the soil as the possibility of its uptake by plant roots that determines the plant reaction to the oxygen conditions in the soil.

In 1952 Lemon and Erickson[419] designed an electrode simulating oxygen uptake by plant roots. The essence of this method is the measurement of the amount of oxygen diffusing to the surface of a platinum electrode where it is reduced electrochemically. The platinum wire is therefore a model of a root absorbing oxygen, and the intensity of oxygen flux to the electrode indicates the maximum amount of oxygen that would be available for a root placed in the same spot as the electrode. It is, therefore, an indirect measure of the potential availability of oxygen for plant roots.

The details of this method are discussed in Chapter 6, so for the present let us consider only the physical conditions under which the measurement is carried out and the analogy to the situation of the root. An important role in this concept, presented in numerous publications,[390,418—420,424,445,709] is played by the presence of water films on the root surface. These films, due to the low diffusion coefficient of gases in water (see Tables 3 and 4), form a significant obstacle in the path of oxygen from soil air to the root.

For the steady-state, in the absence of respiration within the water film, the equation of diffusion, in cylindrical coordinates, has the form:

$$\frac{\partial^2 C}{\partial r^2} + \frac{1}{r}\frac{\partial C}{\partial r} = 0 \tag{125}$$

After integrating this equation for r = R, in the case of an electrode (or root) or radius R surrounded by a water film of thickness d, for boundary conditions $C = C_R$ for $r = R$ and $C = C_{R+d}$ for $r = R + d$, we get an expression for the gradient of oxygen concentration at the surface of the electrode (or root):

$$\frac{\partial C}{\partial r} = \frac{C_{R+d} - C_R}{R \ln\left(1 + \frac{d}{R}\right)} \tag{126}$$

where $C_R$ = oxygen concentration on the surface of the electrode (root) and $C_{R+d}$ = oxygen concentration on the surface of the water film.

Considering that the access of oxygen ($f_r$) to a unit of surface of the electrode (root) is the product of the effective diffusion coefficient in the layer surrounding the root ($D_e$) and the gradient of oxygen concentration, we get:

$$f_r = \frac{D_e\,(C_{R+d} - C_R)}{R \ln\left(1 + \frac{d}{R}\right)} \tag{127}$$

where flow $f_r$ is what is traditionally known as oxygen diffusion rate (ODR) in relation to measurements taken with a Pt electrode.

In the case of a Pt electrode which reduces the oxygen concentration on its surface practically to zero, we can assume that $C_R = 0$, so that:

$$f_r = \frac{D_e\, C_{R+d}}{R \ln\left(1 + \frac{d}{R}\right)} \tag{128}$$

It should be pointed out that the value of $C_{R+d}$ is the oxygen concentration in the soil air multiplied by Bunsen's absorption coefficient (see Table 3). The assumption $C_R = 0$ is also valid, in some situations, for the plant root. As has already been mentioned in Chapter 1, roots can effectively absorb oxygen until its concentration at the root surface falls almost to zero. Such a fall will occur under conditions of restricted flow of oxygen to the root surface.

The effective coefficient of oxygen diffusion $D_e$ in a soil layer saturated with water has never been measured, but assuming that oxygen diffusion in the liquid phase follows a course similar to that in the gaseous phase, it can be assessed from the following equation:

$$D_e = D_w \gamma E_o^{\mu} \tag{129}$$

where $D_w$ = oxygen diffusion coefficient in pure water, $E_o$ = total porosity, and coefficients $\gamma$ and $\mu$ characterizing the soil are identical to those found in Equation (44) describing gas diffusion in the gaseous phase.

The range of $D_e$ values expected is from 0.02 to 0.5 of the values of $D_w$ (cf., Section II.B.1), and therefore, for oxygen, for which $D_w = 2.1 \cdot 10^{-9}$ $m^2s^{-1}$ (temperature 20°C, Table 3), $D_e$ may vary in the soil from $4.2 \cdot 10^{-11}$ to $10^{-9}$ $m^2s^{-1}$.

In practice, the value of ODR is determined on the basis of measurement of the diffusion current intensity, i.e., of the oxygen reduction current on the Pt electrode in a situation where the intensity of that current depends only on the rate of oxygen diffusion to the electrode surface from the surrounding medium. This relationship has the form:

$$ODR = \frac{M\,i}{n\,F\,A} \tag{130}$$

where M = molecular weight of $O_2$ (32 g $mol^{-1}$), i = current in $\mu A$, n = number of $O_2$ equiv. $mol^{-1}$ (=4), F = the Faraday 96,500 C equiv.$^{-1}$, and A = area of the electrode, $m^2$.

After the inclusion of constant values in Equation (130) we get the following formula for ODR:

$$ODR\ [\mu g\ m^{-2}\ s^{-1}] = 8.29 \cdot 10^{-5}\ \frac{i\ [\mu A]}{A\ [m^2]} \tag{131}$$

The ODR index reflects the oxygen availability most comprehensively and this is illustrated in Figure 19. As can be seen, the structural characteristics of the soil determine its total porosity $E_o$, and this combined with a moisture content determines the current air-filled porosity $E_g$ and thus the gas diffusion coefficient D. The latter, together with the respiratory activity q, determines the oxygen concentration distribution C(x) and the redox processes in the soil. The moisture content of the soil determines the effective thickness of the water film surrounding the root, and in combination with the structural characteristics of the soil, the value of the effective diffusion coefficient $D_e$ in the film. This in turn, together with oxygen concentration in the soil air, determines the value of ODR which affects the plants.

The ODR values in soils vary, most frequently, within a range from 0 to 200 $\mu g\ m^{-2}s^{-1}$. They decrease with an increase in soil moisture content,[223,420,469,470] and therefore increase with an increase in the moisture tension and air-filled porosity of the soil. They also decrease with an increase in the soil compaction.[697,700] An example of the relationship of ODR to the moisture tension and compaction of the soil is presented in Figure 20.

The relationship between the ODR and plant response is discussed in Chapter 5.

## E. Diffusion Within Plants

The final stage in the transport of oxygen from the soil to the respiring tissue of the plant root is the radial diffusion of oxygen from the root surface to its center. In turn, part of the carbon dioxide produced in the roots diffuses outwards, in a direction opposite to the radial movement of oxygen, and, part is transported upwards with the transpiration stream and also by longitudinal diffusion to the shoot where it can either be assimilated or released to the atmosphere.

From the point of view of plant adaptation to anoxic conditions, the most significant process is the longitudinal diffusion of oxygen from stems to roots. It is determined by the presence of intercellular spaces filled with air. In the case of mesophytes, the exchange of oxygen and carbon dioxide between the internal air system of the plant and the atmosphere takes place through stomata and lenticells.[25]

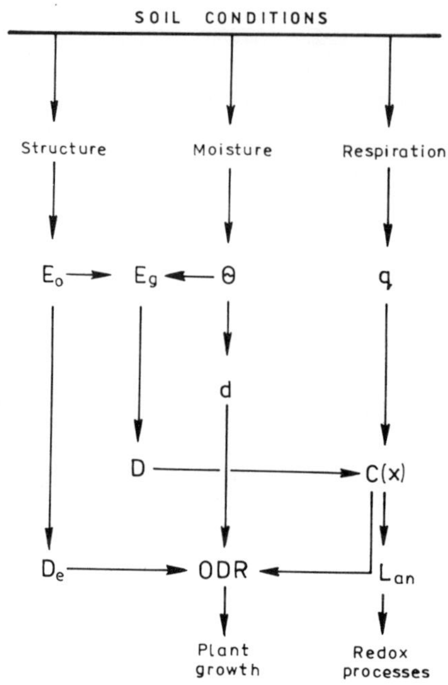


FIGURE 19. A scheme of interrelations between soil parameters influencing soil oxygen status and oxygen availability for plant roots; $E_o$ = total porosity, $E_g$ = air-filled porosity, q = soil respiration, D = coefficient of oxygen diffusion in soil profile (coefficient of macrodiffusion), C(x) = oxygen concentration in soil air, $L_{an}$ = oxygenation depth, d = effective thickness of water film around the root, $D_e$ = effective coefficient of oxygen diffusion within water film surrounding the root (coefficient of microdiffusion). (Modified from Kowalik, P., *Zesz. Nauk. Polit. Gd.*, 15, 3, 1971; Kowalik, P. and Stępniewski, W., *Zesz. Probl. Postepow Nauk. Roln.*, 220, 61, 1979.)

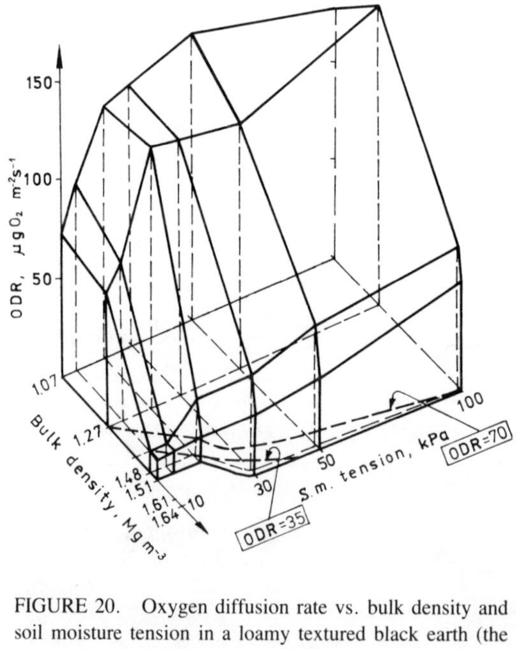


FIGURE 20. Oxygen diffusion rate vs. bulk density and soil moisture tension in a loamy textured black earth (the same soil as in Figures 4 to 6). (Modified from Stępniewski, W., *Pol. J. Soil Sci.*, 13, 3, 1980.)

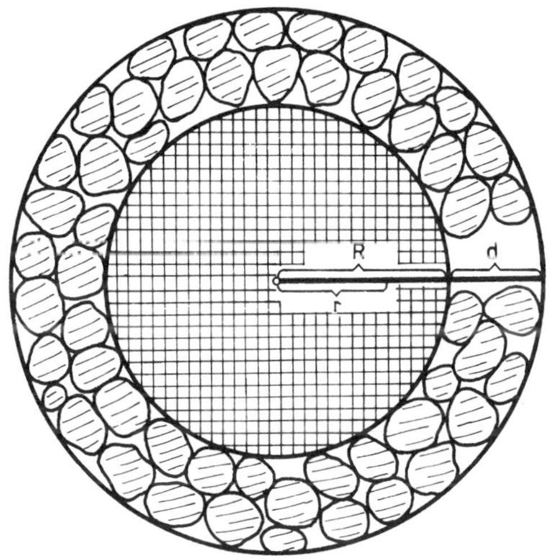


FIGURE 21. Schematic cross-section of a plant root of radius R, surrounded by soil saturated with water of effective thickness d; r = distance of a root element under consideration from the root axis.

A comprehensive review of problems related to oxygen diffusion in the plant has been provided by Armstrong.[25]

The oxidizing power of aquatic plant roots had already been known about a hundred years ago, but the first experimental confirmation of gas diffusion through the roots of mesophytes was obtained only in the 1960s.[39,198]

*1. Radial Diffusion*

Radial diffusion of oxygen from the soil into a root of radius R, having a respiratory activity q and a diffusion coefficient inside the root tissue $D_i$, the root being surrounded by a layer of water saturated soil of effective thickness d with an effective oxygen diffusion coefficient $D_e$ (Figure 21) has been considered by Wanner,[775] and subsequently by Lemon,[418] Lemon and Wiegand,[421] and Wiegand and Lemon.[788] Their reports provided the equations quoted below.

The general equation of diffusion (33), assuming the absence of longitudinal gradients and neglecting the respiratory activity of the soil moisture film surrounding the root, for oxygen in the root tissue in a cylindrical coordinate system for steady-state conditions, has the form:

$$D_i \left[ \frac{d^2C_i}{dr^2} + \frac{1}{r}\frac{dC_i}{dr} \right] - q = 0 \tag{132}$$

and for the soil water film surrounding the root:

$$D_e \left[ \frac{d^2C_e}{dr^2} + \frac{1}{r}\frac{dC_e}{dr} \right] = 0 \tag{133}$$

where $C_i$ and $D_i$ denote the concentration and diffusion coefficient of oxygen in the root, respectively; $C_e$ and $D_e$ denote the same parameters in the soil water film surrounding the root, and q denotes the respiratory activity of the root.

The solution of these equations gives us the following formulas for the oxygen concentration on the root surface $C_R$ when its concentration on the surface of the soil water film surrounding the root, $C_{R+d}$, is known:

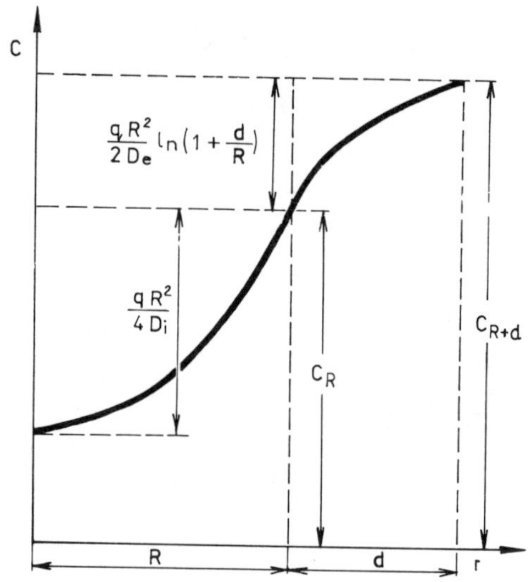


FIGURE 22. Oxygen concentration within the root of a radius R and in surrounding it soil-water layer of a thickness d as a function of the distance from the root axis r calculated from Equation 136.

$$C_R = C_{R+d} - \frac{qR^2}{2D_e} \ln\left(1 + \frac{d}{R}\right) \tag{134}$$

and for the oxygen concentration inside the root:

$$C_i = C_R - \frac{q}{4D_i}(R^2 - r^2) = 0 \tag{135}$$

combining both these equations we get:

$$C_i = C_{R+d} - \frac{qR^2}{2D_e} \ln\left(1 + \frac{d}{R}\right) - \frac{q}{4D_i}(R^2 - r^2) \tag{136}$$

Oxygen concentration distribution both within the root and outside it, calculated on the basis of the above equation, is presented in Figure 22.

It should be emphasized that the above equation is valid only when the respiratory activity of the root does not depend on oxygen concentration, i.e., when $C_R \geqslant C'_R$ (cf. Equation [56]), where $C'_R$ means the critical concentration of oxygen on the root surface (later in the text this is discussed further).

Equation (136) permits calculation of the oxygen concentration $C_{r=0}$ at the center of the root, by substituting $r = 0$; then:

$$C_{r=0} = C_{R+d} - \frac{qR^2}{2D_e} \ln\left(1 + \frac{d}{R}\right) - \frac{qR^2}{4D_i} \tag{137}$$

which means that after the inclusion of Equation (134):

$$C_{r=0} = C_R - \frac{qR^2}{4D_i} \tag{138}$$

If the oxygen concentration at the root axis reaches the critical value $C'_{r=0}$, below which the respiration rate decreases, the critical concentration on the root surface $C'_R$ will be:

$$C'_R = C'_{r=0} + \frac{qR^2}{4D_i} \tag{139}$$

which means it is increased in relation to $C'_{r=o}$ by the concentration drop inside the root.

The critical oxygen concentration at the exterior of the water or soil water film surrounding the root, $C'_{R+d}$, is, of course, $C'_R$ increased by the difference in concentrations across the film ln $(1+d/R)$ $qR/2D_e$ and equals:

$$C'_{R+d} = C'_{r=0} + \frac{qr^2}{4D_i} + \frac{qR^2}{2D_e} \ln\left(1 + \frac{d}{R}\right) \tag{140}$$

Data presented earlier (see Chapter 1) concerning the considerable affinity of cytochrome oxidase to oxygen enable us to assume that with the critical value of oxygen concentration outside the root $C'_{R+d}$ oxygen concentration inside the root is close to zero.

Having considered that $C'_{r=o} = 0$, Equation (140) is simplified to:

$$C'_{R+d} = \frac{qr^2}{4D_i} + \frac{qR^2}{2D_e} \ln\left(1 + \frac{d}{R}\right) \tag{141}$$

and the critical concentration of oxygen on the root surface is determined by the equation:

$$C'_R = \frac{qR^2}{4D_i} \tag{142}$$

The knowledge of oxygen consumption in a unit of its volume q permits the calculation of a unit oxygen flow $f_r$ related to a unit of root surface. If the root has a length l and a radius R:

$$f_r = \frac{\pi R^2 lq}{2 \pi Rl} = \frac{Rq}{2} \tag{143}$$

which, after combinining with Equation (134), gives:

$$f_r = D_e \frac{(C_{R+d} - C_R)}{R \ln\left(1 + \frac{d}{R}\right)} \tag{144}$$

This has been presented by numerous authors.[418—420,708]

It should be emphasized here that during the measurement of the value of flow $f_r$ by means of the polarographic method using a platinum electrode it is assumed that $C_R$ equals zero at a surface of the electrode. In the case of a root this is a simplification, as the critical concentrations of oxygen on the root surface are not lower than 0.02 to 0.03 $m^3m^{-3}$ (cf. Chapter 1).

In view of the analogy of the mathematical description of diffusion flow to electrical current flow where the intensity of diffusive flow corresponds to the intensity of electric current and the difference between concentrations corresponds to the difference in the electric potential, in the design of electrical analogues of diffusion the concepts of diffusive permeability of a medium $P_D$[446] and its reciprocal known as diffusive resistance $R_D$[25] are introduced. These parameters are frequently used in the description of gas exchange in roots and in photosynthesis research.

In the case of uniaxial diffusion, diffusive permeability $P_D$ is equal to a ratio of the diffusion coefficient D to the length of the diffusion path $\Delta x$:

$$P_D = \frac{D}{\Delta x} \tag{145}$$

and is expressed in meters per second.

Fick's Equation (5) for the diffusive flow $f_x$ may, therefore, be written as follows:

$$f_x = P_D \Delta C \tag{146}$$

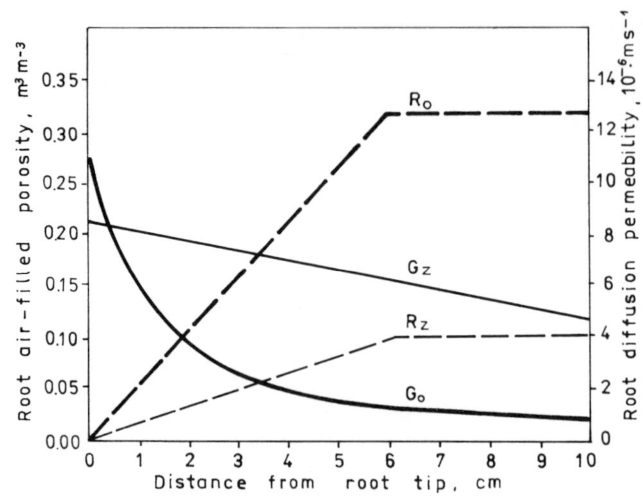


FIGURE 23. Air-filled porosity (R) and radial diffusion permeability (G) of adventitious root of maize (z) and rice (o) as a function of the distance from the root tip. (Data from Luxmoore, R. J., Stolzy, L. H., and Letey, J., *Agron. J.*, 62, 322, 1970.)

where $\Delta C$ = difference in the concentrations over the distance $\Delta x$ and $P_D$ = diffusive permeability of the medium.

Considering that:

$$P_D = \frac{1}{R_D} \tag{147}$$

Fick's equation may be expressed in a form analogous to Ohm's law, that:

$$f_x = \frac{\Delta C}{R_D} \tag{148}$$

where $R_D$ is expressed in seconds per meter.

The use of the concept of diffusion resistance is convenient in that it permits the summation of diffusion resistance values along the diffusion path, in a manner analogous to the summation of electrical resistances. This is important, for example, when determining the diffusion resistance of the whole plant, the complete root system, or individual roots, where we do not have an exact distribution of the diffusion coefficient which surely differs between the root epidermas and the internal tissues. In the literature there is a lack of empirical data concerning the values of the radial diffusion coefficient in the roots.

In roots with no air-filled spaces, the oxygen diffusion coefficient is undoubtedly less than $2.1 \cdot 10^{-9}$ $m^2s^{-1}$ (at 20°C), i.e., below that for pure water, due to obstacles in the form of cell walls. Estimated values of this coefficient adopted by various authors are from $2.2 \cdot 10^{-10}$ to $9 \cdot 10^{-10}$ $m^2s^{-1}$.[421]

As is known, plant root porosity may be considerable and occasionally even reach 0.6 $m^3m^{-3}$.[25] It is difficult, however, to say whether and to what extent it effects an increase in the radial diffusion coefficient in roots because of the lack of data from direct measurements.

Radial diffusive permeability of adventitious roots of maize, as measured by Luxmoore et al.[446] was $4.8 \cdot 10^{-6}$ m $s^{-1}$, and for rice 1 to $11 \cdot 10^{-6}$ m $s^{-1}$, and in both cases decreased with increase in the distance from the root tips (Figure 23). In this figure it can also be seen that, contrary to expectations, an increase in porosity with increase in the distance from the root tips is accompanied, not by an increase in the diffusive permeability, but by its decrease. This probably follows from the aging of the root epidermas and decrease in their permeability.

### Table 7
### ROOT POROSITY

| Plant | Porosity, $m^3m^{-3}$ Abundance of oxygen in soil | Oxygen defficiency in soil | Ref. |
|---|---|---|---|
| Rice | 0.05 | 0.25—0.30 | 39, 230 |
| Maize | 0.06—0.07 | 0.12—0.17 | 487, 824 |
| Festuca | 0.02 | 0.06 | 487 |
| Barley | 0.03 | 0.04 | 824 |
| Sunflower | 0.06 | 0.12 | 824 |
| Wheat | 0.021—0.126 | 0.035—0.145 | 6, 441, 681, 683, 767, 824 |
| Tomato | 0.05 | 0.08 | 824 |

*2. Longitudinal Diffusion*

It is well known that plant roots may contain free spaces filled with air,[11,20,21,25,360] the volume of which, in the case of aquatic plants, is comparable to the air-filled porosity of soils. These are intercellular spaces having continuity along the root down to a distance of 2 to 3 mm from the root tip. They are found in the suberized root tissue known as aerenchyma.[170]

The porosity of mesophyte roots is slight (of the order of several percent of the root volume), however, under conditions of soil inundation with water, even with these plants it is quite common to find the development of adventitious roots containing aerenchyma[172] and having a porosity much greater than in the primary roots.[442] This results in an increase in root porosity under conditions of insufficient oxygen supply from the soil (Table 7).

The existence of a continuous system of air spaces in root tissue makes an internal gas exchange possible both through diffusion and through mass flow. As to the relative contribution of the two mechanisms, in relation to roots all the considerations presented at the beginning of this chapter (Section I.B) for soils are applicable, including the general conclusion that diffusion is the fundamental mechanisms of gas transport. This is true in relation to the transport of oxygen downwards in the roots, but the upward transport of carbon dioxide is also facilitated by the transpiration stream.

The existence of the diffusive transport of oxygen to the roots has been shown experimentally by means of labeled oxygen for English bean,[198] maize,[349,350] rice, and barley.[39,350]

Investigations of the transport of ethylene in rice have shown that it is proportional to the fresh mass of the plant and does not depend on illumination nor on illumination-related changes in the transpiration stream,[416] which indicates that the exchange of gas between the plant shoots and the atmosphere is not a factor limiting the total process of gas transport in the plant. This is supported by the investigations by Stelzer and Läuchli[690] who found that oxygen enters the root system mainly at the stem base. It has also been found that oxygen can diffuse from roots to the surrounding medium in the case of rice,[764,765] bog-bean,[127] maize,[349] barley,[756] and bog plants.[23]

In the case of bog plants, internal gas transport suffices to cover the oxygen requirements of the whole root system, with mesophytes it constitutes only a supplementary factor.

Quantitative determination of the role of internal longitudinal diffusion is complicated by the closing of stomata under conditions of insufficient oxygenation.[337,682,683] However, it follows from the publications quoted previously[416,690] that changes in the diffusive resistance in leaves should not affect the resultant oxygen transport in the roots to any significant degree.

The distribution of oxygen concentration at a depth x in a straight vertical root of constant diameter and length L, under conditions of no radial transport to or from the surrounding medium, assuming that both the respiratory activity q and the oxygen diffusion coefficient

D in the root are constant along the whole length of the root and that q is not a function of oxygen concentration, can be described in a manner analogous to soil (cf. Equation [58]) by the following equation:

$$C = C_o - \frac{q\,(2Lx - x^2)}{2D} \tag{149}$$

where $C_o$, in this case, is the concentration of oxygen at the beginning of the root, i.e., at $x = 0$.

Oxygen concentration at the tip of the root, i.e., at $x = L$ (analogous to the lower boundary of a homogeneous soil profile, see Equation [59]), is

$$C_L = C_o - \frac{qL^2}{2D} \tag{150}$$

If we assume that the growth of roots stops when the critical concentration of oxygen, $C'$ is reached at the root tip, then the limit of root penetration into a soil innundated with water, $L'$, is

$$L' = \sqrt{\frac{2D\,(C_o - C')}{q}} \tag{151}$$

The root oxygenation depth, $L_{an}$, can be calculated by means of Equation (150) substituting values $C_L = 0$ and $L = L_{an}$. Then:

$$L_{an} = \sqrt{\frac{2DC_o}{q}} \tag{152}$$

Replacing the atmospheric concentration of oxygen $C_o$ and the diffusion coefficient in the root D with partial pressure of oxygen in the atmosphere $P_o$ and the diffusion constant in the root K, respectively (see Equations [12] and [14] in this chapter), we get:

$$L_{an} = \sqrt{\frac{2KP_o}{q}} \tag{153}$$

The length $L_{an}$ can represent both the depth at which oxygen concentration in the roots drops to zero and the limit depth of root penetration in an anoxic medium, if the growth of the roots ceases at an oxygen concentration level of zero. A graphic interpretation of this relationship has been presented in Figure 12, where the value $K/K_o$, as long as the diffusion takes place in the gaseous phase, is equivalent to $D/D_o$ and may be interpreted, in relation to the root, as its effective porosity. We can see that when this porosity is below 0.05 $m^3m^{-3}$ at normal respiration values (approximately 100 mg $m^{-3}\,s^{-1}$) in the absense of radial diffusion from the soil the oxygenated section of the roots is only about 7 cm, and therefore longitudinal transport is significant only for the part of the root system near the soil surface.

The literature of the subject does not quote any data concerning the longitudinal diffusion coefficient in the roots, hence it is difficult to use data from Figure 12 in practice. These difficulties are compounded by the simplifying assumptions adopted in the development of the equations in this part of the text. The assumptions referred to are (1) no radial diffusion, (2) constant values of the diffusion coefficient and of the respiratory activity of the root along its whole length, and (3) q independent of oxygen concentration. All these assumptions are not met in reality, as q is greatest at the tip of the root and decreases at oxygen concentrations below the critical value (cf. Chapter 1) and root porosity (and probably also the longitudinal diffusion coefficient) decrease in the direction of the growing tip of the root (which has already been shown in Figure 23).

Calculations made by Luxmoore et al.[443–448] have shown that in the case of maize about 25% of the root requirement for oxygen is met by longitudinal internal transport. The investigations of Greenwood[275] which involve 12 vegetable plants have shown that the internal aeration of their roots is significant for only a 5 cm section of root (measured from the soil surface).

A similar evaluation of the significance of internal transport in mesophytes has been arrived at by Yu et al.,[824] investigating the penetration depth for the roots of five plants in a soil completely saturated with water. Under these conditions the porosity of the roots of maize increased from 0.075 to 0.115 to 0.15 to 0.18 $m^3m^{-3}$ while the depth of root penetration into the soil was 17 cm. For sunflower, the porosity increased from 0.06 to 0.11 $m^3m^{-3}$ and the penetration depth was 15 cm. The porosity of the roots of barley was about 0.035 $m^3m^{-3}$ and did not show any significant changes under the conditions of innundation with water, while their penetration depth was 12 cm. In the case of wheat, the root porosity increased from 0.04 to 0.08 $m^3m^{-3}$ and the penetration depth was 10 cm. Another wheat variety, however, increased its root porosity from 0.06 to 0.15 $m^3m^{-3}$ with a root penetration depth of only 5 cm. Tomato roots increased their porosity from 0.04 to 0.08 $m^3m^{-3}$ with a penetration depth of 6 cm.

The differences in the penetration depth of roots with similar porosity may be the result of differences in both their respiratory activity and in the values of their longitudinal diffusion coefficients. It seems, however, in view of the continuity of the intercellular pores, that the differences in the respiratory activity are probably more significant than in longitudinal diffusion.

## III. SUMMARY

The transport of gases in the soil takes place through both concentration diffusion and mass flow. Both these processes may occur in the soil, in living plant roots, and between the soil and the roots.

The primary mechanism of gas exchange in a soil medium is, under normal field conditions, concentration diffusion. In relation to this there is quite an extensive literature concerning the mathematical modelling of the processes of gas diffusion in the soil, mainly in relation to oxygen. Such models, however, refer to the simplest situations and for the most part have not been verified by experiment.

Difficulties in the modelling of more complex systems closer to natural situations result both from the complexity of the diffusion equations themselves and from the lack of empirical data concerning the relationships between particular soil parameters, e.g., lack of data concerning the diffusion coefficient inside soil aggregates, in the water films surrounding plant roots, and within the roots, or insufficient knowledge of the changes in the respiratory activity in space and time under the effect of various parameters of the soil medium.

The significance of mathematical modelling, emphasized by Smith,[670] in such a complex heterogeneous system as soil, consists not so much in providing accurate quantitative data as in showing which of the variables interacting is of the greatest importance. This enables prediction of the direction of changes both between different soils and in a single soil with variable external factors.

Relatively less is known concerning the transport of gases within the plant and its contribution towards satisfying the total oxygen requirements of the plant roots, as well as its significance for plant survival through periods of anoxic conditions. There is, therefore, need for more comprehensive and more accurate studies concerning the balance of gaseous components of the soil medium, taking account of the gaps in our present knowledge.

Soil oxygenation is determined by several parameters, the mutual relationships of which are presented in Figure 19. Thus the structure and moisture content determine the content of soil air; in turn, the content of soil air affects the gas diffusion coefficient, and therefore the possibility of gas exchange with the atmosphere. The processes of gas transport in the soil together with the processes of gas evolution and absorption by the soil determine the composition of soil air.

## SYMBOLS AND UNITS

$C$, $C_{O_2}$ = oxygen concentration in soil air ($m^3m^{-3}$, mol $m^{-3}$ or g $m^{-3}$).
$C_o$ = oxygen concentration in atmospheric air.
$C(x)$, $C_1(x)$, $C_2(x)$, $C_k(x)$ = oxygen concentration in soil air at a depth x, in layers 1,2,k, respectively.
$C_1$, $C_2$, $C_k$, $C_i$, and $C_n$ = oxygen concentration in soil air of a first, second, kth, ith, and nth layer of soil profile, respectively.
$C_{Hk}$, $C_{Hn}$, $C_L$ = oxygen concentration at a depth $H_k$, $H_n$, L, respectively.
$C_R$ = oxygen concentrations on the surface of root (electrode or aggregate).
$C_{R+d}$ = oxygen concentration on the surface of water film surrounding the root (electrode).
$C_i$ = oxygen concentration inside the root.
$C_e$ = oxygen concentration in the water film surrounding the root.
$C_r$ = oxygen concentration at a distance r from aggregate center.
$C'$ = critical $O_2$ concentration in soil.
$C'_R$ = critical $O_2$ concentration on aggregate surface.
$D$ = gas diffusion coefficient in soil ($m^2s^{-1}$).
$D_o$ = gas diffusion coefficient in atmospheric air (Tables 3 and 4, this chapter and Table 2, Chapter 3).
$D/D_o$ = relative coefficient of gas diffusion in soil, dimensionless.
$D_e$ = effective gas diffusion coefficient in liquid film surrounding the root.
$D_c$ = gas diffusion coefficient in soil aggregates.
$D_i$ = coefficient of radial gas diffusion in root tissue.
$D_w$ = oxygen diffusion coefficient in water (Table 3).
$D^*$ = $D/E_g$ = gas diffusion coefficient within air-filled soil pores = apparent diffusivity.
$D_G$ = $D/(E_g + \alpha_B\theta)$
$D(x)$ = gas diffusion coefficient at a depth x.
$D_1$, $D_2$, $D_k$, $D_i$, $D_n$ = oxygen diffusion coefficient in the soil layer indicated by the subscript.
$d$ = effective thickness of the moisture film surrounding the root.
$E_o$ = total soil porosity ($m^3m^{-3}$).
$E_g$ = air-filled porosity in soil ($m^3m^{-3}$).
$E_i$ = interaggregate soil porosity.
$E_g(x)$ = air-filled porosity at a depth x.
$E_{go}$ = air-filled porosity at the soil surface.
$f_x$ = diffusive oxygen flux along x axis (mol $m^{-2}s^{-1}$, m $s^{-1}$, or kg $m^{-2}s^{-1}$).
$f_r$ = radial oxygen flux.
$G$ = gas content in an unit soil volume (mol $m^{-3}$).
$K(K_{O_2}, K_{CO_2})$ = gas diffusion constant (for $O_2$, $CO_2$) (mol $m^{-3}s^{-1}Pa^{-1}$).
$K_o$ = gas diffusion constant in the atmospheric air.
$K_w$ = gas diffusion constant in water.
$k$ = air permeability of soil ($m^2$).
$L$ = depth of the soil profile.
$L_{an}$ = depth of oxidized zone or depth of anoxic zone appearance.
$P(P_{CO_2}. P_{O_2})$ = partial pressure ($CO_2$, $O_2$) (Pa).

| | | |
|---|---|---|
| $P_D$ | = | diffusive permeability = $D/\Delta x$ ($m\ s^{-1}$). |
| $q$ | = | soil respiration rate ($mg\ O_2\ m^{-3}s^{-1}$ or $mol\ m^{-3}s^{-1}$). |
| $q_o$ | = | respiration rate at soil surface. |
| $q(x)$ | = | respiration rate at a depth x. |
| $q^*$ | = | $q/E_g$ = apparent respiration. |
| $q_1, q_2, q_i, q_k, q_n$ | = | respiration rate in the soil layers indicated by the subscripts. |
| $q'$ | = | the critical value of respiration rate. |
| $R$ | = | radius of a capillary (electrode, aggregate). |
| $R'$ | = | the critical value of soil aggregate radius. |
| $R_D$ | = | diffusive resistance = $\Delta x/D$ ($s\ m^{-1}$). |
| Re | = | Reynold's number. |
| $r$ | = | distance from the center of root (aggregate, electrode). |
| $t$ | = | time (s). |

Other symbols are explained in the text of this Chapter.

Chapter 3

# SOIL AIR

## I. AIR CONTENT IN SOIL

Soil air fills that part of the soil volume that is not occupied by water. The air content in the soil ($E_g$), often called the air-filled porosity of the soil, is therefore equal to the difference between the total porosity of the soil ($E_o$) and its current moisture content by volume ($\theta$):

$$E_g = E_o - \theta = E_o - \rho W \tag{1}$$

where $\rho$ = soil bulk density in Mg $m^{-3}$ and W = moisture content by weight in kg $kg^{-1}$.

In the case of a nonswelling soil, the air content decreases in a linear manner with an increase in moisture content both by weight and by volume. In a swelling soil the relationship becomes nonlinear.

Figure 1 presents the relationship of the air-filled porosity to both of its fundamental determinants, i.e., the soil moisture tension and soil bulk density. Curves corresponding to soil samples of the lowest compaction ($\rho$ = 1.07 and 1.27 Mg $m^{-3}$) show a continuous increase with moisture tension, which is typical of nonswelling soils. In samples displaying shrinking the increase in air-filled porosity with the dewatering of the sample may be slowed down, and even a decrease in the air-filled porosity may occur (see curves for compacted samples of $\rho$ = 1.48 to 1.62 and 1.51 to 1.64 Mg $m^{-3}$ which shrank at moisture tension values within the range from 30 to 40 kPa).

The content of air in a soil profile is subjected to dynamic changes depending on the changes in the total porosity and soil moisture content, although it does not usually exceed 0.4 $m^3m^{-3}$. The extent of these changes, in the course of year, for a soddy-podzolic soil, is presented in Figure 2.

The amount of air in the soil determines not only the reserve of oxygen present but also, and primarily, the value of the oxygen diffusion coefficient, and it therefore affects the rate of gas exchange. In turn, the rate of gas exchange in the soil, together with the rate of soil respiration, determine the composition of the soil air.

## II. SOIL AIR COMPOSITION

The basic components of the soil air are nitrogen, oxygen, carbon dioxide, and water vapor (see Table 1). The last component occurs in the soil in amounts corresponding to 93 to 100% of the saturated vapor pressure. It causes the dilution of the other gaseous components of atmospheric air. Moreover, the vapor pressure at temperatures within the range from 0 to 30°C is from 0.6 to 4.2 kPa, which does not exceed 4.2% of the total pressure of atmospheric air. Hence, this component is most frequently neglected in the investigations of soil air.

The basic components of soil air are unevenly distributed between the liquid and gaseous phases of the soil. Carbon dioxide, distributed in similar quantities in both phases (Figure 3), is in contrast here to oxygen and nitrogen which are characterized by very low solubility in water. It can be seen in Figure 3 that almost all of the soil oxygen is contained in the gaseous phase of the soil, while the content of carbon dioxide in the soil does not change significantly with change in the proportion of the liquid and gaseous phases at temperatures in the region of 20°C.

In comparison to atmospheric air, soil air contains less oxygen (the quantity of which usually varies around 0.1 to 0.2 $m^3m^{-3}$, though it can even drop to zero) and more carbon

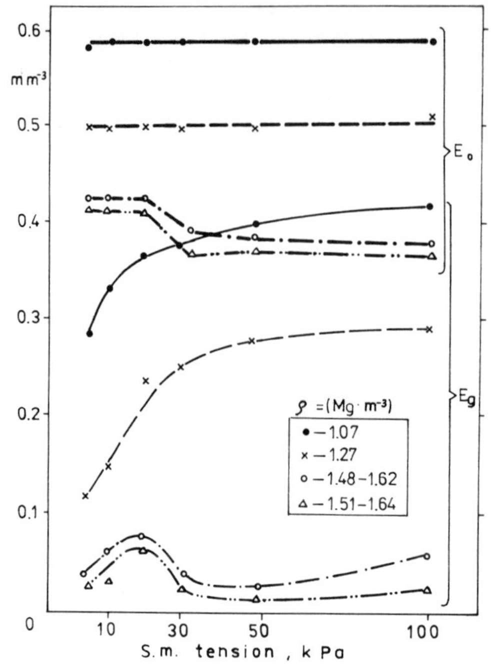


FIGURE 1. Total $E_o$ and air-filled porosity $E_g$ of a loamy textured black earth at different bulk densities (the same soil as in Figures 4 to 6 and 20, Chapter 2). (Modified from Stępniewski, W., *Pol. J. Soil Sci.*, 14, 15, 1981.)

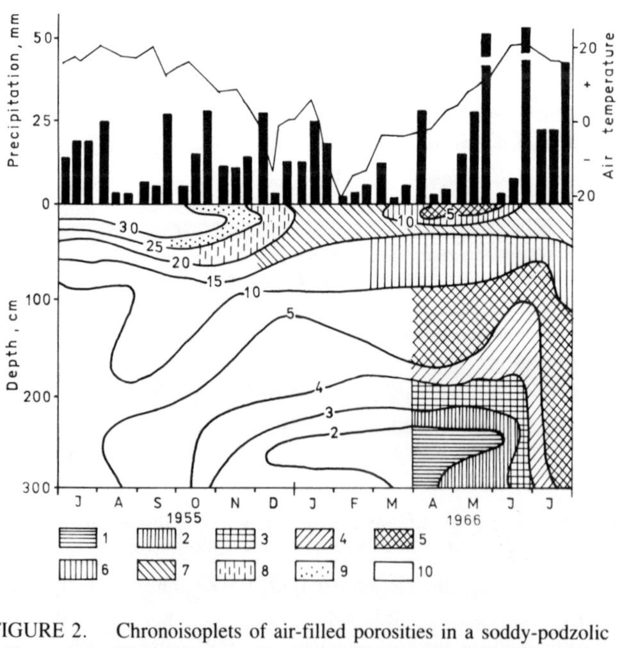


FIGURE 2. Chronoisoplets of air-filled porosities in a soddy-podzolic soil cropped with grasses in connection with air temperature and precipitations. Air-filled porosity in $m^3m^{-3}$: 1 $<$ 0.02, 2: 0.02 to 0.03, 3: 0.03 to 0.04, 4: 0.04 to 0.05, 5: 0.05 to 0.10, 6: 0.10 to 0.15, 7: 0.15 to 0.20, 8: 0.20 to 0.25, 9: 0.25 to 0.30, 10: $>$ 0.30. (Modified from Nikolaeva, I. N., *Air Conditions of Soddy-Podzolic Soils*, Izd-vo "Kolos", Moscow, 1970.)

## Table 1
## VOLUMETRIC COMPOSITION OF DRY AND MOIST ATMOSPHERIC AIR OF DIFFERENT TEMPERATURES

| | | Conc $m^3m^{-3}$ | | | |
|---|---|---|---|---|---|
| | | Air saturated with water vapor at a temp of | | | |
| Component | Dry air | 0°C | 10°C | 20°C | 30°C |
| $N_2$ | 0.7809 | 0.7762 | 0.7715 | 0.7629 | 0.7481 |
| $O_2$ | 0.2095 | 0.2082 | 0.2082 | 0.2047 | 0.2007 |
| $CO_2$ | 0.0003 | 0.0003 | 0.0003 | 0.0003 | 0.0003 |
| Ar + other inert gases | 0.0093 | 0.0092 | 0.0092 | 0.0091 | 0.0089 |
| Water vapor | 0.0000 | 0.0060 | 0.0120 | 0.0230 | 0.0420 |

Data from *Handbook of Physicochemistry* (Polish), Wyd. Naukowo-Techniczne, Warsaw, 1974.

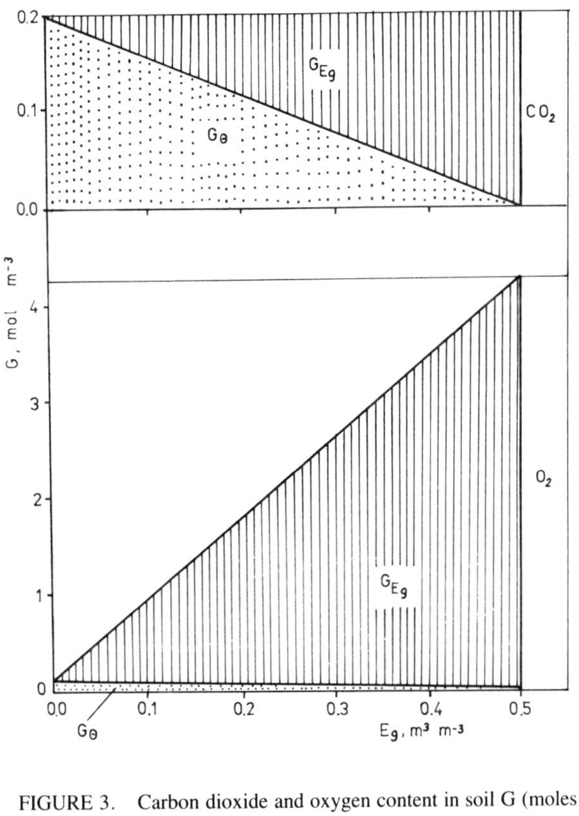


FIGURE 3. Carbon dioxide and oxygen content in soil G (moles per cubic meter of soil) as a function of air-filled porosity $E_g$, as divided between liquid ($G_\theta$) and gas ($G_{E_g}$) soil phases, assuming $CO_2$ concentration in soil air 0.01 $m^3m^{-3}$, and $O_2$ concentration 0.203 $m^3m^{-3}$. Calculations performed for total soil porosity $E_o$ = 0.5 $m^3m^{-3}$, assuming equilibrium between liquid and gas phase at 20°C.

**Table 2**
**PROPERTIES OF GASES BEING COMPONENTS OF SOIL AIR**

| Component | Mean free path × $10^{-8}$ m (at 100 kPa)[a] | $V_m$ $dm^3$ NTP[b] | Density kg $m^{-3}$ NTP[b] | Solubility in water $\alpha_B$ 0°C[b] | Solubility in water $\alpha_B$ 20°C[b] | Diffusion coeff In air $D_o$ $10^{-5}$ $m^2s^{-1}$ NTP[b] | Diffusion coeff In water $D_w$ $10^{-9}$ $m^2s^{-1}$ 25°C, NP[b] |
|---|---|---|---|---|---|---|---|
| $N_2$ | 8.50 | 22.40 | 1.251 | 0.0235 | | 1.81 | 1.9 |
| $O_2$ | 9.05 | 22.39 | 1.429 | 0.0489 | 0.0333 | 1.78 | 2.5 |
| $CO_2$ | 5.56 | 22.26 | 1.977 | 1.713 | 0.942 | 1.39 | 1.96 |
| $CH_4$ | | 22.36 | 0.717 | 0.0556 | 0.0331 | | |
| $NH_3$ | 5.92 | 22.08 | 0.771 | 1 110.0 | | 1.98 | 2.0 |
| $H_2S$ | | 22.14 | 1.539 | 4.670 | 2.582 | | |
| $N_2O$ | | 22.25 | 1.978 | | 0.629 | 1.43 | |
| $H_2$ | | 22.43 | 0.080 | 0.0215 | | 6.34 | 5.85 |
| $H_2O$ vapor | | 23.45 | 0.768 | | | 2.82 | |
| $C_2H_4$ | | 22.24 | 1.261 | 0.226 | 0.122 | 1.37 | |
| Ar | | 22.39 | 1.784 | | | | |

*Note:* NTP = normal conditions, pressure 101.3 kPa, temp 0°C; NP = normal pressure 101.3 kPa; $V_m$ = molar vol.

[a] Data from Stolzy, L. H., in *The Plant Root and its Environment*, Carson, E. W., Ed., University Press of Virginia, Charlottesville, Va., 1974, 335.

[b] Data from *Handbook of Physicochemistry* (Polish), Wyd. Naukowo-Techniczne, Warsaw, 1974.

dioxide (the content of which varies usually from 0.001 to 0.05 $m^3m^{-3}$ and sometimes may even exceed 0.1 $m^3m^{-3}$).

The content of argon and other inert gases remains practically unchanged, while the content of nitrogen varies within a range of several percent as a result of changes in the sum of oxygen and carbon dioxide. In exceptional cases, under anoxic conditions, with extremely high concentrations of carbon dioxide and in the presence of methane, the content of nitrogen may even drop to 0.4 to 0.3 $m^3m^{-3}$.[183]

The sum of the oxygen and carbon dioxide, under conditions of oxic metabolism, i.e., when the respiration quotient is close to 1, is about 0.2 $m^3m^{-3}$ or somewhat less due to solubility of carbon dioxide in water; it can exceed this value under conditions of partial anoxia.[183,190]

In certain cases, when the soil oxygenation is insufficient, other gases may occur in the soil air, such as $CH_4$, $H_2S$, $N_2O$, $C_2H_4$, and $H_2$, the presence of any of these, despite their low concentrations, is an important indicator of the status of the soil.

Some characteristics of the components of soil air are presented in Table 2, and more detailed information concerning oxygen and carbon dioxide can be found in Table 3.

## A. Oxygen

Soil oxygen is contained primarily in the gaseous phase of the soil, and only to a small extent in the soil water (Table 3, Chapter 2; Table 3, this chapter; Figure 3). Exact values of the oxygen concentration in air and water at various temperatures are presented in Table 3. Runkles et al.[620] state in addition that some soil clay minerals, when dry, can absorb certain amounts of oxygen.

With an increase in the soil moisture content the adsorption of oxygen on minerals decreases so that the amount of it adsorbed by the water-soil system is equivalent to the

## Table 3
## OXYGEN CONCENTRATION AT NP AND DIFFERENT TEMPERATURES AND PARTIAL PRESSURE IN WATER SATURATED ATMOSPHERIC AIR AND IN WATER EQUILIBRATED WITH MOIST AIR

| | | Conc | | | | |
|---|---|---|---|---|---|---|
| | | In moist air | | | In water | |
| Temp °C | Partial pressure in moist air kPa | $m^3m^{-3}$ | $g\ m^{-3}$ | $mol\ m^{-3}$ | $g\ m^{-3}$ | $mol\ m^{-3}$ |
| 0 | 21.10 | 0.2082 | 297.5 | 9.30 | 14.6 | 0.456 |
| 5 | 21.03 | 0.2076 | 291.3 | 9.09 | 12.6 | 0.397 |
| 10 | 20.97 | 0.2070 | 285.4 | 8.92 | 11.3 | 0.352 |
| 15 | 20.86 | 0.2060 | 279.0 | 8.72 | 10.0 | 0.315 |
| 20 | 20.74 | 0.2047 | 272.5 | 8.52 | 9.13 | 0.284 |
| 25 | 20.56 | 0.2030 | 265.7 | 8.30 | 8.22 | 0.257 |
| 30 | 20.33 | 0.2007 | 258.4 | 8.16 | 7.52 | 0.233 |

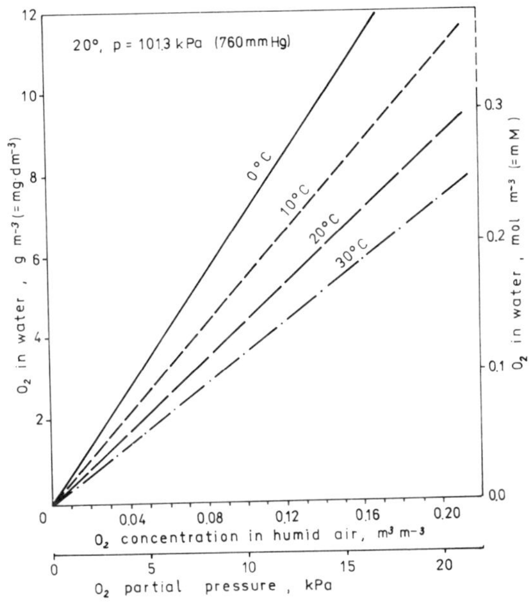


FIGURE 4. Equilibrium oxygen concentration in water as a function of oxygen concentration in air saturated with water vapor, at normal pressure (101.3 KPa), at different temperatures.

theoretical solubility of oxygen in the water phase only, which permits the use of values compiled in Table 3, Chapter 2, Table 3 and Figure 4, this chapter. However, a reservation should be made here that the values refer to pure water, while in the soil we deal with a solution in which oxygen solubility is still lower due to the so-called "salting out" effect, and also due to the presence of other gases.[456]

Variability of oxygen solubility in the soil solution under the effect of the factors enumerated has not been investigated more closely in view of the low solubility of oxygen and therefore the insignificant role dissolved oxygen plays in the oxygen balance of the soil.

The oxygen content in well-aerated cultivated soils does not usually fall below 0.15 $m^3m^{-3}$.[69,212,231,270,526,627,693,694] This is exemplified in Figure 5 where even at a depth of 3 m the oxygen concentration is still considerable.

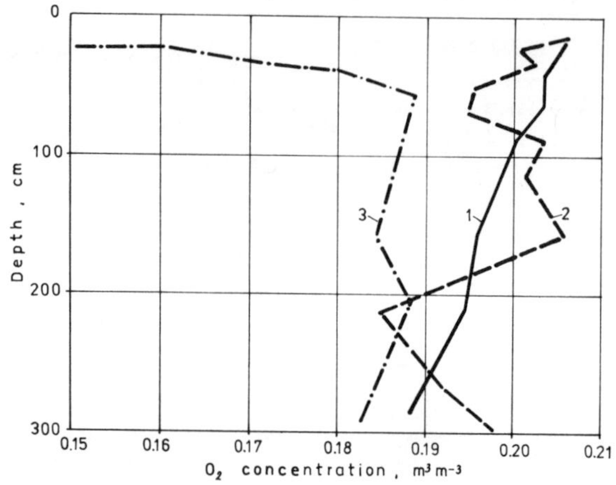


FIGURE 5. Examples of oxygen distribution in soddy-podzolic soil; (1) 4.08.1955, (2) 2.10.1955, (3) 18.04.1956. (Modified from Nikolaeva, I. N., *Air Conditions of Soddy-Podzolic Soils*, Izd-vo "Kolos", Moscow, 1970.)

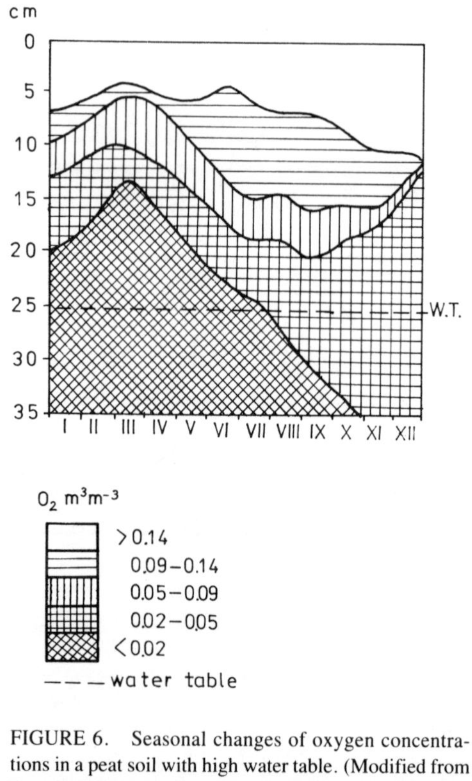


FIGURE 6. Seasonal changes of oxygen concentrations in a peat soil with high water table. (Modified from Boggie, R., *Plant Soil,* 48, 447, 1977.)

Oxygen concentration in the soil usually decreases with depth in the soil profile. It also changes in the vegetation season with a tendency for a minimum to occur during the summer, related to the occurrence of higher temperatures. In poorly drained soils, oxygen concentration may be considerably less,[63,221,224,542,548] and under especially adverse conditions may even fall to zero (Figure 6). This happens particularly in the case of soil zones saturated with water. Drastic anoxibiosis occurs when soil is treated with slurry[405,701] or liquid waste.[734]

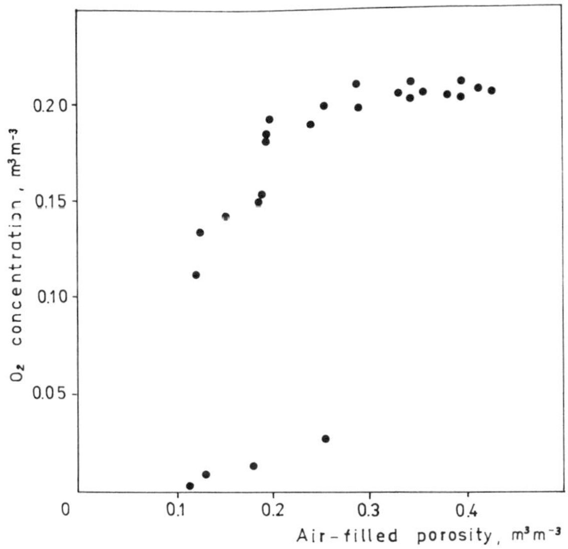


FIGURE 7. Oxygen concentration in air of a meadow loamy-silt soil at a depth of 7.8 cm as related to air-filled porosity. (Modified from Grable, A. R. and Siemer, E. G., *Soil Sci. Soc. Am. Proc.*, 32, 180, 1968.)

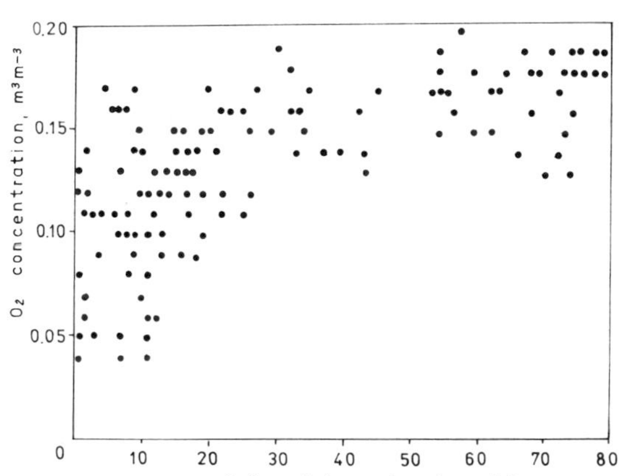


FIGURE 8. Oxygen concentration in the air of a silty clay loam soil as a function of its moisture tension. (Modified from Meek, B. D., Donovan, T. J., and Graham, L. E., *Soil Sci. Soc. Am. J.*, 44, 433, 1980.)

In any given soil the basic physical parameter, determining its oxygen content, is its air-filled porosity, which is decisive for the rate of gas exchange. With a decrease in this porosity, oxygen concentration in the soil also decreases. At a certain critical value of air-filled porosity, oxygen concentration drops to zero (e.g., Figure 7). This critical value of the air-filled porosity of soil may vary within quite a broad range. In the case presented in Figure 7 it is from 0.12 to 0.20 $m^3m^{-3}$, while Flühler,[207] found a value of 0.07 $m^3m^{-3}$.

Oxygen concentration in the soil is affected by all those factors and measures that affect its diffusion coefficient and all those that affect the soil respiratory activity. The first of these include primarily the soil moisture content and the soil structure in its broadest sense. The effect of moisture is presented in Figure 8, where for the soil in question, at a moisture

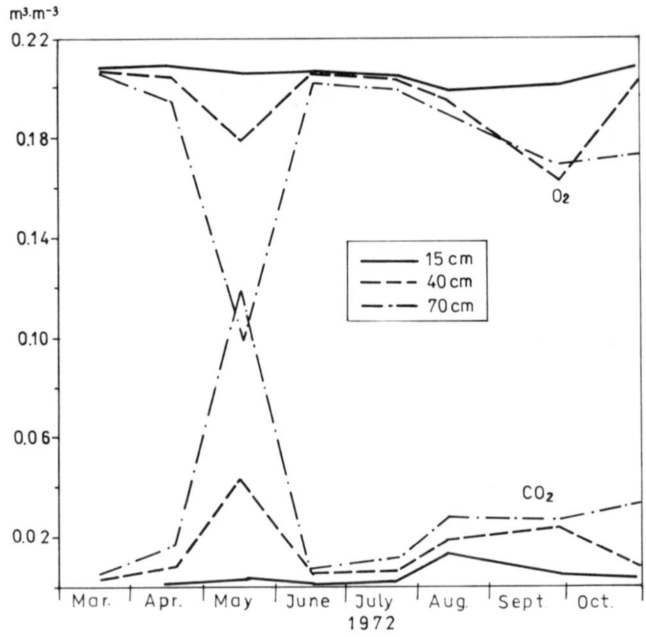


FIGURE 9. Oxygen and carbon dioxide content in the air of a pseudo-podzolic loamy sand soil. (Modified from Stępniewski, W., *Zesz. Probl. Postepow Nauk. Roln.*, 197, 203, 1977.)

tension of about 30 kPa, there is a distinct decrease in the oxygen content related to some discontinuity in the soil pores occurring at this moisture level. The value of this threshold moisture tension will be different for various soils, and even for the same soil at different levels of compaction. Therefore, all irrigation measures are liable to decrease the oxygen concentration in the soil air. A particular drastic effect is that of surface flooding of the soil with water. This may result in total exhaustion of the oxygen reserves in the soil in as little as 24 hr, particularly in the summer.[738] The other major factor hampering gas exchange in the soil is its compaction.[190]

The factors decreasing oxygen concentration in the soil air through increasing the soil respiratory activity include the application of manure,[260] green fertilizers,[190] and slurry.[405,701]

Factors increasing the oxygen content in the soil air include all measures which loosen the soil.[542,548] In a zero-tillage of a clay soil an increase in the oxygen concentration was observed in comparison to the same soil when tilled.[165] This was explained by the formation of continuous channels between the arable layer and the subsoil. However, the same authors, in another paper,[92] reported an opposite effect.

Stępniewski[694] has noted an advantageous affect of increased mineral fertilization on the oxygen content in the soil during the vegetation period, which he relates to the increase in evapotranspiration and therefore decrease in the soil moisture content.

The role of oxygen in the processes occurring in the soil, as well as the plant response to oxygen, are discussed in subsequent chapters.

### B. Carbon Dioxide

The content of carbon dioxide in the soil varies dynamically in a manner opposite to that of oxygen content. In the case of well-aerated soils its increase is approximately equal to the decrease in the oxygen content (e.g., Figure 9). This has been considered in detail by Wood and Greenwood.[807] In such soils the content of carbon dioxide does not usually exceed some few percent of the soil air volume[12,212,213,219,231,270,526,570,571,627] even at depths of several meters.[476,583]

In soils with a very deep ground water table, carbon dioxide concentration frequently reaches its maximum in the middle part of the profile.[583] In soil profiles of very shallow ground water tables these concentrations usually increase with depth, reaching values of 0.1 to 0.2 $m^3m^{-3}$ at the ground water level.[221,389] With these higher concentrations of carbon dioxide, the sum of $CO_2$ and $O_2$ may, volumetrically, be less than 0.21 $m^3m^{-3}$,[69] as a result of the greater water solubility of carbon dioxide in comparison to oxygen (cf. Tables 3 and 4, Chapter 2). Similar values may occur after the soil has been innundated with water for a few days.[738]

Maximum concentrations of carbon dioxide in the soil air occur during periods of increased moisture content and soil temperature. Carbon dioxide content in the soil air is affected by all those factors that affect the oxygen content, but here they act in an opposite direction. Thus an increase in carbon dioxide is caused by irrigation, the presence of plant cover,[571,627] fertilization with fresh plant material,[190] manure,[627] and particularly liquid wastes[734] or slurries.[183,405,701] In extreme situations, carbon dioxide concentrations in the soil may exceed 0.2 $m^3m^{-3}$,[183] which is the result of increased respiration quotient under anoxic conditions.

The carbon dioxide content in soil air is lowered by measures which loosen the soil.[571]

The comparison of the gradients of partial pressures of oxygen and carbon dioxide carried out by Greenwood[277] is of interest for (as it is shown, under conditions of sufficient oxygen availability) the amount of carbon dioxide evolved in equal, by volume (and in moles), to the amount of oxygen consumed (respiration quotient equal to one), and therefore oxygen flow is equal to carbon dioxide flow. As follows from Equation (13) in Chapter 2, the partial pressure gradients of oxygen and carbon dioxide will be in inverse proportion to the diffusion constants, that is

$$\text{grad } P_{CO_2} \cdot K_{CO_2} = \text{grad } P_{O_2} \cdot K_{O_2} \tag{2}$$

and therefore:

$$\frac{\text{grad } P_{CO_2}}{\text{grad } P_{O_2}} = \frac{K_{O_2}}{K_{CO_2}} \tag{3}$$

In the situation in which diffusion takes place in the gaseous phase of the soil, the ratio of the diffusion constants $K_{O_2}$ to $K_{CO_2}$ will be the same as in the atmospheric air (see Tables 3 and 4, Chapter 2), i.e., 1.25 at a temperature of 20°C. Hence the gradient of partial pressure of carbon dioxide, and therefore of its concentration in the soil air expressed as a fraction of volume or as the number of moles per unit of volume, will be 1.25 times greater than for oxygen. Therefore at any given depth in the soil, omitting the $CO_2$ present in the atmospheric air where its concentration, in relation to that in the soil air, can be considered negligible, we shall get:

$$P_{CO_2} = 1.25 \ (20.7 \text{ kPa} - P_{O_2}) \tag{4}$$

or

$$C_{CO_2} = 1.25 \ (0.203 - C_{O_2}) \tag{5}$$

whence it follows that the sum of the concentrations of the two gases, under conditions of sufficient oxygenation, should only slightly exceed the value of oxygen concentration in the atmosphere.

In a situation in which diffusion takes place only in the liquid phase of the soil (soil in a state of saturation or close to saturation with water), the ratio of $K_{O_2}/K_{CO_2}$ is equal to the ratio of diffusion constants of these gases in water and at a temperature of 20°C equals 0.0416 (see Tables 3 and 4, Chapter 2). Therefore the gradient $P_{CO_2}$ is 24 times smaller than the gradient $P_{O_2}$, as the diffusion of carbon dioxide in water, at the same gradient of partial pressure, takes place 24 times faster than the diffusion of oxygen. Hence, at the same depth in the profile:

$$P_{CO_2} = 0.0416\ (20.7\ \text{kPa} - P_{O_2}) \tag{6}$$

These relationships have been verified experimentally by Greenwood[277] and provide an explanation for the observation that the sum of $O_2$ and $CO_2$ in the soil air decreases at higher soil moisture contents. At the same time, however, under these conditions anoxic areas begin to appear, the respiration quotient in which is greater than 1, which results in an increase in the carbon dioxide concentration in relation to the value calculated on the basis of Equation (6).

The effect of carbon dioxide on plants is discussed in Chapter 5.

### C. Nitrous Oxide

The first report on the presence of nitrous oxide ($N_2O$) in the soil air under field conditions was published by Burford and Millington.[93] They noted the occurrence of nitrous oxide in a red-brown soil profile at depths down to 90 cm in amounts of up to 100 $cm^3m^{-3}$, with the maximum content at a depth of 30 to 60 cm.

Further investigations[92,167,168] have shown that the concentration of nitrous oxide may be greater than 100 $cm^3m^{-3}$, and can sometimes reach 1000 and, in individual samples, even 6500 $cm^3m^{-3}$.[167] It should be emphasized that nitrous oxide concentration in soil air is characterized by great variability, so that many authors give the cumulative frequency of concentration distribution without giving mean values.[208,209,211] The scatter of results was so broad that no significant negative correlation was observed between the concentration of nitrous oxide and the concentration of oxygen, though such a tendency seems to exist.[92,167]

The content of nitrous oxide in the soil increases with increase in the content of nitrates and of moisture.[208,209] In the case of innundation with water, the content of nitrous oxide shows an initial rapid increase and, having reached a maximum (after a period of several days), decreases.[738]

It has been observed that in clay soils the concentration of nitrous oxide is many times greater with zero-tillage than in tilled soil.[92]

There was a wide-spread opinion that nitrous oxide is formed only in the process of denitrification. This process occurs in the soil when the oxygen concentration drops to practically zero.[209] Hence the presence of nitrous oxide in the soil air has been considered to be an indicator of the presence of anoxic pockets in the soil in which denitrification took place.

The view connecting the occurrence of nitrous oxide only with the process of denitrification has recently been questioned. It has been shown that considerable quantities of nitrous oxide can be also generated in the process of nitrification.[74,75,665] Therefore the use of nitrous oxide as an index of the presence of anoxic areas in the soil is unjustified.

It should be pointed out that $N_2O$ is decomposed microbiologically to nitrogen; the reaction being inhibited by the presence of 0.001 to 0.05 $m^3m^{-3}$ acetylene in the soil air.[37,627,822]

There have been no reports regarding any possible physiological function of nitrous oxide in relation to plant roots.

### D. Ethylene

Of the various organic components of the soil air, ethylene[2] (which is a plant hormone) deserves particular attention as even in low concentrations it affects plants.

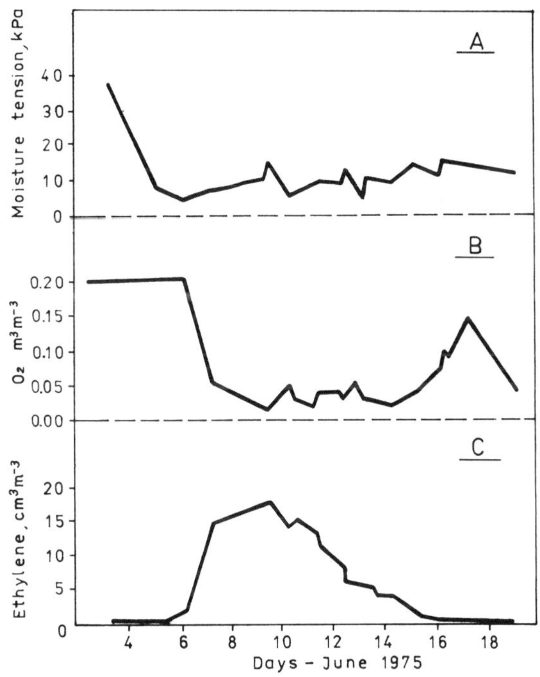


FIGURE 10. Changes of soil moisture tension (A), oxygen concentration (B), and ethylene content (C) in the air of a sandy loam soil irrigated in summer. (Modified from Campbell, R. B. and Moreau, R. A., *Am. Pot. J.*, 56, 199, 1979.)

The presence of ethylene in the soil has been detected by the application of gas chromatography.[676]

It should be emphasized that ethylene concentration in the soil does not reach high values, and even with soil incubation in closed containers it does not exceed 40 $cm^3m^{-3}$.[303,662,676] Concentrations met under field conditions are generally from 0.01 to 10 $cm^3m^{-3}$. The occurrence of concentrations higher than 1 $cm^3m^{-3}$ is connected with periods of excessive moisture contents,[100,168,172,672] application of slurries,[90] and increased soil temperature,[672] which are usually accompanied by a decrease in the oxygen concentration (Figure 10). The maximum value encountered under field conditions was 18.4 $cm^3m^{-3}$.[100]

In the case of soil innundation with water, ethylene concentration reaches a maximum value of the order of several cubic centimeters per cubic meter, within a period of several days.[100,334,738]

An attempt has been made to explain the fact that higher concentrations of ethylene do not occur in the soil by assuming the simultaneous formation in the soil of some volatile component restricting the evolution of ethylene.[674] The increase in ethylene concentration observed during period of increased soil moisture content coupled with a decrease in oxygen concentration, does not necessarily mean its increased evolution under these conditions as the cause of ethylene accumulation in the soil may lie in its restricted diffusion to the atmosphere.

It should be noted that the formation and fate of ethylene in the soil are not fully explained. What is known, however, is that its source can be both the soil itself and living plant roots.

Soil ethylene is of microbiological origin,[303,674] and so far it has been established that ethylene may be metabolite of fungi (e.g., *Mucor hiemalis,* or *Penicillium* spp.[124,156,330,450-452]),and also of spore forming bacteria.[662,663]

Undoubtedly, ethylene evolves from the soil under anoxic conditions which may be the result of both the saturation of the soil with water,[303,820] replacement of soil air with nitrogen,[126,248,249,820] and also of self-exhaustion of oxygen reserves during incubation of soil in a closed container.[662]

Smith and Restall[674] observed the evolution of ethylene in a water suspension of soil through which pure nitrogen or its mixture with air containing 0.01 $m^3m^{-3}$ of oxygen was bubbled. Higher concentrations of oxygen limited the evolution of ethylene.

The intensity of ethylene production in the soil increases with increase in the organic matter content,[248] and also under the effect of organic amendments in the form of plant residues[249,303,820] and other organic substances such as ethanol[249] or glucose.[820] The previous history of the soil also has an effect on the production of ethylene. Thus an increase in its evolution has been observed in a soil incubated in an atmosphere of nitrogen as a result of its previous drying.[248]

Soil pH may be another factor affecting the evolution of ethylene in the soil.[248]

Apart from investigations relating the evolution of ethylene under anoxic conditions, there are reports that ethylene may also be formed under oxic conditions.[249,820] Lynch and Harper[452] ascribe the formation of ethylene under anoxic conditions to the activation of the substrates necessary for its biosynthesis rather than to the existence of an anoxic pathway for the synthesis of this gas. These authors have determined that the fungus *M. hiemalis* grew better and produced more ethylene in the presence of oxygen, and that a decrease in the rate of ethylene evolution occurred at oxygen concentrations below 0.025 $m^3m^{-3}$.

The production of ethylene in living plant roots is also related to the status of root oxygenation, although the conditions of the course of this process are not fully understood.

Generally, it is known that ethylene is formed in plant tissues under oxic conditions.[429,460] This is also true for the roots, as the presence of ethylene was observed, in amounts approximately 0.9 $cm^3m^{-3}$, in the roots of intact maize plants grown in an aerated nutrient solution.[171,336] In addition, Konings and Jackson[376] observed the evolution of ethylene from undisturbed root systems of rice, mustard, and three varieties of tomato cut off from the shoots and incubated in a humid atmosphere containing 0.21 $m^3m^{-3}$ of oxygen. The production of ethylene in this case was positively correlated to the rate of root elongation, thus rice roots, evolving ethylene in the amount of $1.5 \cdot 10^{-3}$ $mm^3$ (g fresh mass)$^{-1}$ $hr^{-1}$, elongated at a rate of about 20 mm $day^{-1}$, while the roots of mustard, producing ethylene at a rate of $6.5 \cdot 10^{-3}$ $mm^3$ (g of fresh mass)$^{-1}$ $hr^{-1}$, elongated by about 65 mm $day^{-1}$.

On the other hand, other authors[171,336] observed that the ethylene concentration in the roots of plants grown in a nonaerated solution (up to 1.9 $cm^3m^{-3}$) was twice as much as when grown in an aerated one (0.9 $cm^3m^{-3}$). Also Kawase[361] reported an increase in the ethylene concentration in the roots of garden plants (chrysanthemum, sunflower, tomato, and radish) under the effect of flooding the roots in the soil with water.

The examples that reported of increase in the ethylene concentration in roots due to lack of aeration of the nutrient solution or flooding the soil with water cannot constitute a proof of an increase in the production of ethylene by the roots under anoxic conditions, as these conditions cannot be considered anoxic. In the nonaerated nutrient solution oxygen concentration, as a result of internal oxygen transport through the plant was equivalent to approximately 0.012 $m^3m^{-3}$ in the air.[171,336] In the case of the waterlogged garden plants mentioned,[361] there must also have been certain amount of oxygen in the plant roots from the internal transport.

In turn, under anoxic conditions (in an atmosphere of nitrogen), no evolution of ethylene was observed during the incubation of excised roots.[171,337]

On the basis of the foregoing it can be stated that ethylene in the roots is only produced in the presence of oxygen, although the required oxygen content is low. The lower limit of oxygen concentration, as well as its optimum concentration for ethylene production, have not yet been determined.

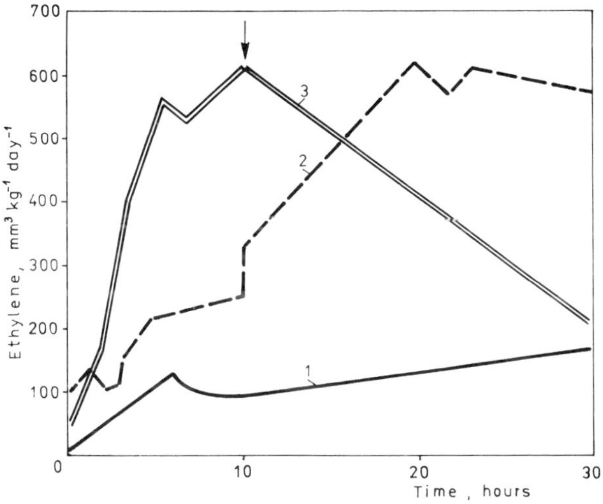


FIGURE 11. Ethylene evolution rate from roots of *Vicia faba* L. as affected by facing a physical barrier; (1) control, (2) with physical barrier, (3) physical barrier removed after 10 hr (as indicated by the arrow). (Modified from Kays, S. J., Nicklow, C. W., and Simons, D. H., *Plant Soil,* 40, 565, 1974.)

Moreover, it has been observed that waterlogging stimulates the evolution of ethylene in the shoots.[72,334,361] Another factor stimulating the production of ethylene in plant roots is a mechanical obstacle in the way of root growth,[363] as shown in Figure 11.

Bradford and Yang[72] explain the process of ethylene formation in the plant by the presence of a precursor of ethylene-ACC (1-aminocyclopropane-1-carboxylic acid) which is formed in the roots under anoxic conditions and then is transported through the xylem to the shoots where it is transformed into ethylene.

A third factor affecting the concentration of ethylene in the soil air, apart from the intensity of production and the rate of its escape from the soil medium, is the phenomenon of ethylene degradation in the soil. Little is known of this phenomenon so far, and the infrequent reports on the subject are contradictory. Cornforth[126] found that under oxic conditions ethylene in two soils was subject to microbial degradation. The degradation took place 50 times faster than its production in these soils under anoxic conditions. The degradation rate was practically constant at oxygen concentrations within the range from 0.21 to 0.005 $cm^3m^{-3}$ and ceased at oxygen concentrations below 0.002 $m^3m^{-3}$.

In turn, Yoshida and Suzuki,[820] while investigating the soils of the rice fields in the Philippines, observed that the ability to degrade ethylene was shown only by the soil of a field which had been submerged for a period of 9 years. This ability disappeared with an even partial drying of the soil. Unflooded soils or soils flooded only periodically did not degrade ethylene.

## E. Methane and Other Hydrocarbons

Under anoxic conditions, methane appears in the soil and in the presence of considerable quantities of organic matter in the soil, may reach concentration of several and even several tens percent.[183,701,821]

Smith and Restall[674] and Goodlass and Smith,[249] who incubated soils under anoxic conditions, also observed occurrence of trace amounts of higher hydrocarbons, such as propane, ethane, and butane.

## III. SUMMARY

Oxygen is the component of soil air which is of the greatest biological significance. Its concentration in the soil, depending on the conditions, is subject to considerable variations, from 0.00 to 0.21 $m^3m^{-3}$. The availability of this component for plants depends not only on its concentration in the soil air, but also on the ability of the roots to take it up, which is affected both by the arrangement of the soil particles and by the soil moisture content. The measure of this availability is the ODR.

Another biologically significant component of soil air is carbon dioxide. Its concentration in the soil air does not usually exceed a few percent by volume. Under conditions of soil saturation with water, its concentration is in the region of 0.1 $m^3m^{-3}$ and, in exceptional cases only, exceeds 0.2 $m^3m^{-3}$.

Apart from oxygen and carbon dioxide, there are also other gaseous components of the soil air, though present only in much lower quantities, e.g., nitrous oxide (from several to several tens and, occasionally, even to several thousand cubic centimeters per cubic meter), ethylene (0.01 to 20 $cm^3m^{-3}$), methane (reaching under an extremely anoxic conditions concentration of several percent), and some other hydrocarbons. Among these latter components ethylene has received the greatest attention, due to its high physiological activity even at concentrations below 1 $cm^3m^{-3}$. The origin of ethylene in the soil is not fully understood, but its presence is considered to be a proof of the existence of anoxic pockets in the soil. Similar interpretations of the presence of nitrous oxide in the soil air as an indicator of the presence of denitrification, and therefore also anoxic conditions, is less justified due to its formation (as has been established recently) in the process of nitrification.

The presence of methane and other saturated hydrocarbons in the soil is related to extreme anoxic conditions.

Chapter 4

# BIOLOGICAL AND CHEMICAL PROCESSES RELATED TO SOIL AIR

Soil air, particularly oxygen as its most important component, significantly affects biological and chemical transformations that occur in soil. Under anoxic conditions, caused for instance by soil submergence, changes in the microbial population are induced which bring about changes in the respiration of the soil, its enzyme activity, and redox potential. The consequence of oxygen deficiency in soil are also variations of soil reaction and chemical transformations of many soil components, both organic and inorganic, leading to characteristic morphological features in the soil profile.

In this chapter all these topics are treated with the exception of redox transformations of soil organic matter and of soil minerals which are beyond the scope of this book.

## I. MICROBIAL POPULATION, RESPIRATION, AND ENZYME ACTIVITY

A decrease in soil oxygen concentration brings about a lessening of oxygen consumption which was discussed in Chapter 1. Simultaneously, a change in the ratio of aerobic to anaerobic microorganisms, in favor of the latter, occurs.[367]

Immediately after submergence the number of aerobes and facultative anaerobes increases, reaches a maximum after 1 to 2 days and then sharply decreases.[546] This is shown, as an example, in Figure 1.*

Simultaneously the number of obligatory anaerobes increases reaching its maximum after 5 to 8 days of submergence and subsequently slowly decreases. Thus in the course of several days of submergence the soil, normally inhabited by aerobic bacteria, actinomyces, and fungi, becomes dominated by anaerobic bacteria (Figure 2) and algae. The latter, due to their ability to assimilate carbon dioxide with evolution of oxygen, cause stagnant surface water to become oversaturated with oxygen as compared to saturation with atmospheric air. Among the numerous algae present in flooded soil, those of particular importance are the blue-green algae able to fix molecular nitrogen.

The changes in activity of particular groups of microbes are accompanied by changes in enzyme activity in the soil. These changes, however, have been little studied. In the pertinent literature[107,241,612,729] the authors dealt mainly with dehydrogenase. The results of these studies are divergent and do not lead to compatible conclusions.

Thus Tate[729] observed a decrease in dehydrogenase activity during the first 5 days of soil submergence followed by a subsequent increase until the soil dried out. On the other hand Černa[107] reported a 2.5-fold increase in dehydrogenase activity in a soil brought to full water capacity. Moreover, Ross[612] found that dehydrogenase activity obtained during anoxic incubation is higher as compared to incubation in the presence of oxygen.

The only data concerning dehydrogenase and catalase activities as related to soil oxygen availability, expressed by ODR values, were reported by Gliński et al.[241] Their results are presented in Figure 3. As can be seen, a linear decrease of activity of dehydrogenase and an increase in activity of catalase with increase of ODR was observed throughout the entire range studied, from 10 to 60 $\mu g\ m^{-2}s^{-1}$.

## II. REDOX POTENTIAL AND ELECTRON ACTIVITY IN SOIL

In soil, as in living organisms, numerous redox processes, i.e., reactions connected with electron transfer, take place. The source of electrons in soil is organic matter while their

* Figures and tables for this chapter appear at the end of the text.

acceptors are oxidized inorganic compounds. These reactions proceed in the presence of enzymes produced by soil microbes.

The driving force of any chemical reaction is a tendency to decrease free energy of the system until, at equilibrium, the sum of the free energies of the products is equal to that of the remaining substrates. For a reversible redox reaction this driving force can be expressed in joules or in volts.

The change of free energy $\Delta G$ for the reversible reaction:

$$Ox + ne + mH^+ \rightleftharpoons Red \qquad (1)$$

is given by:

$$\Delta G = G_o + RT \ln \frac{[Red]}{[Ox]} - mRT \ln [H^+] \qquad (2)$$

where [Ox] and [Red] are the activities of the oxidized and reduced forms, respectively; $\Delta G_o$ is the change in free energy when [Ox] = [Red] and $[H^+] = 1$ mol $dm^{-3}$; R is the universal gas constant; and T is the absolute temperature.

This can be expressed in terms of potential using the relationship $\Delta G = - nFE$, viz.:

$$E = E_o + \frac{RT}{nF} \ln \frac{[Ox]}{[Red]} + \frac{m}{n}\frac{RT}{F} \ln [H^+] \qquad (3)$$

where E is the electromotive force of the reaction in volts, $E_o$ is the reaction voltage when [Ox] = [Red] and $[H^+] = 1$ mol $dm^{-3}$, and F is the Faraday.

If E is measured with respect to the standard hydrogen electrode it is denoted by Eh, Equation (3) then becomes:

$$Eh = E_o + \frac{RT}{nF} \ln \frac{[Ox]}{[Red]} + \frac{m}{n}\frac{RT}{F} \ln [H^+] \qquad (4)$$

or

$$Eh = E_o + \frac{2.303\ RT}{nF} \log \frac{[Ox]}{[Red]} - \frac{2.303\ RT}{F}\frac{m}{n} pH \qquad (5)$$

which is known as the Nernst equation.

This relationship can also be described by using the pe value which is defined as the negative logarithm of electron activity (in mol $dm^{-3}$)[313,562,720]

$$pe = - \log [e] \qquad (6)$$

Treating electrons as normal reactants in Equation (1) the equilibrium constant K can be expressed as follows:

$$K = \frac{[Red]}{[Ox]\,[e]^n\,[H^+]^m} \qquad (7)$$

The pe value is then:

$$pe = \frac{1}{n} \log K + \frac{1}{n} \log \frac{[Ox]}{[Red]} - \frac{m}{n} pH \qquad (8)$$

Denoting 1/n log K as $pe_o$, being the negative logarithm of electron activity when [Ox] = [Red] and pH = 0 we obtain:

$$pe = pe_o + \frac{1}{n} \log \frac{[Ox]}{[Red]} - \frac{m}{n} pH \qquad (9)$$

Comparison of Equations (9) and (5) shows that:

$$pe = \frac{Eh \cdot F}{2.303\ RT} \tag{10}$$

and

$$pe_o = \frac{E_o \cdot F}{2.303\ RT} \tag{11}$$

The factor 2.303 RT/F is 0.0591 V at 25°C and for this temperature

$$pe = \frac{Eh}{0.059} \tag{12}$$

Thus descriptions of a redox system in terms of redox potential (Equation [5]) and pe (Equation [9]) are equivalent to each other due to relations (10) to (12). In this book redox potential will be preferred as this value is taken directly from the measurement.

Equations (5) and (9) say that high Eh or pe values are connected with higher activities of oxidizing forms while low Eh and pe values characterize reduced conditions. These equations concern single redox couples, each having a characteristic value of $E_o$ or $pe_o$. The values of Eh and pe at [Ox] = [Red] and pH = 7 are denoted as $E_o^7$ and $pe_o^7$, respectively.

Examples of the more important redox systems, which are likely to operate in soil, with their $E_o$, $E_o^7$, $pe_o$, $pe_o^7$ values (after Ponnamperuma[562]) are listed below. (*Note*: pressure of gases $P_{O_2}$ and $P_{CO_2}$ is in kPa and due to this the free term in Equations (14) and (24), being the Eh value at a gas partial pressure of 1 kPa and pH = 0, is not equal to $E_o$ corresponding to the gas partial pressure of 100 kPa).

| Reactions and equations for Eh | $E_o$(V) | $E_o^7$(V) | $pe_o$ | $pe_o^7$ | |
|---|---|---|---|---|---|
| $O_2 + 4H^+ + 4e \rightleftharpoons 2H_2O$ | 1.229 | 0.812 | 20.80 | 13.80 | (13) |
| $Eh = 1.259 + 0.0148 \log P_{O_2} - 0.0591\ pH$ | | | | | (14) |
| $NO_3^- + 2H^+ + 2e \rightleftharpoons NO_2^- + H_2O$ | 0.834 | 0.420 | 14.11 | 7.11 | (15) |
| $Eh = 0.834 + 0.0295 \log \frac{[NO_3^-]}{[NO_2^-]} - 0.0591\ pH$ | | | | | (16) |
| $MnO_2 + 4H^+ + 2e \rightleftharpoons Mn^{++} + 2H_2O$ | 1.229 | 0.401 | 20.80 | 6.80 | (17) |
| $Eh = 1.229 - 0.0295 \log [Mn^{++}] - 0.118\ pH$ | | | | | (18) |
| $Fe(OH)_3 + 3H^+ + e \rightleftharpoons Fe^{++} + 3H_2O$ | 1.057 | −0.185 | 17.87 | −3.13 | (19) |
| $Eh = 1.057 - 0.059 \log [Fe^{++}] - 0.177\ pH$ | | | | | (20) |
| $SO_4^{--} + 10\ H^+ + 8e \rightleftharpoons H_2S + 4H_2O$ | 0.303 | −0.214 | 5.12 | −3.63 | (21) |
| $Eh = 0.303 + 0.0074 \log \frac{[SO_4^{--}]}{[H_2S]} - 0.0739\ pH$ | | | | | |
| $2H^+ + 2e \rightleftharpoons H_2$ | 0.000 | −0.414 | 0.00 | −7.00 | (22) |

$$\text{Eh} = 0.059 - 0.0295 \log P_{H2} - \tag{23}$$

$$0.059 \text{ pH} \tag{24}$$

As can be seen from Equations (13) to (23), in a soil solution many redox systems are involved and the resulting potential is called a "mixed potential".[64] The relation of this "mixed potential" to activities of particular redox components has not yet been established.

The range of Eh changes in soil is from 800 to −400 mV; the related pe range being from 14 to −7. The upper Eh and pe values correspond to equilibrium of the soil solution with atmospheric oxygen (Equation [14]) while the lowest correspond to equilibrium with gaseous hydrogen at atmospheric pressure (Equation [24]).

It should be mentioned that in the past another measure of oxidizing intensity called rH was used. This index was thought to comprise both Eh and pH effects. It was defined as the negative logarithm of a hypothetical hydrogen pressure (in bars) corresponding to given Eh and pH conditions.

Rearranging Equation (24) we obtain:

$$\text{rH} = \frac{\text{Eh}}{0.029} + 2\text{pH} \tag{25}$$

or

$$\text{rH} = 2\text{pe} + 2\text{pH} \tag{26}$$

It was later realized that in the case when oxidized or reduced forms are basic or acidic, the system does not resemble a simple hydrogen electrode and respective dissociation constants should be included in the formula.[313] Due to this the rH concept has been discarded by its proposer in whose opinion "rH has become an unmitigated nuisance"[116] although subsequent misuse of rH has been made by other workers.

As the reduction proceeds in the soil, a decrease in redox potential is observed; its value being governed by consecutive redox systems dominating with respect to concentration.

The redox potential in a "normally wetted" nonflooded soil is governed by the oxygen concentration as shown in Figure 4.

It can be seen that Eh decreases distinctly at oxygen concentrations below 0.05 $m^3m^{-3}$. The range of the Eh changes directly connected with oxygen concentration does not exceed 150 mV, and even at zero oxygen concentration Eh is about 300 mV.[185,264,743]

It should be noted that the redox potential of the $O_2/H_2O$ system (cf. Equation [14]) depends more on pH (59 mV per pH unit) than on partial pressure of oxygen (15 mV per tenfold change in $O_2$ pressure).

The course of the curves in Figure 4 also shows that they do not conform to the Nernstian equation, which confirms that these are "mixed" potentials. Equation (14) predicts, at pH 7 and oxygen partial pressure of 20 kPa to 0.2 kPa, Eh values in the range from 865 to 820 mV, i.e., far above those in Figure 4. Moreover, the slope of the curves is higher than theoretically predicted. In addition it varies between the soils.

The relationship of Eh to ODR is similar to that of Eh to $O_2$ concentration, and is shown in Figure 5. In the arable loess soil incubated in the laboratory for 8 days at oxygen concentrations which ranged from 0.00 to 0.21 $m^3m^{-3}$, the redox potential varied from 300 to 600 mV for the entire ODR range. However, in the peat soil subjected to prolonged reduction processes in the field, the Eh values were much lower and approached −200 mV at the lowest ODR values.

When soil is flooded, the exhaustion of oxygen usually occurs within several hours and further decrease of Eh below 300 mV is caused by soil reduction and subsequent involvement of other redox systems. It has been shown that reduction in soil is a microbial process and does not occur when the soil is sterilized. It should be stressed that in the case of submergence of air dry soil the drop of Eh is extremely rapid, and its value may reach −400 mV within 1 day.[815]

As can be seen from Figure 6, submergence of the soil causes its redox potential to decrease gradually until a certain fairly stable level is reached. The rate of the decrease and the minimum value of Eh depend on the intensity of reduction being related mainly to temperature and the amount of easily decomposable organic substrate and on the amount of bioreducible oxidized inorganic compounds which are electron acceptors (nitrates, manganese III and IV compounds, and ferric oxides).

Thus, incorporation of easily decomposible organic material and increase in temperature accelerate the reduction rate and the decrease in soil redox potential. On the other hand the presence of bioreducible inorganic compounds maintains Eh at a fairly constant level, e.g., in the case of nitrates at about 100 to 200 mV (cf. Figure 7). Further decrease in Eh occurs after exhaustion of these forms.[32,88]

When the submerged soil is subjected to drying out, regardless of the degree of reduction, a rapid increase of redox potential to the initial level is observed.[472,550,585]

In field conditions Eh undergoes dynamic changes, their amplitude and frequency being related to all the factors affecting it directly or indirectly viz. moisture content, respiration rate, temperature and air-filled porosity. An example of Eh changes in a soil irrigated with potato processing water is presented in Figure 8. It can be noted that in the case of surface irrigation variations of Eh were highest in the top layer of the soil.

An example of the dependence of Eh on the air-filled porosity of the soil is shown in Figure 9. A sharp drop of Eh is observed at a certain critical air-filled porosity value, where the pores become discontinuous. This critical value depends on soil structure and is higher for aggregates than for soil of natural structure. This statement is in accordance with the observation that the effectiveness of air-filled pores for mass flow of gases is lower in loose soils than in compacted ones (cf. Chapter 2, Section I.A.3).

In the case of prolonged soil submergence high Eh gradients occur at very shallow depths which is given in Figure 10. As can be seen, the depth of oxygenation corresponding to Eh $\geqslant$ 300 mV is from 14 mm for unamended soil to less than 1 mm in soil amended with 2% of rice straw. These distances of oxygen diffusion are close to those calculated for saturated soil (cf. Figure 12, Chapter 2). It should be pointed out that in paddy soils Eh in the root zone is higher than that in the bulk soil.[825]

In order to compare Eh values of different soils a correction to a common pH basis (usually to pH 7) is frequently made as follows:

$$Eh_7 = Eh - 0.059\,(7 - pH) \tag{27}$$

This correction is based on the tacit assumption that the Eh change is always 59 mV per unit of pH, which is not always true (cf. Equations [18], [20], and [22]). This objection has been supported by the results of Jugsujinda and Patrick[355] and Moraghan and Patrick[493] who observed, in a reduced soil incubated at controlled pH values, differences of Eh exceeding 120 mV per pH unit.

## III. SOIL REACTION

Processes of reduction in soil are accompanied by changes of reaction towards neutral. In other words, the pH of acid soils increases while that of alkaline soils decreases, stabilizing after several weeks within the range 6 to 7.5. This has been confirmed both under conditions of soil submergence[565] (cf. Figure 11) and during incubation at various oxygen concentrations[240] (Figure 12).

The changes of soil pH during 30 days of incubation in flooded conditions, as a function of initial pH (in oxygenated status) are presented in Figure 13. These changes, as can be seen, do not usually exceed 2 pH units. Some authors[546,562] reported higher pH changes (up to 3 units) induced by soil redox processes; this is also shown for one soil in Figure 11.

The increase in pH of acid soils on submergence is explained by $H^+$ uptake in most of the reduction processes (cf. Equations [13] to [24]). The decrease in pH of alkaline soils is caused by accumulation of $CO_2$ under anoxia.[562]

It has been found that after prolonged submergence the pH of soils is governed by the partial pressure of carbon dioxide. For various soils after 15 weeks of submergence, this relationship was[565]

$$pH = 7.17 - 0.36 \log P_{CO_2} \tag{28}$$

where $P_{CO_2}$ is in kPa. The pH of these soils is very sensitive to $P_{CO_2}$ and removal of $CO_2$ may raise pH by 2 units.[562]

For particular soils the following relationships have been found:[563]

for ferruginous soils:

$$pH = 7.28 - 0.58 \log P_{CO_2} \tag{29}$$

for calcareous soils:

$$pH = 7.33 - 0.66 \log P_{CO_2} \tag{30}$$

and for sodic soils:

$$pH = 9.85 + \log (\text{alkalinity}) - \log P_{CO_2} \tag{31}$$

where $P_{CO_2}$ is always in kPa.

## IV. SPECIFIC CONDUCTANCE AND IONIC STRENGTH OF SOIL

Submergence of the soil results in an increase of ion concentration in the soil solution during the first several weeks after which it gradually decreases. This is accompanied by parallel changes of specific conductance and ionic strength.[546,562]

A close relationship has been found between the specific conductance of the soil solution and its ionic strength viz. for ionic strength up to 0.05 *M*, its value (mol $dm^{-3}$) is numerically equal to 16 times the specific conductance (mmhos $cm^{-1}$) at 25°C.[566]

The initial increase in specific conductance occurs due to liberation of $Mn^{++}$ and $Fe^{++}$ ions, accumulation of $NH_4^+$, $HCO_3^-$, and $RCOO^-$ and removal of $Ca^{++}$ from the exchange complex by $Fe^{++}$. In addition in calcareous soils dissolution of carbonates in the presence of $CO_2$ and organic acids may be involved.

The subsequent decrease of specific conductance and thus ionic strength is explained by reprecipitation of $Fe^{++}$ and $Mn^{++}$ in the form of $Fe_3O_4{\cdot}nH_2O$ and $MnCO_3$.

The maximum values of specific conductance in most flooded soils are 2 to 4 mmhos $cm^{-1}$. The value of 4 mmhos $cm^{-1}$ is, according to Ponnamperuma,[562] critical for rice growth. This and higher values of conductance are met in saline soils rich in organic matter, and in acid sulfate soils.

## V. NITROGEN TRANSFORMATIONS

Most of the processes contributing to the nitrogen pool in soil and to its chemical changes are related to the oxygen status of the soil. They are denitrification, nitrification, processes of microbial fixation of atmospheric nitrogen, and ammonification.

### A. Denitrification

When soil is flooded or is close to water saturation the nitrate content decreases sharply with time.[557,738] This phenomenon is caused by nitrate reduction. According to Fewson and Nicholas[199] two kinds of nitrate reduction can be distinguished. One is $NO_3^-$ assimilation whereby $NO_3^-$ is reduced to $NH_4^+$ with subsequent incorporation into the cell, and the other is $NO_3^-$ dissimilation or $NO_3^-$ respiration, whereby $NO_3^-$ is a terminal electron acceptor in the absence of $O_2$. In the latter situation $NO_2^-$, $N_2O$, $N_2$, and $NH_4^+$, which are the reduction products are excreted by the microbes.

A special kind of nitrate respiration in which the end-products are gaseous ($N_2O$, $N_2$, NO) is called denitrification. It is performed by many groups of microbes, mainly by heterotrophic bacteria, such as[464,660] *Pseudomonas, Achromobacter, Bacillus, Chromobacterium, Mycoplana, Serratio, Vibrio,* and by some autotrophic bacteria, such as chemoautotrophic *Thiobacillus denitrificans* oxidizing sulfur compounds and *Micrococcus denitrificans* oxidizing hydrogen, as well as by some soil fungi (e.g., *Fusarium oxysporum, F. solani*[65]) and symbiotic *Rhizobia*.[332,826]

Denitrification is a strictly anoxic process and its appearance in soil indicates that at least part of the soil volume is deprived of oxygen. This process can be observed even when there is some oxygen content in the soil air; this is due to the presence of anoxic microsites inside soil crumbs or aggregates.

The rate of denitrification, high in water saturated soil, decreases sharply in a narrow air-filled porosity interval with the appearance of continuous air-filled pores increasing the gas diffusion coefficient (Figure 14).

According to Patrick[539] the reduction of nitrate occurs at a redox potential between 300 and 200 mV. During this process the rate of redox potential decrease is related to the rate of $NO_3^-$ depletion.[557] When nitrates or nitrites are presented in soil the redox potential is about 100 to 200 mV. The rate of denitrification depends on pH and temperature.[34,210,562,753]

According to Ponnamperuma[562] in short term experiments lasting a few hours or days, the rate of denitrification increases with temperature up to 60°C. However, over a period of 2 weeks the overall rate is almost unchanged within the range 15 to 40°C. Denitrification of nitrates ceases at a temperature of about 5°C, while the denitrification of nitrites is still performed. As Ponnamperuma[563] pointed out, the denitrification rate depends linearly (first order kinetics) on nitrate concentrations and is slower in acid soils than in neutral ones.

Soil denitrification is of great interest in connection with gaseous losses of nitrogen fertilizers. Due to this, much attention has been paid to the $N_2O$ content in soil air and its flow to the atmosphere.[92,163,166,211,608,609,626] Losses of nitrogen in the form of $N_2O$ may be up to 1.2 kg N $ha^{-1}$ $day^{-1}$.[92,166,205,626]

A more precise insight into gaseous nitrogen losses due to denitrification can be obtained by application of the acetylene inhibition method.[628] It has recently been found that acetylene at concentrations of 0.001 to 0.05 $m^3m^{-3}$ in the soil air inhibits nitrification and microbial conversion of $N_2O$ to $N_2$ without affecting reduction of nitrate to $N_2O$; thus all the denitrified nitrogen can be measured as $N_2O$.[37,626,822]

### B. Nitrification

The opposite process to microbial nitrate and nitrite reduction is nitrification, i.e., the microbial oxidation of ammonia to nitrate. Nitrification consists of two steps. The first step comprises conversion of ammonia to nitrite and is brought about by the group of autotrophic bacteria called *Nitrosomonas*. The second step, i.e., conversion of nitrite to nitrate, is performed by a collective group of autotrophic bacteria called *Nitrobacter*. The process of nitrification is also brought about by certain heterotrophic fungi, e.g., *Aspergillus flavus,* some species of *Penicillium,* and by other bacteria (e.g., *Arthrobacter*.[305])

Nitrification is a typical oxic process positively correlated with both the oxygen concentration in the soil air and with the soil ODR; this is shown in Figure 15.

Recent work has shown that $N_2O$ can also be evolved from well-oxygenated soils, due to nitrification.[74,75]

It should be mentioned that nitrite, which is the intermediate product of both nitrification and denitrification, is readily decomposed chemically in soils, with the evolution of gaseous $N_2O$, NO, and $N_2$.[665,754] Because of this it never accumulates in soils and even in flooded soils its concentration rarely exceeds 3 mg $kg^{-1}$.[562] The relative contribution of chemical and microbial processes to gaseous nitrogen losses has not been fully determined.

### C. Nitrogen Fixation

There are two kinds of microorganisms able to fix atmospheric nitrogen in the soil: the symbiotic and free-living nonsymbiotic.

The most important, from the agricultural point of view, is the fixation of nitrogen by symbiotic *Rhizobia* living in the nodules of leguminous plants.

The best known nonsymbiotic nitrogen fixers are *Clostridium, Azotobacter, Beijerinkia, Derxia,* and *Azotomonas*. In addition some blue-green algae, several groups of fungi, some actinomyces, and many other bacteria possess this ability. Among these fixers there are both aerobic and anaerobic organisms. Thus symbiotic *Rhizobia,* and nonsymbiotic nitrogen fixers such as *Azotobacter, Arthrobacter, Beijerinkia, Derxia, Azotomonas,* and blue-green algae belong to aerobic organisms while *Clostridia* are strict anaerobes.

It has been shown that a low oxygen concentration in the rhizosphere leads to a decrease in the nitrogen fixing activity of detached nodules[49,148,531] and intact nodulated roots.[135,519] An example of the dependence of nitrogen fixation on oxygen pressure in the case of intact soybean nodules is presented in Figure 16.

In long-term experiments an interesting phenomenon was noted, that nitrogen fixation by intact soybean nodules could adapt to altered oxygen pressure.[134] Nitrogen fixation occurs also in flooded soils due to the existence of anoxic microorganisms, such as *Clostridia,* as well as oxic blue-green algae in the surface layer of these soils.[112,514,586]

According to Ponnamperuma[562] nitrogen fixation in rice soils is enhanced in the root zone of rice owing to internal transport of oxygen and nitrogen through the plants to the rhizosphere. Rice plants increase the nitrogen fixation rate due to presence of aerobic nitrogen fixing bacteria on their roots. It was found that nitrogen fixing bacteria from the rhizosphere are microaerophiles[777] for which the optimum oxygen partial pressure for nitrogen fixation is about 4 kPa.[431]

### D. Ammonification

Ammonification is a process of ammonia release from organic matter and is accomplished by a wide group of microorganisms, of which fungi are normally the most important.

Ammonification comprises many types of reactions including both oxidative and reductive processes.[305] Ponnamperuma[562] states that although oxic ammonification may be more rapid than anoxic, the overall rate of ammonia release is higher in submerged soils due to the decreased rate of the process of its immobilization.

## VI. MANGANESE TRANSFORMATIONS

Manganese (Mn) in soil is considered to exist in four forms: water-soluble, exchangeable, reducible, and residual.[256,547] The relative contribution of these forms depends on the soil oxygen status.

Investigations of Mn transformations and also those of iron and nitrogen in soils have been performed in several ways. Most frequently it was an incubation of a flooded soil.[537,743] In some cases a soil suspension was incubated at controlled pH[493] or Eh[538,743] or with control of both Eh and pH.[119,256] Lately a method of controlling Eh in packed soil cores has been developed.[544] Another approach is incubation of soil samples at established moisture and oxygen contents.[692] These methodical differences must be kept in mind when transferring the results obtained in the laboratory to field conditions.

During flooding of soil the water-soluble fraction of Mn initially increases with time, and after reaching a maximum (after one to several weeks) decreases (cf. Figure 17). According to Ponnamperuma[563] in acid soils high in Mn and in organic matter, the concentration of water-soluble Mn increases up to 90 mg $kg^{-1}$ and then shows a rapid decline to a fairly stable level of about 10 mg $kg^{-1}$.

Soils high in Mn and low in organic matter also give high maxima but they are late and broad. In alkaline soils and in soils low in Mn, water-soluble Mn rarely exceeds 10 mg $kg^{-1}$ at any stage of submergence.

Ponnamperuma[562] and Ponnamperuma et al.[564] suggested that the decrease of $Mn^{++}$ concentration after the peak is due to precipitation of $MnCO_3$. In these conditions the $Mn^{++}$ concentration conforms to:[537,562]

$$pH + \frac{1}{2} \log [Mn^{++}] + \frac{1}{2} \log P_{CO_2} = Q \qquad (32)$$

where Q is from 4 to 4.4, while the theoretical value for $MnCO_3$ in equilibrium with $Mn^{++}$, $H_2O$, and $CO_2$ should be 4.06.[562]

Reduction of Mn is performed mainly by heterotrophic facultative anaerobes, while its oxidation is exclusively chemical. The reduction of Mn in soil takes place simultaneously with that of nitrate in an Eh range from 400 to 200 mV[82,83,256,544,743] at pH close to neutral. In acid conditions (pH 5) reduction of Mn is performed even at a redox potential as high as +500 mV.[256]

It was shown that coincidence of nitrate and Mn reduction normally observed in soils is not an obligatory phenomenon. In the case of bacteria strains not possessing nitrate reductase, nitrates remained unaffected even at Eh levels of −300 mV at pH 4.7 while the reduction of Mn proceeded normally.[497,524]

The relationships between water-soluble plus exchangeable and reducible Mn fractions are presented in Figure 18. At the Eh values allowing Mn to be reduced, a decrease in reducible Mn was observed to the advantage of its exchangeable and water-soluble fractions.

When the soil is flooded or deprived of oxygen a gradual increase in the reduced forms of Mn occurs when Eh is in the range between 400 and 200 mV[547] (Figure 19). In prolonged flooding after the bioreducible Mn has disappeared, a decrease in the extractable Mn can be observed due to its reprecipitation in the form of carbonates and sulfides.

It was found that during incubation of a soil suspension at controlled Eh-pH conditions the activity of Mn in solution increased with the decrease of Eh and pH values.[256] This relation is described by the equation:

$$Eh = 0.061 + 0.276\, pMn^{++} - 0.158\, pH \ (R^2 = 0.896) \qquad (33)$$

or

$$pMn^{++} = -0.221 + 3.623\, Eh + 0.512\, pH \qquad (34)$$

where $pMn^{++} = -\log [Mn^{++}]$

The empirical Equation (33) does not correspond to the Nernstian relation of any of the numerous known redox systems of Mn.[256,563,564,810]

Incubation of loess soil at moisture tensions of 10 to 100 kPa brings about an increase in the amount of reduced Mn with time at lower oxygen concentrations (below 0.1 $m^3m^{-3}$). An example of this relationship is shown in Figure 20.

The relation of $Mn^{++}$ to Eh and pH for the same soil is presented in Figure 21. The changes of Eh from an initial value of about 500 mV to about 350 mV and of pH from an initial value of about 6 to about 8 occurring during incubation were accompanied by an apparent increase of $Mn^{++}$ content at Eh levels below 400 mV.

## VII. IRON TRANSFORMATIONS

In the case of iron, as in the case of manganese, the following four forms can be distinguished: water-soluble, exchangeable, reducible, and residual. The relative contribution of these forms is determined by the redox conditions in the soil.

In water-soluble fraction 10 to 30% of $Fe^{++}$ was found to be present as organic complexes.[563]

Iron transformations in soil are both chemical and microbial; iron reduction being predominantly microbial while during oxidation, chemical processes are of equal or even greater importance.

Iron reduction is performed by heterotrophic facultative anaerobes (e.g., *Bacillus polymyxa, Enterobacter aerogens*) and by obligatory anaerobes such as *Clostridia*.[294,496,498,521,522] The ability to reduce iron is also possessed by *Escherichia freundii, Aerobacter aerogens, B. circulans, B. cereus, E. coli, Serratio marcescens, B. pumilis, B. sphaericus,* and *Pseudomonas aeruginosa*.[464,523]

It has been shown, in addition, that amorphous ferric compounds are reduced in preference to the crystalline forms.[498,499,524]

The Eh values reported for iron reduction in soil usually vary from 400 to 120 mV.[82,83,258,538,544,692,728] Gotoh and Patrick,[257] during the incubation of soil suspensions under controlled Eh-pH conditions, found that the critical Eh value for iron reduction in soil was highly dependent on pH; being 300 mV at pH 5 and $-100$ mV at pH 8. This is shown in Figure 22.

Iron reduction follows the reduction of nitrates and manganese. This sequence, although always observed during incubation of soil with the original microbial population, can not be treated as obligatory.

Munch and Ottow[497] showed that reduction of iron by *Clostridium butyricum*, which is incapable of using nitrate as an electron acceptor during respiration, is not inhibited by the presence of nitrates. Nitrate was not attacked through the entire incubation period even in completely reducing conditions (Eh $= -300$ mV at pH 4.5 to 4.8).[497,525] On the other hand the presence of $MnO_2$, which is utilized as an electron acceptor by the *Clostridium* strain, inhibited iron reduction. In the case of another strain of bacteria, *B. polymyxa,* capable of utilizing both nitrates and $MnO_2$ as well as iron oxides as electron acceptors, iron reduction was inhibited by the presence of nitrate and/or $MnO_2$. The suppressing effect of nitrate on iron reduction by bacteria possessing nitrate reductase (capable of performing nitrate respiration) has also been reported in other papers.[294,521,522,524]

These results, and also later studies[499] with separation of microbial cells from the Fe(III)-oxides, confirm an opinion that Fe(III)-oxides are used directly as hydrogen acceptors by microbes rather than that they are transformed at low Eh and/or by certain "reducing" metabolites. Nevertheless, in natural conditions Eh and pH values are the best indicators of soil redox status and are commonly used to characterize it.

When the soil is submerged the content of water-soluble iron increases, reaches a peak, and declines (Figure 23). For acid soils, rich in organic matter and in active iron, peak values as high as 600 mg $kg^{-1}$ are reached within 2 to 4 weeks of flooding, and then decline to level of 50 to 100 mg $kg^{-1}$. In soils high in organic matter and low in active iron, high water-soluble iron content may persist for several months. In neutral and calcareous soils water-soluble iron usually does not exceed 20 mg $kg^{-1}$. Low temperature retards the peak and broadens the area under it.[562,563]

During the last few years some workers[252,537,562,563,566,809] have attempted to describe the changes in concentrations of $Fe^{++}$ in submerged soils quantitatively in terms of Eh, pH, and $Fe^{++}$ activities. They have shown that in such soils the following redox equilibria are involved:

$$Fe(OH)_3 + e + 3H^+ \rightarrow Fe^{++} + 3H_2O \quad (35)$$

for which

$$Eh = 1.057 + 0.059\ pFe^{++} - 0.177\ pH \quad (36)$$

or

$$pe = 17.9 + pFe^{++} - 3\ pH \quad (37)$$

$$3Fe(OH)_3 + e + H^+ \rightarrow Fe_3(OH)_8 + H_2O \quad (38)$$

for which

$$Eh = 0.43 - 0.059\ pH \quad (39)$$

or

$$pe = 7.3 - pH \quad (40)$$

$$Fe_3(OH)_8 + 8H^+ + 2e \rightarrow Fe^{++} + 8H_2O \quad (41)$$

for which

$$Eh = 1.373 + 0.0885\ pFe^{++} - 0.236\ pH \quad (42)$$

or

$$pe = 23.3 + 1.5\ pFe^{++} - 4\ pH \quad (43)$$

It was found that these three systems operate after the peak of water-soluble iron in a wide range of salt concentrations.[537] It was suggested that in the first stage of submergence the $Fe(OH)_3/Fe^{++}$ system should be active, but this has not been clearly confirmed.[562]

Patrick[540] pointed out that inorganic redox systems can be described in both intensity and capacity terms; the intensity factor determining the relative ease of the reduction, whereas the capacity factor denotes the amount of the redox system undergoing reduction. The intensity factor can be expressed by the free energy of the reaction, or by Eh, or pe. The capacity factor of the redox system can be expressed, for instance, in terms of its oxygen equivalent.

The reaction of inorganic bioreducible oxidants with glucose and their oxygen equivalents are presented in Table 1. It is shown that, e.g., 6.98 mg $Fe^{+++}$ $kg^{-1}$, is equivalent to 1 mg $O_2$ $kg^{-1}$ while as expressed on molecular basis 2 mol of $Fe_2O_3$ are equivalent to 1 mol of oxygen.

Among the redox systems operating in soils the ferric oxyhydroxide system usually has a greater equivalent than all the other redox systems together. The capacity factors for oxygen (about 10 mg $kg^{-1}$ of the soil) and nitrate (from 0 to more than 100 mg $kg^{-1}$) systems are usually low in the soils, while the manganese oxide system is intermediate (the bioreducible manganese content varying from less than 100 mg $kg^{-1}$ to over 1000 mg $kg^{-1}$ corresponding to 30 and 300 mg $kg^{-1}$ of oxygen, respectively).

In some soils the amount of bioreducible iron is sometimes over 6000 mg $kg^{-1}$,[257,529,539,540] which is equivalent to about 850 mg $O_2$ $kg^{-1}$.

The rate of iron reduction in soil depends on Eh, pH, temperature, and organic matter content. Reduction of iron and manganese (cf. Figures 18 and 22) and metal oxides in general is favored by hydrogen ions while nonmetal redox systems, such as nitrate or sulfate, are inhibited by low pH.[539]

However, in natural conditions when initially acid soil is subjected to anoxia accompanied by decrease of Eh and increase of pH, another pattern is recorded (Figure 24), similar to that in the case of manganese (cf. Figure 21).

An opposite process to iron reduction, i.e., its oxidation, occurs in the presence of oxygen and is performed in both biochemical and chemical ways. Chemical oxidation of iron is favored by high pH. The ability to oxidize iron is possessed by several groups of bacteria. The best known are obligatory autotrophs (*Galionella ferruginea, Thiobacillus ferrooxidans*, and *Ferrobacillus ferrooxidans*) and some facultative autotrophs (*Leptotrix, Crenotrix*, and *Cladotrix*).[464]

Iron transformations play an important role in soil-forming processes[375,574] and especially in gleyzation and ferrolysis (cf. Section XIII.A).

## VIII. SULFUR TRANSFORMATIONS

During prolonged flooding, microbial reduction of sulfates occurs in soils rich in organic matter and the end products of the reduction are sulfides. It takes place after the reduction of nitrates, manganese, and iron has been completed, under the extreme reducing conditions. The reported redox potential values at which sulfate reduction begins vary from 40 to $-200$ mV.[82,83,120,186,375,574]

Addition to the soil of oxidants such as $KNO_3$, $MnO_2$, ferricitrophosphate, and $FePO_4$ delays reduction of sulfates and the appearance of sulfides; the effect being most pronounced in the case of $KNO_3$ due to its high solubility in water.[186]

There are only a few groups of bacteria performing the reduction of sulfates to sulfides; obligate anaerobic *Desulfovibrio* sp. being the most important. This ability is also possessed by *Clostridium nigrificans*, *B. megatherium*, and *Pseudomonas zielinskii*.[464,568] There are also numerous bacteria as well as actinomyces and fungi able to transform partly reduced sulfur compounds such as thiosulfates, sulfites, etc., to sulfides.

The appearance of sulfide ions brings about major changes in soils due to the precipitation of sulfides of heavy metals.

In submerged alkaline and neutral soils sulfate up to 1500 mg $kg^{-1}$ may be entirely reduced within a few weeks of flooding. Acid soils first show an increase in water soluble $SO_4^{--}$ after flooding (due to release of $SO_4^{--}$ sorbed by clay and hydrous oxides of iron and aluminium following increase in pH) then a slow decrease lasting several months.[562]

Reduction of sulfate follows first order reaction kinetics; the velocity constant being several hundred times higher for alkaline soils than for acid soils. Due to this, acid sulfate soils undergo slow reduction which is accelerated by liming.

The appearance of sulfide ions $S^{--}$ or $H_2S$ causes precipitation of insoluble sulfides, mainly FeS; although other heavy metal sulfides such as CuS, PbS, and CdS may be involved. Due to this the concentration of $H_2S$ in the solution of submerged soils rarely exceeds 0.1 mg $kg^{-1}$.[562]

Reduction of sulfate in rice soils may cause insufficient sulfur supply, immobilization of zinc and copper and, in soils low in iron, toxic effects of $H_2S$ may appear.

When oxygen enters the reduced soil a reverse process of microbial oxidation of sulfides to sulfate is brought about. The best known bacteria engaged in this process are *Thiobacillus, Beggiotoa, Thiotrix, Thioploca, Chromatium, Thiospirillum, Thiodictyon*, and *Thiocapsa*.[247,464]

## IX. SOLUBILITY OF PHOSPHORUS

Although phosphorus can exist in different valence states the process of its reduction in soil is very rare.[541,562] The main phosphorus transformations under anoxic conditions are the changes of its solubility due to transformation of other elements such as iron and manganese and changes in soil reaction.

Flooding the soil increases the availability of both native and added P, whether judged by solubility in different extractants or uptake by rice plants.[371,543,562,576,782]

Due to this a negative correlation between Eh and P solubility has been recorded.[538,635,782] An example of such a relationship is shown in Figure 25.

Obviously the relationship between extractable phosphorus and Eh is an indirect one and is caused by reduction of iron and changes in pH.

According to Stumm and Morgan[720] the increase in water soluble P in submerged acid soils results from: 1° hydrolysis of Al and Fe(III) phosphates, 2° release of P held by anion exchange on clay and hydrous oxides of Fe(III) and Al, and 3° reduction of Fe(III) to Fe(II) and liberation of sorbed or chemically bonded P; the first two reactions being brought about by pH increase during reduction.

Increase in P solubility in alkaline soil is a consequence of the decrease in soil pH on flooding and the increase of solubility of hydroxyapatite. However, incubation of soil suspension at controlled pH caused an increase of concentration of native P in solution at all the pH levels studied (5, 6.5, and 8).[319]

Quantitative studies of phosphate equilibria in flooded soils are difficult because of the sorption, the high contribution of organic P in a water-soluble fraction, the presence of phosphorus in colloids, the formation of water-soluble complexes with Ca, Mg, Al, and Fe, and the differentiated nature of solid phosphorus containing compounds.

## X. SOLUBILITY OF OTHER ELEMENTS

Besides the elements discussed above which undergo direct redox transformations in soil, the solubility of many other elements may change during anoxia due to secondary effects. These effects comprise changes in solubility, precipitation, adsorption, chelating, and pH changes which induce shifts in ion equilibria.

Among these elements heavy metals are of particular interest. Their ionic equilibria in different soil redox conditions are very complex[84,302,310,312,348,432,466,584,658,687,720] and the trend of solubility changes is not easy to predict. The empirical data from incubation show that solubility of Cu, Zn, Ni, Co, Cr, Pb, Mo, and V usually increases in anoxic conditions.[234,311,312,505,546,563,580,687,805] However, under extremely reduced conditions characterized by negative Eh values a sharp decrease in extractable Cd, Cu, Zn, and Pb was recorded (after Yamane and Okazaki[811]), which was caused by precipitation of sulfides.

Regarding other elements, the solubility of silicon generally increases while that of boron decreases on soil submergence.[563]

## XI. HYDROGEN TRANSFORMATIONS

Hydrogen formation in soils is observed at Eh below − 150 mV.[726] It may appear several days after flooding and disappear some days after reaching peak value. Hydrogen production in submerged soils is favored by the addition of organic matter.[492,814] The opposite process to hydrogen formation is its oxidation to water by autotrophic bacteria of genera *Hydrogenomonas* and also by *Desulfovibrio desulfuricans* and *Micrococcus denitrificans*.[247,464]

Hydrogen appeared in flooded soils amended with organic matter within 3 days after flooding, reached its maximum 10 days after flooding, and disappeared entirely 21 days after flooding.[492] Small amounts of hydrogen (up to 5%) are usually present in gas extracted from paddy fields.[808,821]

Microbial oxidation of hydrogen in soil is related to oxygen content according to the Michaelis-Menten relation (Equation [14] from Chapter 1) and shows a curvilinear relation with a maximum with respect to soil moisture content.[123] It was found that hydrogen normally fixed in the soil[430,647,648] may be evolved to atmosphere in the presence of leguminous plants.[121,122] It was shown[353] that N-fixing blue-green algae also evolve hydrogen in the presence of light.

Nodules of leguminous plants also contribute to the hydrogen budget in soil. They are able both to produce it[196,197,415,618,642] and to fix it.[22,157,158]

## XII. FORMATION OF CHARACTERISTIC ORGANIC COMPOUNDS

### A. Gases

The soil oxygen status is connected with formation of some volatile organic compounds; the main ones being ethylene and methane.

Ethylene formation in soil, although intensively studied, is still obscure and no definite Eh and pH conditions can be given as yet. Detailed discussion of ethylene formation in soils has been presented in Chapter 3.

In extremely reduced conditions, methane formation is observed both in flooded soils[492,726,814] and in soils of relatively low moisture content incubated under anoxic conditions.[242,243,693] Its content in field conditions has been reviewed in Chapter 3. It was found that methane in flooded soils appeared from several days to several weeks after flooding and was accumulating throughout the incubation periods. Organic matter amendment stimulated methane formation in an alkaline soil and suppressed it in an acid soil.[492] Methane production takes place during methane fermentation performed by anaerobic bacteria such as *Methanobacterium, Methanobacillus, Methanosarcina,* and *Methanococcus.*[464] Takai and Kamura[726] ascribe methane formation in flooded soils to Eh values from $-150$ to $-220$ mV. On the other hand Gliński and Stępniewski[243] observed methane formation during anoxic incubation and nonflooded soils at an Eh as high as 220 mV. In the presence of oxygen methane is easily oxidized by bacteria: *Bacillus methanicus, Methanomonas methanooxidans, Methylococcus capsualtus,* and *Pseudomonas methanica.*[316,317]

### B. Organic Acids and Other Organic Compounds

Generally, anoxic conditions slow down decomposition of organic matter in soil.

A specific feature of anoxic decomposition of organic matter is the accumulation of organic acids which are intermediate compounds in the process of transformation of organic matter to gases. This accumulation is especially abundant at low temperatures due to the retardation of organic acid decomposition. The organic acids are mainly acetic, butyric and also, in smaller amounts, formic, propionic, lactic, and valeric acids.[726,811,814]

The pattern of production of organic acids in curvilinear with a peak value of (10 to 50 m*M*) which appears after a period of several days to several weeks of submergence.[110,253-255,489,490,563,812-814] The amount of the acid and time of its maximum production depend on soil, temperature, and organic matter.

Among other organic substances present in flooded soils amino acids, phenolic substances, amines, alcohols, aldehydes, mercaptans, and esters have been reported.[10]

Some of the organic compounds formed during anoxia may adversely affect plant roots (cf. Chapter 5).

# XIII. SOIL AERATION AND PEDOGENESIS

## A. Hydromorphic Soils

Under certain conditions redox processes dominate other soil-forming factors and determine the main soil properties. These conditions are connected with permanent or seasonal submergence of soils.[562]

Soils may be overwetted with rainwater, flooded naturally or artificially, waterlogged due to impermeability of the soil material, or may have an impervious layer or a high water table. In these conditions hydromorphic soils are formed which belong to the Histosols and the suborder Aqu in the U.S. nomenclature[680] or Gleysols and Planosols in the F.A.O./U.N.E.S.C.O. nomenclature[176].

The high moisture content of these soils causes permanent or temporary anoxic conditions, which in mineral soils initiate gleyzation or gleying processes.[639] As a result of these processes, a characteristic color develops in the soil profile: olive gray, blue, or bluish green. This color is due to the presence of different forms of reduced iron.

The form of gley depends on the kind of soil wetting. Seasonal rainwater or waterlogging, when there is insufficient water outflow and percolation, causes partial anoxibiosis in the upper part of soil profile with the appearance of iron and manganese accumulation flecks and blotches, from millimeter to coin size, which are called mottles. They may be scarce or numerous, faint or prominent, spherical or irregular. In medium-textured soils the mottles, after many cycles of drying and wetting, harden to brownish-black concretions rich in Fe, Al, Ti, and Mn.[115,684]

The processes described above take place in moderately humid climates and form so called Pseudogleys (the surface-water gley).

In conditions of high and permanent moistening of deeper soil layers continuous gleyzation occurs, which is connected with advanced anoxibiosis. This phenomenon leads to the development of Gley soils, which are mineral or organo-mineral. They are situated in flat land or in depressions at a high ground water level. Gley soils are rich in total Fe and preserve a high proportion of $Fe^{++}$ in the reduction (*G*) horizon because of the nearly neutral soil reaction.

Another soil forming process connected with permanent high water saturation and anoxia is the bog-forming process. It consists of the accumulation of organic matter. In conditions of partial anoxia a mud is accumulated while fully anoxic conditions lead to the formation of peat. Lowering the ground water level in bog soils cause oxic processes to proceed which result in an intensive mineralization of organic matter. As a result postbog Muck soils or Black earths are formed.

The phenomenon of pseudogley or gley often occrs in Podzols or Podzoluvisols. It forms gley spots or permanent gley layers. In these soils the partial anoxibiosis, mainly in the coniferous forest litter after a long period of rainfall, causes mobilization of different compounds, which are easily displaced down the soil profile before being deposited. This results in the formation of the bleached horizon rich in quartz and poor in iron and clay which is characteristic of podzolic soils. This process has been called podzolization.

A process similar in effect may occur in surface-water gley soils.[77] This process is called ferrolysis (from ferrous and lysis; clay decomposition based on the alternate reduction and oxidation of iron). Ferrolysis, it has been suggested, is the main soil-forming process in seasonally wet, nearly level terrace and river plain landforms in the humid monsoon climate of Southeast Asia. This process consists of

1. Reduction of iron to ferrous ions
2. Displacement of the exchangeable cations from particles by $Fe^{++}$ and $Mn^{++}$
3. pH Rise from 4 to 5 to 6 to 7

4. Reoxidation of iron with formation of $H^+$ upon entry of oxygen to the profile
5. Displacement of the exchangeable ferrous iron by $H^+$ and corrosion of octahedral layers of the clay minerals at their edges

In this process, clay minerals are destroyed and basic cations are leached down the soil profile. It results in the development of albic horizons low in clay and of silica cementation in the underlying horizons.

### B. Paddy Soils

The so called ''paddy soils'', which are managed in a special way for the wet cultivation of rice, have a specific place in soil classification. Their management includes (1) leveling the land and construction of levees to impound water, (2) puddling, i.e., plowing and harrowing the water-saturated soil, (3) maintenance of 5 to 10 cm of standing water during the 4 to 5 months of rice growth, (4) draining and drying the fields at harvest, (5) reflooding after an interval of a few weeks to 8 months.[562]

These soils comprise almost all major groups of soils, occurring in land that is used for growing rice and whose surface is submerged for several months during growing season. They are generally located on relatively imperfectly drained land, naturally wet and with the water table near the soil surface. Their profile is characterized by (1) a partly oxidized *A* horizon high in organic matter, (2) a mottled zone in which oxidation and reduction alternate, and (3) a permanently reduced gley zone.

In paddy soils oxygen is rapidly consumed as a result of (1) microbial respiration, (2) chemical oxidation of $Fe^{++}$ and $Mn^{++}$, (3) microbial oxidation of $NH_4^+$ and carbon, and (4) oxidation of sulfides.[587]

The rapid disappearence of oxygen from the flooded soil is accompanied by an increase of other gases, in proportions 0.01 to 0.20 $m^3m^{-3}$ $CO_2$, 0.1 to 0.95 $m^3m^{-3}$ $N_2$, 0.15 to 0.75 $m^3m^{-3}$ $CH_4$, and 0.00 to 0.1 $m^3m^{-3}$ $H_2$.[546] This variation is due to differences in the duration of flooding, soil, and other environmental conditions, microbial inhabitants, and the nature of the organic and inorganic substances available as substrate.

The entire profile of a flooded soil is not uniformly low in oxygen. Normally the oxygen content is relatively high in a layer which may extend to 1 cm below the soil-water interface. Flood water always contains some dissolved oxygen which is available for various biological processes. Beneath this thin oxygenated layer, the oxygen content drops off sharply and with increased depth becomes virtually nonexistent.[546] An example of a flooded soil profile, indicating the concentration of dissolved oxygen at different depths, is shown in Figure 26.

All other processes connected with anoxia due to excess of water in soil, discussed in this chapter, are also relevant to paddy soils.

The reader interested in pursuing the subject of paddy soils further is referred to the excellent review on rice soils provided by Ponnamperuma.[562] A great deal of information regarding these soils is also contained in the *Proceedings of the Symposium on Paddy Soils* organized in China in 1980[572] and in the *Transactions of the 12th International Congress of Soil Science* in New Delhi, India, 1982.[736] Much interesting data on the genesis and use of hydromorphic soils can be found in the *Transactions of Commission V and VI of the International Soil Science Society on Pseudogley and Gley.*[574]

## XIV. SUMMARY

Soil oxygen status affects soil processes by governing the soil metabolism. It influences the composition of the microbial population and the transformations of many organic and inorganic compounds occurring in soils, which in turn result in changes of redox potential and the reaction of the soil.

Although redox potential is not a cause of most of these transformations (except chemical oxidation of $Mn^{++}$ or $Fe^{++}$) but their consequence, it is the best indicator for evaluating the degree of reduction. However, a quantitative prediction of soluble forms of particular nutrients in soil on the basis of Eh and pH is not as yet possible.

Much attention has been paid to chemical transformations in flooded soils. It has been well established that with the disappearance of oxygen in soil reduction of manganese, oxides and nitrates takes place first followed by reduction of iron. During prolonged flooding leading to the reduction of all bioreducible manganese and iron accompanied by a decrease of Eh to negative values, reduction of sulfates to sulfides and methane fermentation take place.

The sequence of the main redox processes in soil together with corresponding Eh intervals is summarized in Table 2. The protective role of iron oxides consists in both delaying reduction of sulfates and in decreasing $S^{--}$ concentrations (toxic for plant roots) by precipitation of FeS.

The redox transformations are accompanied by numerous changes which are still obscure; these are changes in ionic equilibria, sorption, precipitation, and ion exchange.

All these redox processes result in an increase of solubility of some elements, e.g., iron, manganese, phosphorus, silicon, molybdenum and decrease in the case of others, e.g., boron. The behavior of copper, zinc, lead, and cadmium is more complex; they may be released on soil reduction but precipitated later.

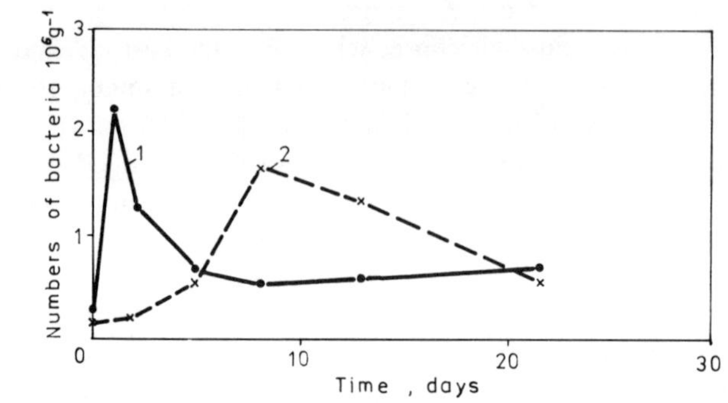


FIGURE 1. Changes in microbial population after flooding a paddy soil incubated at 35°C; (1) aerobic bacteria, (2) anaerobic bacteria. (Modified from Takai, Y. and Kamura, T., *Folia Microbiol.*, 11, 304, 1966.)

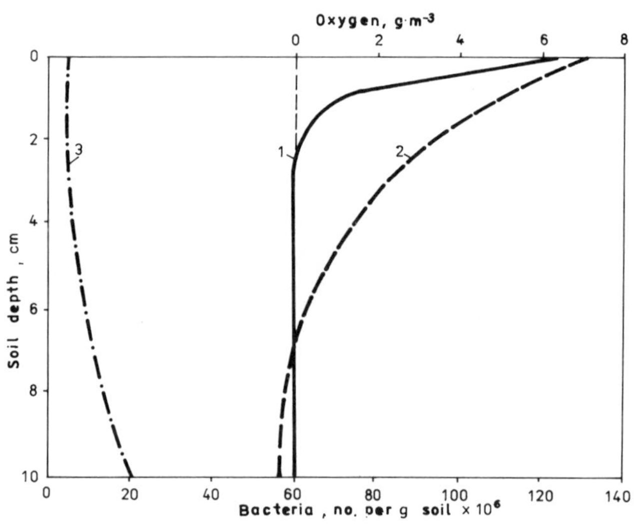


FIGURE 2. Distribution of oxygen concentration in soil water (1), aerobic (2), and anaerobic (3) bacteria in a flooded soil profile. (Based on Patrick, W. H. and Mikkelsen, D. S., in *Fertilizer Technology and Use*, 2nd ed., Soil Science Society of America, Madison, Wisc., 1971, 187.)

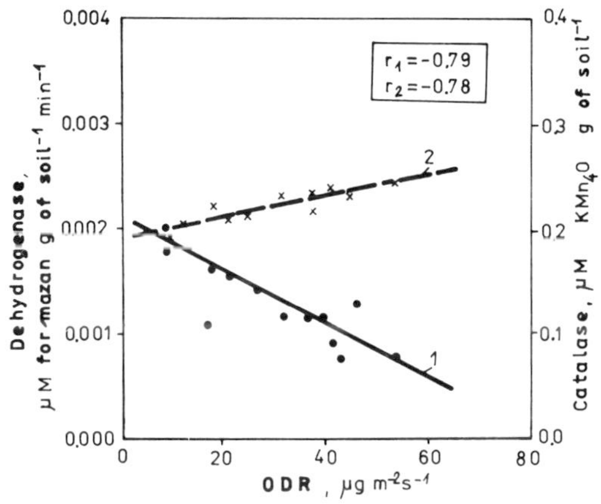


FIGURE 3. Dehydrogenase (1) and catalase (2) activities in a loamy soil ODR. (Data from Gliński, J., Stępniewski, W., and Kasiak, A., *Roczn. Glebozn.*, 34, 53, 1983.)

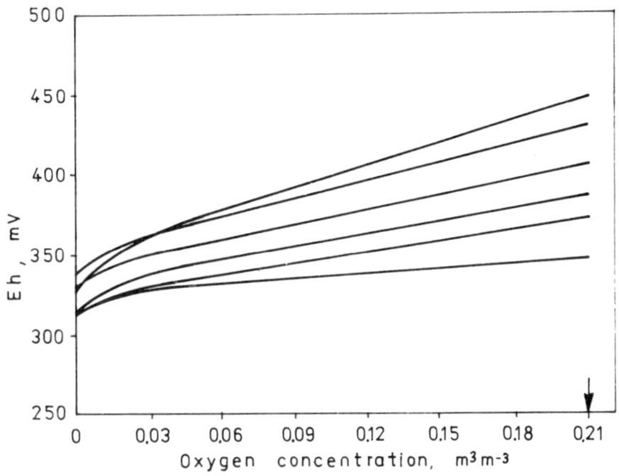


FIGURE 4. Redox potential of six different soils of moisture tension 33 kPa as a function of oxygen concentration in soil air. (Modified from Turner, F. T. and Patrick, W. H., in *Trans. 9th Int. Congr. Soil Sci.*, Adelaide, Australia, 1968, 55.)

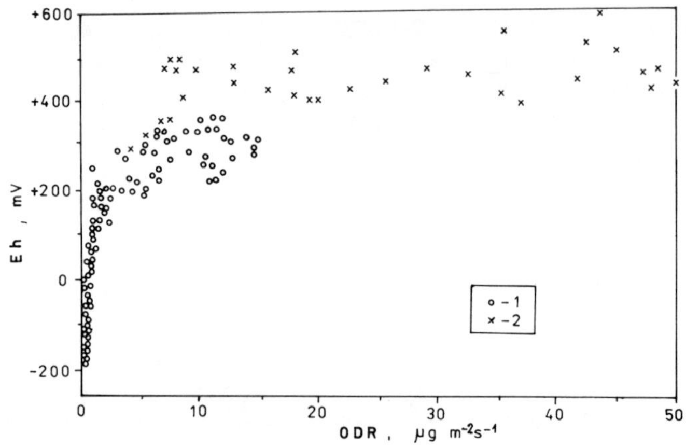


FIGURE 5. Redox potential as a function of ODR in (1) Scottish blanked-bog peat soil (field measurement), (2) ploughed layer of loess soils incubated at moisture tension 10 kPa during 8 days, at 20°C, at oxygen concentration varying from 0.00 to 0.21 $m^3m^{-3}$. (Data from Armstrong, W., *J. Soil Sci.*, 18, 27, 1967; Gliński, J., Stępniewski, W., Łabuda, S., and Przywara, G., *Roezn. Glebozn.*, in press.)

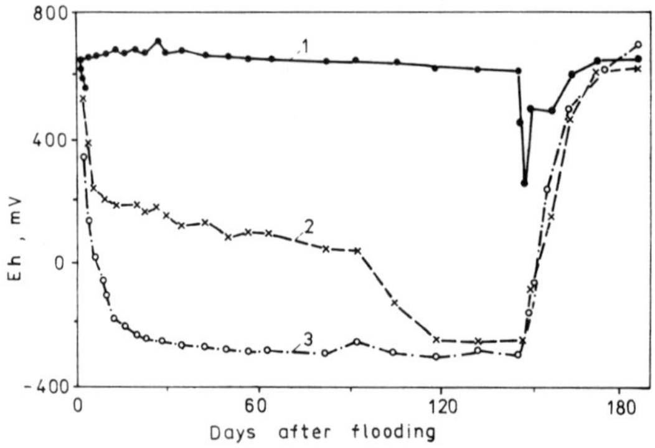


FIGURE 6. Effects of flooding and organic matter treatments on Eh of a clay soil; (1) control, (2) flooded, (3) flooded, with organic matter. (Modified from Willett, I. R., Divisional Rep. No. 31, CSIRO, 1978.)

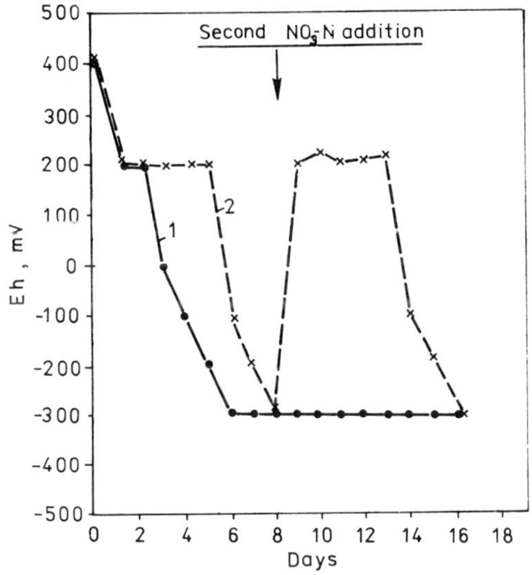


FIGURE 7. Changes of redox potential of a silty clay loam (Ap horizon) after saturation with water; (1) untreated soil, (2) soil with nitrate addition. (Modified from Bailey, L. D. and Beauchamp, E. G., *Can. J. Soil Sci.*, 51, 51, 1971.)

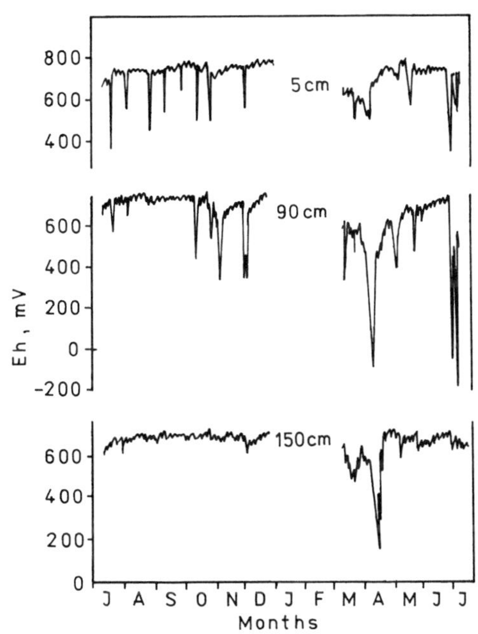


FIGURE 8. Redox potential in a sandy loam soil irrigated in field conditions with potato processing waste water and waste water mixed with canal water; 210 cm waste water during a year. (Modified from Smith, J. H., Gilbert, R. G., and Miller, J. B., *J. Environ. Qual.*, 7, 571, 1978.)

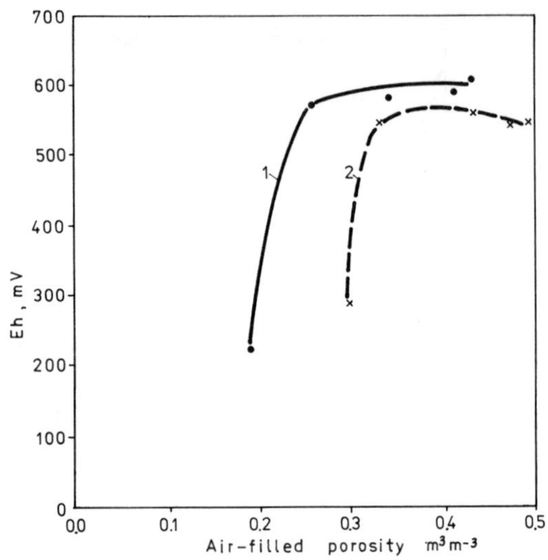


FIGURE 9. Redox potential vs. air-filled porosity in a loamy soil (1) and in its 2 to 3 mm aggregate fraction (2). (Modified from Grable, A. R. and Siemer, E. G., *Soil Sci. Soc. Am. Proc.*, 32, 180, 1968.)

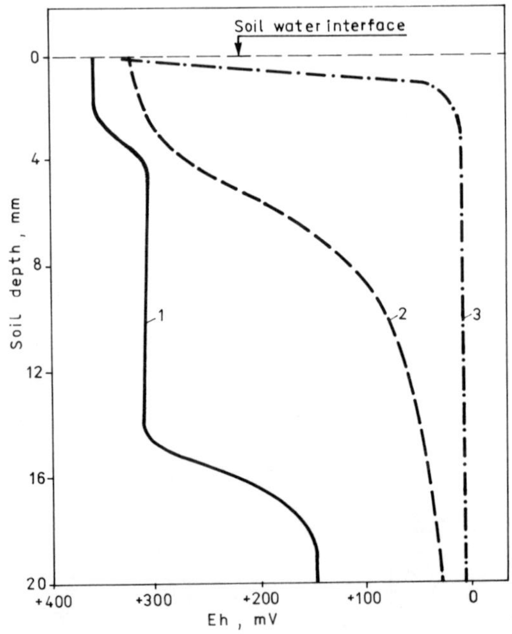


FIGURE 10. Distribution of redox potential in a paddy silt-loam soil after 3 weeks of flooding with a 7.6 cm water layer without (1) and with addition of 0.5% (2) and 2% (3) of finally ground rice straw. (Modified from Engler, R. M. and Patrick, W. H., *J. Environ. Qual.*, 3, 409, 1974.)

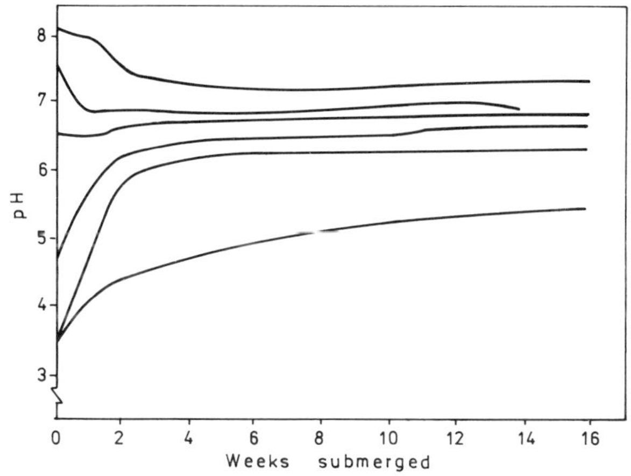


FIGURE 11. Changes of pH in different soils after flooding with water. (Data from Ponnamperuma, F. N., *Adv. Agron.*, 24, 29, 1972; Ponnamperuma, F. N., Martinez, E., and Loy, T., *Soil Sci.*, 101, 421, 1966.)

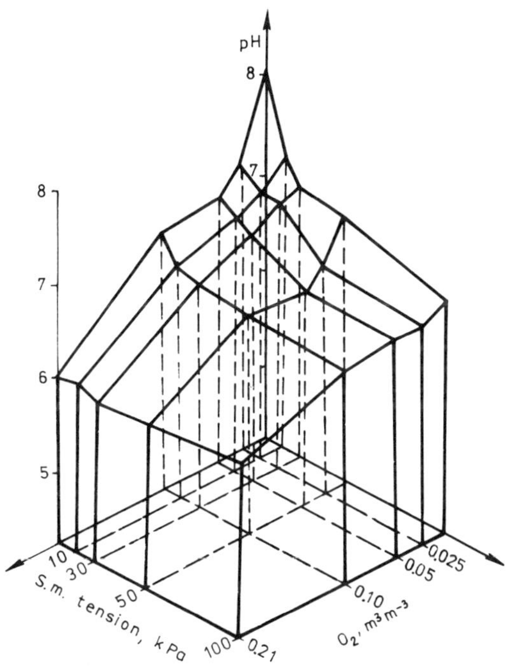


FIGURE 12. pH of a brown loess soil (measured *in situ* by placing the glass electrode bulb directly into the soil after 3 weeks of incubation at 20°C, at indicated soil moisture tension and oxygen concentration in soil air.

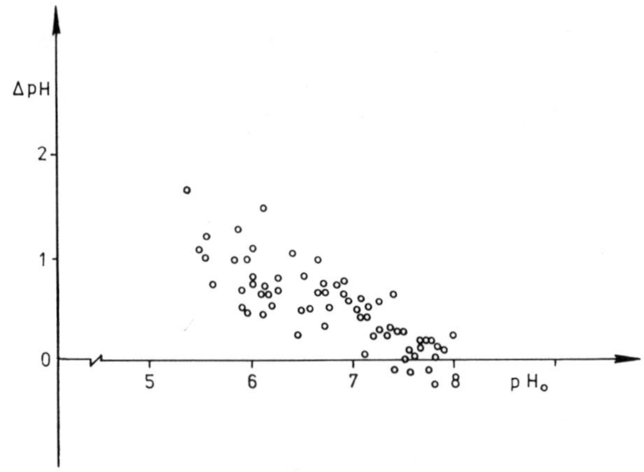


FIGURE 13. Changes *in situ* measured pH (ΔpH = pH final − pH initial as a function of initial pH values ($pH_o$) in 66 different soils after 30 days of flooding at 20°C.

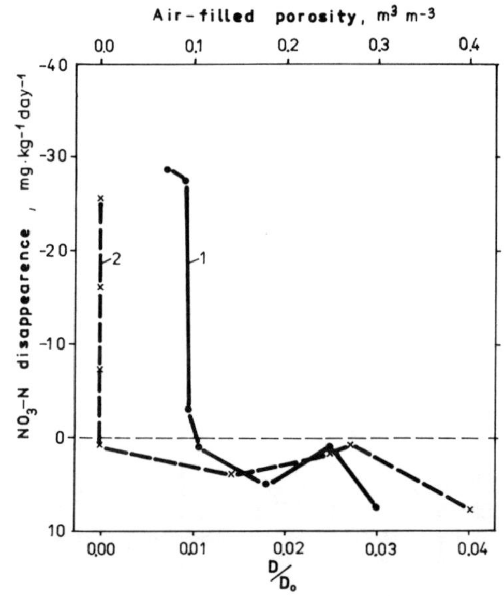


FIGURE 14. Nitrate reduction rate as related to air-filled porosity (1) and relative oxygen diffusion coefficient $D/D_o$ (2) in Crevasse loamy sand. (Data from Pilot, L. and Patrick, W. H., *Soil Sci.*, 114, 312, 1972.)

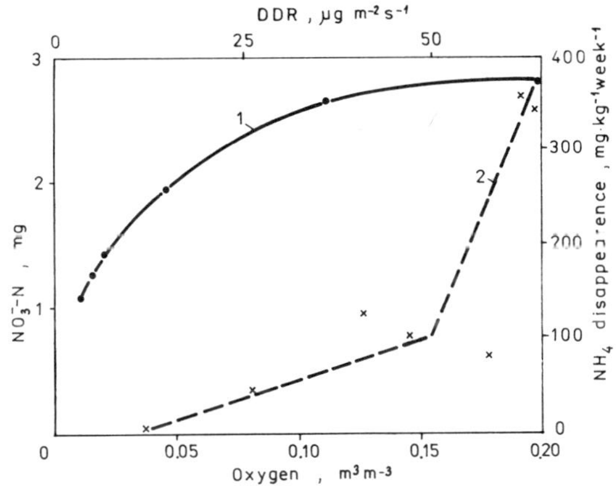


FIGURE 15. Nitrate produced during 21 days of incubation of soil in atmosphere containing different oxygen concentrations (1) and maximum nitrification rates estimated from ammonium disappearence as related to ODR (2).

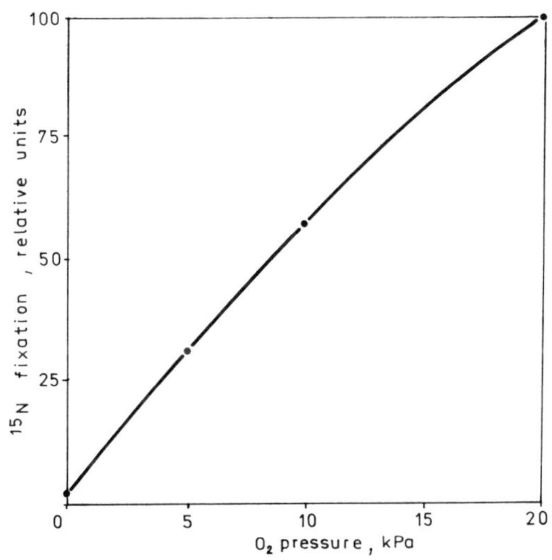


FIGURE 16. Effect of oxygen partial pressure on $^{15}N$ incorporation into the intact soybean nodules exposed to $^{15}N_2$ at nitrogen partial pressure 10 kPa and He as a supplementary gas. (Data from Ohyama, T. and Kumazawa, K., *Soil Sci. Plant Nutr.*, 26, 321, 1980.)

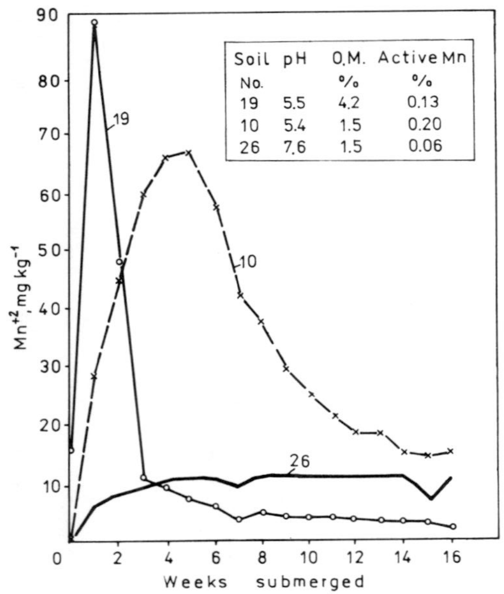


FIGURE 17. Kinetics of water-soluble $Mn^{++}$ in three submerged soils differentiated with respect to pH, organic matter, and active mangenese content. (Modified from Ponnamperuma, F. N., in *Proc. Symp. Paddy Soils China,* Nanjing, China, 1980, Institut Soil Sci., Academic Sinica, Ed., Science Press Beijing-Springer Verlag, Berlin, 1981, 59.)

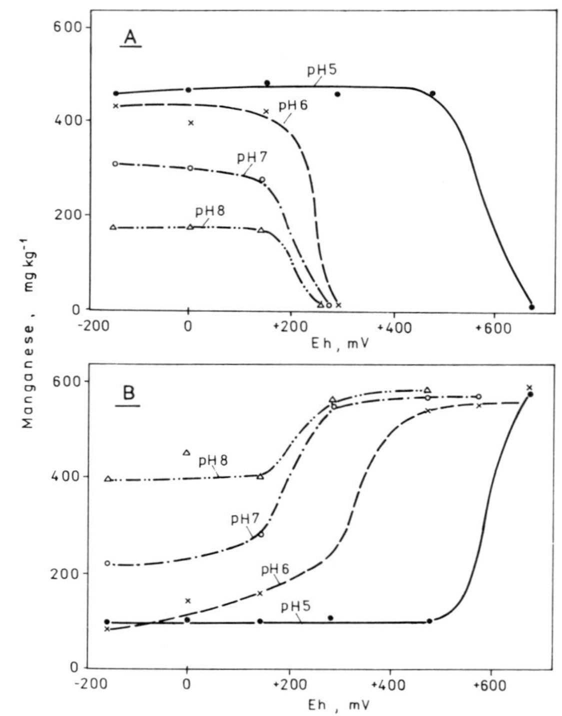


FIGURE 18. Distribution of soil manganese at different Eh-pH combinations between water soluble and exchangeable fraction (A) and reducible fraction (B). Soil suspensions were incubated at constant pH and Eh for 2 to 3 weeks. (Data from Gotoh, S. and Patrick, W. H., *Soil Sci. Soc. Am. Proc.*, 36, 738, 1972.)

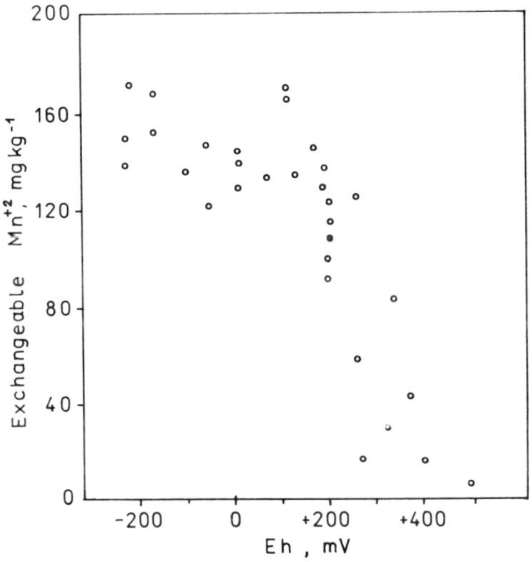


FIGURE 19. Exchangeable Mn (sodium acetate) in loamy sand soil as related to its Eh. The packed soil cores were incubated at different established Eh values. (Modified from Patrick, W. H. and Hendersen, R. E., *Soil Sci. Soc. Am. J.*, 45, 35, 1981.)

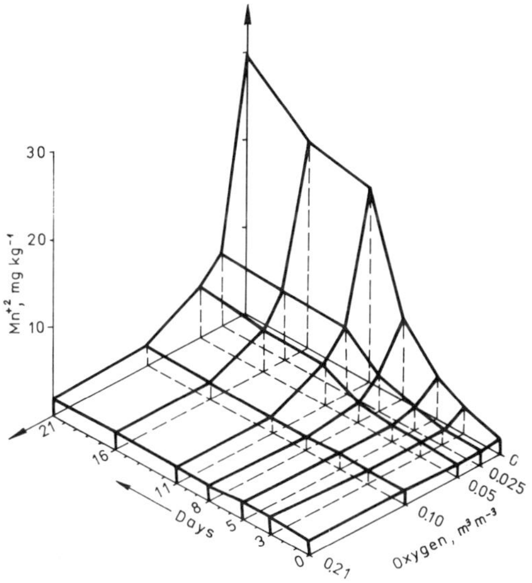


FIGURE 20. Changes of $Mn^{++}$ (1 *M* ammonia acetate fraction) in a loess soil during 3 weeks incubation in the stream of different $O_2$-$N_2$ mixtures at a soil moisture tension 10 kPa.

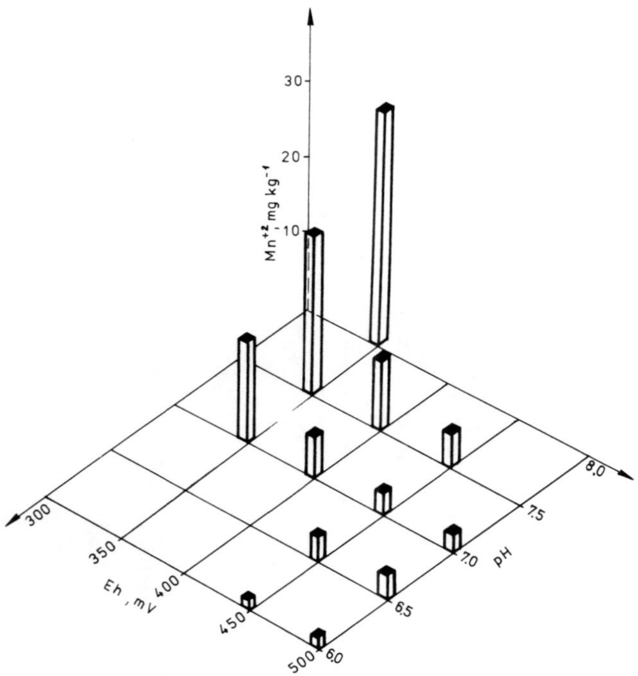


FIGURE 21. Dependence of $Mn^{++}$ (1 *M* ammonia acetate fraction) on Eh and pH (measured *in situ* by placing the glass electrode bulb directly into the soil) in a loess soil of initial pH 6 and Eh 500 mV. The data present changes occurring during 3 weeks incubation of soil samples of different moisture tension (10 to 100 kPa) and at oxygen concentration range from 0 to 0.2 $m^3m^{-3}$ in a stream of its mixture with nitrogen.

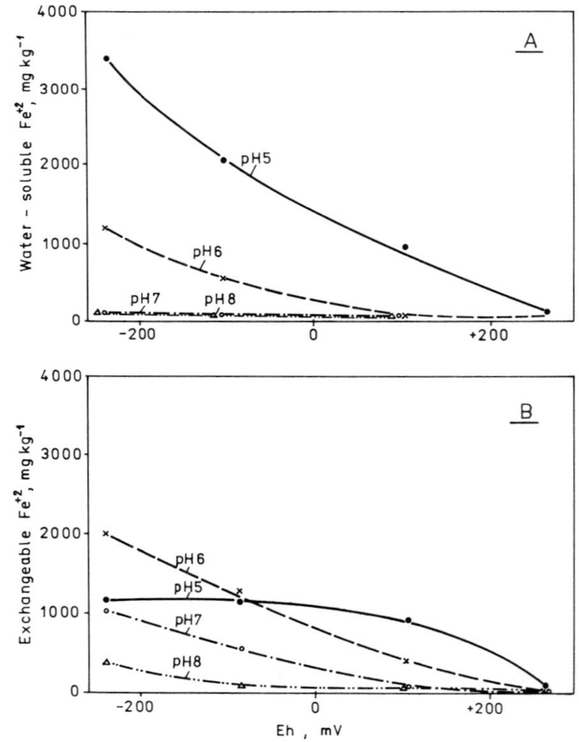


FIGURE 22. Water soluble (A) and exchangeable (*M* $NH_4OAc$ extractable minus water soluble) (B) iron fractions at various Eh-pH combinations. Soil suspensions were incubated at established pH and Eh for 2 to 3 weeks. (Data from Gotoh, S. and Patrick, W. H., *Soil Sci. Soc. Am. Proc.*, 38, 66, 1974.)

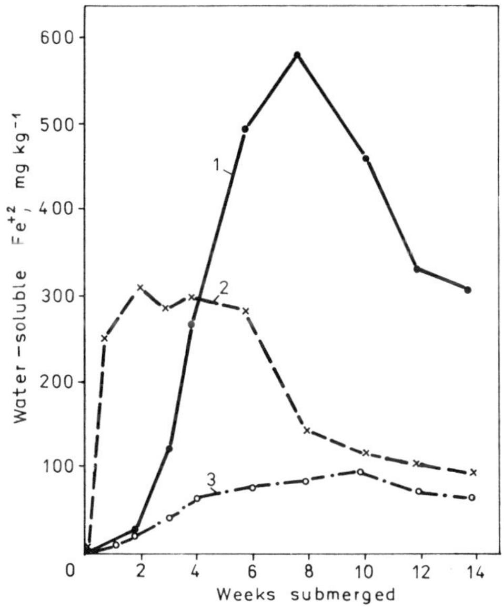


FIGURE 23. Changes of water-soluble iron in three soils: (1) Louisiana clay, (2) Casiguran sandy loam, (3) Maahas clay (after flooding at a temperature of 20°C). (Modified from Ponnamperuma, F. N., in *Proc. Symp. Paddy Soils China,* Nanjing, China, 1980, Institut Soil Sci., Academia Sinica, Ed., Science Press Beijing-Springer Verlag, Berlin, 1981, 59.)

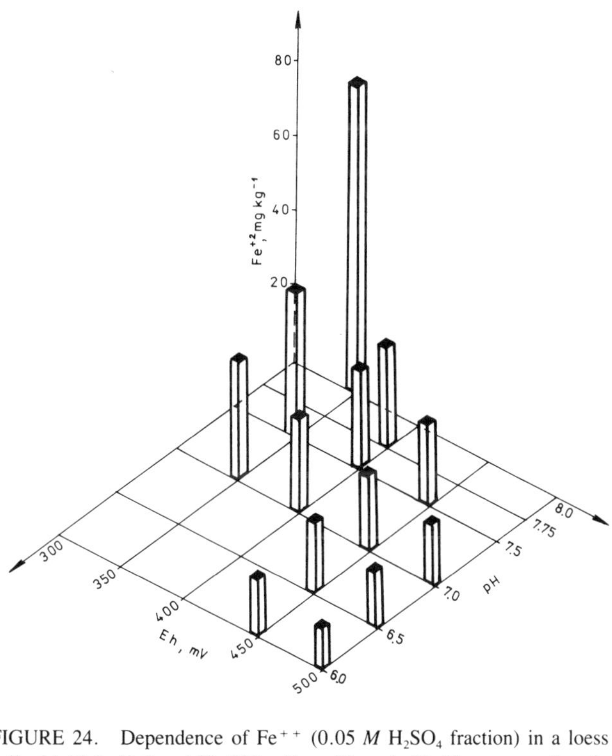


FIGURE 24. Dependence of $Fe^{++}$ (0.05 *M* $H_2SO_4$ fraction) in a loess soil of initial pH 6 and Eh 500 mV on Eh and pH. The same soil as in Figure 21.

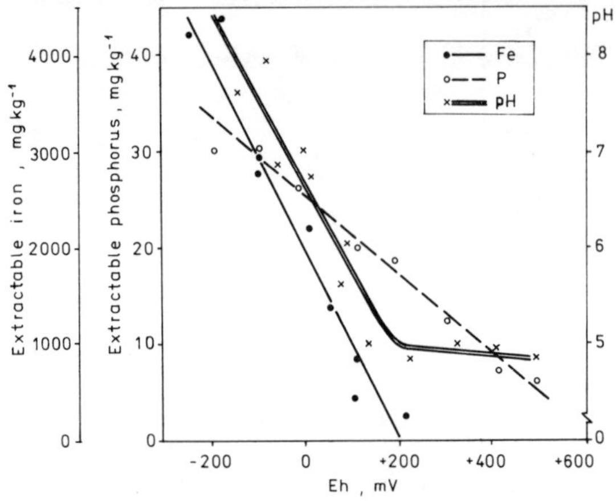


FIGURE 25. Dependence of pH, extractable iron (1 *M* sodium acetate of pH 2.8 fraction) and phosphorus (1 *M* HCl + 0.03 *M* $NH_4F$) on Eh obtained during 4 months incubation of a flooded Crowley silt loam soil. (Data from Patrick, W. H., in *Trans. 8th Int. Congr. Soil Sci.*, Bucharest, Romania, 1964, 605.)

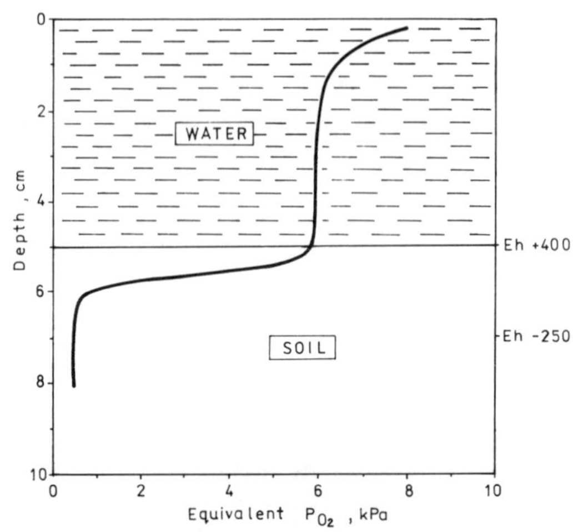


FIGURE 26. Oxygen profile in a flooded soil. (Modified from Patrick, W. H. and Mikkelsen, D. S., in *Fertilizer Technology and Use*, 2nd ed., Soil Science Society of America, Madison, Wisc., 1971, 187.)

**Table 1**
**REACTION OF INORGANIC OXIDANTS WITH GLUCOSE AND THEIR $O_2$ EQUIVALENTS REFERRED TO PURE COMPONENTS**

| Oxidant | Reaction | $O_2$ Equivalent Mole basis | $O_2$ Equivalent mg $kg^{-1}$ basis |
|---|---|---|---|
| $O_2$ | $C_6H_{12}O_6 + 6O_2 \rightarrow 6CO_2 + 6H_2O$ | 1 | 1 |
| $NO_3^-$ | $5C_6H_{12}O_6 + 24NO_3^- + 24H^+ \rightarrow 30\ CO_2 + 12N_2 + 42H_2O$ | 0.8 | 0.35 |
| $MnO_2$ | $C_6H_{12}O_6 + 12MnO_2 + 24H^+ \rightarrow 6CO_2 + 12Mn^{++} + 18H_2O$ | 2 | 3.43 |
| $Fe_2O_3$ | $C_6H_{12}O_6 + 12Fe_2O_3 + 48H^+ \rightarrow 6CO_2 + 24Fe^{++} + 30H_2O$ | 2 | 6.98 |

Modified from Patrick, W. H., in *Proc. Symp. Paddy Soils China*, Nanjing, China, 1980, Institut Soil Sci., Academia Sinica, Ed., Science Press Beijing - Springer Verlag, Berlin, 1981, 107.

**Table 2**
**SEQUENCE OF REDOX PROCESSES IN SOILS**

| Succession of processes | Eh (mV) at the beginning of the process |
|---|---|
| Respiration: $O_2 \rightarrow H_2O$ | >400 |
| $NO_3^-$-reduction | +600 to +300 |
| Formation of $Mn^{++}$ | +500 to +220 |
| Production of $Fe^{++}$ | +400 to +180 |
| $S^{--}$-accumulation | +100 to −200 |
| $CH_4$ formation | −150 to −280 |

Data from Ottow, J. C. G., in *Proc. Symp. Paddy Soils China*, Nanjing, China, 1980, Institut Soil Sci., Academica Sinica, Ed., Science Press Beijing-Springer Verlag, Berlin, 1981, 330.

# Chapter 5

# SOIL AIR AND PLANT RESPONSE

The influence of soil air on plants is complex. It comprises direct and indirect effects. The former are connected with the physiological effects of oxygen and carbon dioxide which are basic components of soil air. The latter are connected with the numerous changes occurring in the soil under the influence of oxygen and carbon dioxide (cf. Chapter 4). None of these changes is without an effect on the plants.

The indirect effects comprise the influence of both $O_2$ and $CO_2$ on redox processes in the soil, on its acidity and on the living organisms within it. All these factors influence the availability of nutrients for the plants as well as their well-being in general. A scheme of these mutual interrelations is presented in Figure 1.

The extreme case in the soil-air to plant relation is that in which the soil is submerged and the soil air is almost completely replaced by water. Such conditions may be permanent (e.g., in bog or marsh soils), intermittant (e.g., in rice soils), or sporadic and of short duration.

As to the plant's tolerance to soil submergence, we have a whole spectrum of plants — from hygrophites adapted to permanent flooding (e.g., rice) to mesophytes among which significant differentiation exists. Hence, among these plants there are those moderately tolerant to flooding (most grasses and cultivated plants), as well as those very sensitive to it so that after only a few days of flooding they die away (e.g., tobacco, peas). Xerophytes have been investigated to only a limited extent. Even if their root system is sensitive to oxygen deficiency, the fact is that some of them can live with a reduced root system, being capable of increasing their resistance to all stresses, and among them to an oxygen deficiency stress.

Obviously, plant tolerance to root flooding is relative because it also depends on other factors. One of them is the stage of plant development. Thus in the case of cereal plants (wheat, oats, and barley) we find they are most sensitive to soil flooding in the stem elongation stage and much less so in the earlier stages.

Among the environmental factors the most important is the soil temperature, the effect of which is shown in Figure 2.

The influence of flooding on plant roots is complex and can comprise all the factors presented in Figure 1. The general opinion is that oxygen deficiency is the main factor influencing the plant, at least during the first days of flooding. It was found,[382] however, that damage to plants (e.g., tobacco, tomato, sunflower) was greater in the case of plants growing on soils than in the case of those growing on pure sand. This implies the existence of some indirect effects of a reduced soil environment.

In this chapter both the direct and the indirect effects of soil air on plants will be discussed.

## I. DIRECT EFFECTS OF SOIL AIR ON PLANTS

It is difficult to separate the direct effects of soil air on plants from the indirect ones caused by the soil processes. It is especially difficult in the case of soil compaction and/or its flooding with water or sewage. The effects of soil compaction include, apart from all aspects connected with oxygenation, a mechanical factor, which seriously restricts root penetration and root growth. These problems are treated in the reviews by Cannell[101] and Chancellor.[108]

Most of the negative phenomena connected with deficient aeration are caused solely and directly by lack of oxygen for metabolic processes. Indirect effects appear only when there is a long-lasting shortage of oxygen in the soil.

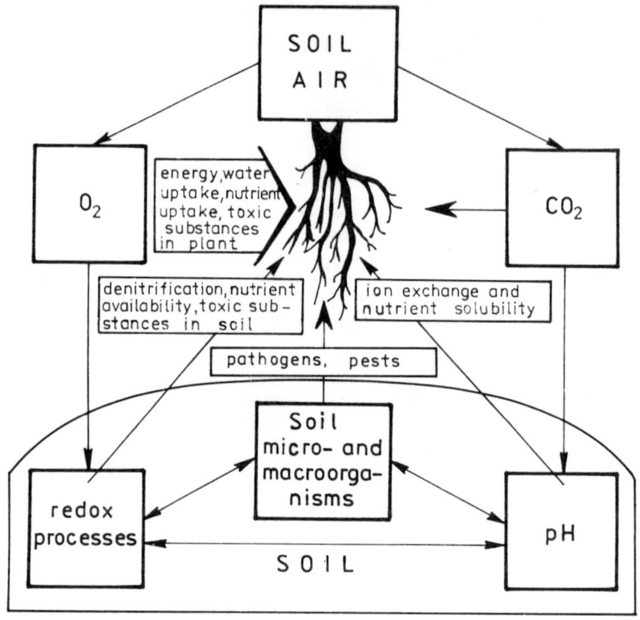


FIGURE 1. A scheme of direct and indirect effects of soil air on plants.

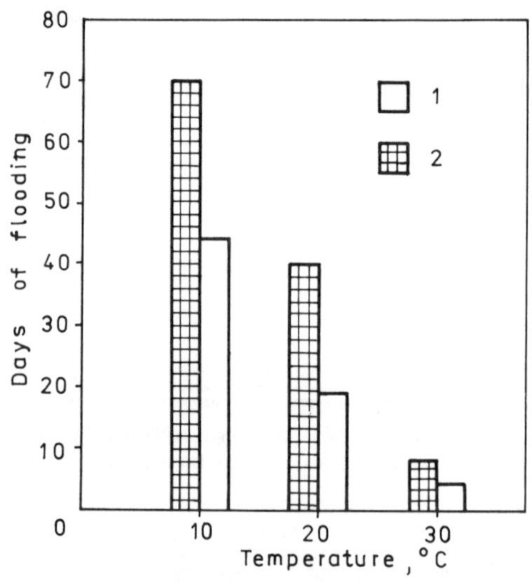


FIGURE 2. Flooding time causing death of (1) 50% and (2) 100% of Merion Kentucky bluegrass plants at three temperatures: 10, 20, and 30°C. (Based on data from Beard, J. B. and Martin, D. P., *Agron. J.*, 62, 257, 1970.)

The term "incorrect oxygenation" is often thought to be synonymous with "oxygen deficiency". However, there are hints in the literature, suggesting that the effect of surplus of oxygen at the plant roots is also deleterious. This is referred to as supraoptimal oxygenation.

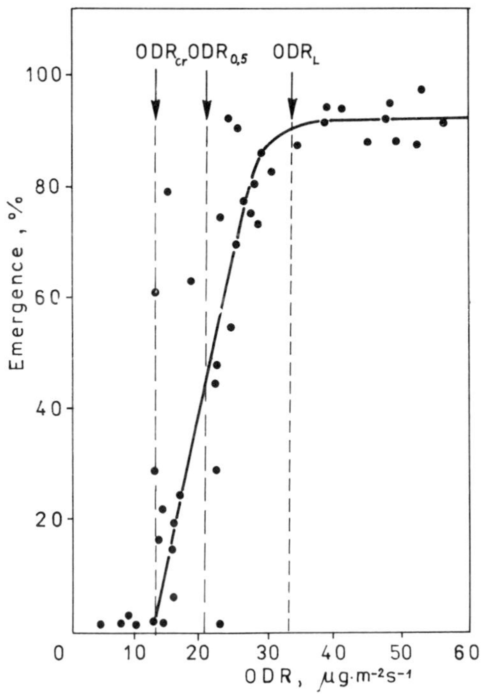


FIGURE 3. Final emergence of oats vs. ODR. ODR$_{cr}$, ODR$_l$, ODR$_{0.5}$ — critical, limiting, and half emergence values of ODR, respectively. Explanations in text. (Modified from Glínski, J., Stępniewski, W., Łabuda, S., and Przywara, G., *Rocz. Glebozn.*, 35, 3, 1984.)

## A. Plant Response to Oxygen Deficiency

### *1. Seedling Emergence*

As mentioned in Chapter 2, the oxygen concentration was used as an index of soil oxygenation in earlier papers.[50,261] Later, oxygen concentration was replaced by a more reliable index of oxygen availability to plant roots — the oxygen diffusion rate (ODR) (cf. Chapter 2, Section II.D).

In examining the influence of oxygen on the initial stage of plant development we must distinguish between germination and emergence.

Many plants, e.g., wheat,[5,150,238,356] lettuce, marigold, carrot, onion, and portulaca[611,652] germinate well even at a low oxygen concentration. Some plants, such as rice, cucumber, and cockscomb[611,652] preserve their germination ability even in completely anoxic conditions. There are also plants showing increase of germination ability with increase of oxygen concentration throughout practically the whole range of concentrations found in soil air. Plants in this category include grasses such as *Phalaris tuberosa, Oryzopsis holciformis, Agropyrum desertorum, A. elongatum,*[150] and tomato.[611,652]

The emergence of plants is preceded by a considerable development of the root system. For this reason the effects of soil oxygen on seedling emergence are better described in terms of ODR than in terms of the oxygen concentration in the soil air.

A typical course of the relationship between plant emergence and ODR is shown in Figure 3. Starting with the highest values of ODR, initially we observe a plateau range, where the differentiation in emergence is not significant. Then, after reaching the so-called limiting value, ODR$_l$, we observe a significant decrease in final emergence in comparison with the germination capacity. Thereafter there is a rapid linear decline in the number of emerging

**Table 1**
**THRESHOLD ODR VALUES FOR EMERGENCE OF SOME CROP PLANTS**

| Plant | $ODR_l$ | $ODR_{0.5}$ | $ODR_{cr}$ |
|---|---|---|---|
| Barley | 25 | 17 | 8 |
| Oats | 30 | 20 | 12 |
| Beans | 33 | 25 | 12 |
| Wheat | 40 | — | 8 |
| Flax | 40 | 25 | 13 |
| Maize | 40 | 27 | 16 |
| Tomato | 40 | 30 | 25 |
| Sugar beet | 50[a] | 30[a] | 13[a] |
| Rye | 50 | 33 | 12 |

*Note:* $ODR_l$, limiting value; $ODR_{0.5}$, half emergence value corresponding to 50% of emergence; $ODR_{cr}$, critical value.

[a] Threshold values concerning emergence but not further survival of the plants.

Data from Glínski, J., Stępniewski, W., Łabuda, S., and Przywara, G., *Roczn. Glebozn.*, 35, 3, 1984.

plants as the ODR decreases; this number falling to zero at the critical ODR value, $ODR_{cr}$ (the level making emergence impossible.) Between critical and limiting value there is the half-emergence value denoted as $ODR_{0.5}$ (the value at which the number of emerging seedlings reaches 50% of the seed germination capacity).

Glínski et al.[236-239,244,245] carried out laboratory experiments on the influence of oxygen availability in different soils on the emergence of nine cultivated plants. The results obtained show the typical course of the relationship between seedling emergence and ODR presented in Figure 3 and allow one to rank the plants investigated according to their increasing oxygen requirements during emergence (Table 1).

When the criteria mentioned above concerning the limiting, half emergence, and critical ODR values are applied to consideration of the results published by other authors, it is not always possible to distinguish all the three threshold values because the experiments were carried out in too narrow range of ODR or because of a lack of data about the germination capacity of the seeds studied.

For maize, apart from the values given in Table 1, there are also data obtained by Singh and Ghildyal[659] and Wengel.[783] The results of Singh and Ghildyal[659] suggest values of $ODR_{0.5}$ equal 65 $\mu g\ m^{-2}s^{-1}$, $ODR_l$ equal 100 $\mu g\ m^{-2}\ s^{-1}$, and $ODR_{cr}$ equal $<15$ $\mu g\ m^{-2}s^{-1}$. According to data of Wengel[783] $ODR_l$ is 25 $\mu g\ m^{-2}s^{-1}$, $ODR_{0.5}$ is 12 $\mu g\ m^{-2}s^{-1}$, and $ODR_{cr}$ below 7 $\mu g\ m^{-2}s^{-1}$. Hanks and Thorp[296] obtained data for wheat indicating $ODR_l$ is 135 $\mu g\ m^{-2}s^{-1}$, $ODR_{0.5}$ is 100 $\mu g\ m^{-2}s^{-1}$, and $ODR_{cr}$ is 80 $\mu g\ m^{-2}s^{-1}$.

From the data of Erickson and Van Doren[193] it is possible to obtain the following threshold values of ODR: for one variety of sugar beet $ODR_l$ 55 $\mu g\ m^{-2}s^{-1}$, $ODR_{0.5}$ 25 $\mu g\ m^{-2}s^{-1}$, and $ODR_{cr}$ below 15 $\mu g\ m^{-2}s^{-1}$. For the other sugar beet variety $ODR_{cr}$ is the same as for the former, but emergence increased linearly when ODR increased reaching 50% at ODR 65 $\mu g\ m^{-2}s^{-1}$. The lack of data concerning the germination capacity of seeds in this case, makes the determinations of $ODR_l$ and $ODR_{0.5}$ impossible.

The ODR values determined by Erickson and Van Doren[193] for potato are 100, 70, and 40 $\mu\ m^{-2}s^{-1}$ for the limiting, half emergence, and critical values, respectively. From the data concerning pea it is possible to estimate only $ODR_{cr}$, which is somewhere between 30 and 60 $\mu g\ m^{-2}s^{-1}$.

Hughes et al.[326] found the emergence of two grasses did not depend on ODR in the range from 16 to 150 $\mu g\ m^{-2}s^{-1}$.

Comparison of the threshold values obtained by different authors for the same plants (sugar beet, wheat, and maize) shows that they are not in agreement. Only in the case of sugar beet the results obtained by Glínski et al.[239] and by Erickson and Van Doren,[193] for one variety, are similar.

The cause of this discrepancy may be sought in intervarietal differences in oxygen requirements of the plants which have been observed, e.g., by Erickson and Van Doren.[193] The extremely high threshold values of ODR obtained by Hanks and Thorp[296] for wheat may result from the superposition of soil oxygenation and compaction effects.

*2. Root Respiration, Anoxic Metabolites in Plants, and Their Physiological Effects*

***a. Oxygen Uptake by Roots***

Root respiration is the first process to be restricted by oxygen deficiency. Other disturbances of the vital functions of plants, such as growth, water, and nutrient uptake, are the consequence of the respiration disturbances. The influence of the oxygen concentration in the soil on root respiration has already been discussed in Chapter 1.

The most credible view seems to be that presented by Armstrong,[25] Armstrong and Gaynard,[26] and Greenwood[273] which is that root respiration does not decrease unless the oxygen partial pressure on the root surface is lower than 1 to 2 kPa. This is in agreement with physiological theory, as it has been established that cytochrome oxidase has a high affinity for oxygen.

On this basis, the higher values of critical oxygen concentration found in the literature (cf. Chapter 1, Section II.B.1.a) must be the result of the presence of water films on the root surfaces or the result of the diffusive resistance of the plant tissue.

Nonetheless, the phenomenon of decreasing root respiration at a soil air oxygen pressure of a few kPa does in fact occur.[646,723]

The above picture is, however, a considerable simplification because conditions of full oxygenation or of complete anoxia hardly ever, if ever, occur within plant roots. This is especially underlined by Crawford[131] who draws attention to the fact that root merystemes display some anoxic metabolism even in contact with atmospheric air. This is confirmed by the high respiratory quotients in merystemes and the presence of ethanol in them.[131]

When the oxygen concentration in the root environment decreases the zone of anoxic metabolism localized in the merysteme expands to the other parts of the root system although rarely, if ever, occupying it entirely because its upper parts are supplied with oxygen through internal diffusion from the shoots.

Thus the physiological role of the decline in oxygen uptake by the root system as a whole, observed below the critical value of oxygen concentration in the surroundings, may be of greater importance, than is implied by the relative value of this decline as these anoxic areas, although limited in size, are located in the parts which are most important for root functioning.

It must be stressed that after a short anaerobic period which does not injure the roots permanently, an increased oxygen uptake by the roots appears (the so called "respiration rebound"). This phenomenon has been observed both in excised roots[827] and in intact roots.[178,795]

***b. Ethanol Accumulation in Plants***

Exclusion of oxygen from root metabolism causes a change of this metabolism towards fermentation. A few types of fermentation are possible which is shown in Figure 4. In the case of plants the most common is alcohol fermentation.[129,131,132,364]

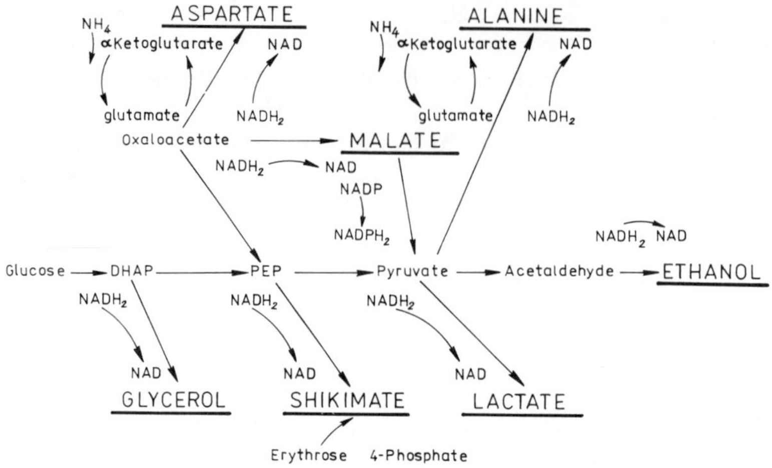


FIGURE 4. Diagram illustrating the various means of proton disposal and the range of end-products of glycolysis found in plants capable of enduring prolonged periods of partial or total anoxia. (Modified from Crawford, R. M. M., in *Nitrogen in the Environment*, Vol. 1, Nielsen, D. R. and McDonald, J. G., Eds., Academic Press, New York, 1978, 427.)

Ethanol is found in the root merystemes, even if they are in contact with atmospheric air. In this case its concentration rapidly decreases with increase in the distance from the root cap.[130] Ethanol occurrence is also common in germinating seeds.[131,132]

Ethanol formation in germinating seeds, as well as in roots when in anoxic conditions, is connected with the plant's capability of surviving periods of flooding. It was found that in water-saturated soaking seeds of flood-tolerant plants, ethanol accumulates in much smaller amounts than in similarly saturated seeds of plants which are not resistant to flooding. This is shown in Figure 5.

As can be seen, flood-tolerant plants also show lower respiratory intensity, and a smaller Pasteur effect, which according to Crawford,[130] is one of the hazards of plant survival under anoxia. Roots of many marsh plants tolerant of flooding, and seeds tolerant of soaking under anoxia do not exhibit the Pasteur effect which is typical for intolerant species.

Floods cause a much greater increase in ethanol accumulation in the case of intolerant plants (2 to 5 by a factor) than in tolerant plants (in which the increase is by a factor of less than 2) and in some tolerant species even a drop in ethanol content is observed.[131]

Similar relations were found for alcohol dehydrogenase which is positively correlated with ethanol content in the plant.[131]

In the case of plants which are intolerant of flooding (e.g., pea) a connection exists between the ethanol concentration in the seedlings and the proportion of plants emerging.[40] This is shown in Figure 6. Ethanol accumulation to the extent of 40 $\mu M$ per seedling causes some reduction in emergence, which at a concentration of 60 $\mu M$ is already down to the level of 50% and disappears entirely at an ethanol concentration of 80 $\mu M$. The authors, however, draw attention to the fact that this relationship is not necessarily the cause of seedling mortality.

The ethanol formation in plant roots is connected with environmental conditions. Fulton and Erickson[217] stated, that ethanol accumulation in flooded tomato plants was closely correlated with the ODR values in the soil, which is shown in Figure 7. Further investigations of the relationship between ethanol accumulation in the plant and soil conditions showed that ethanol concentration is higher in plants subjected to sudden anoxic stress than in plants grown for a long time in conditions of insufficient oxygenation, and moreover, depends on the physiological age of the plant.[29]

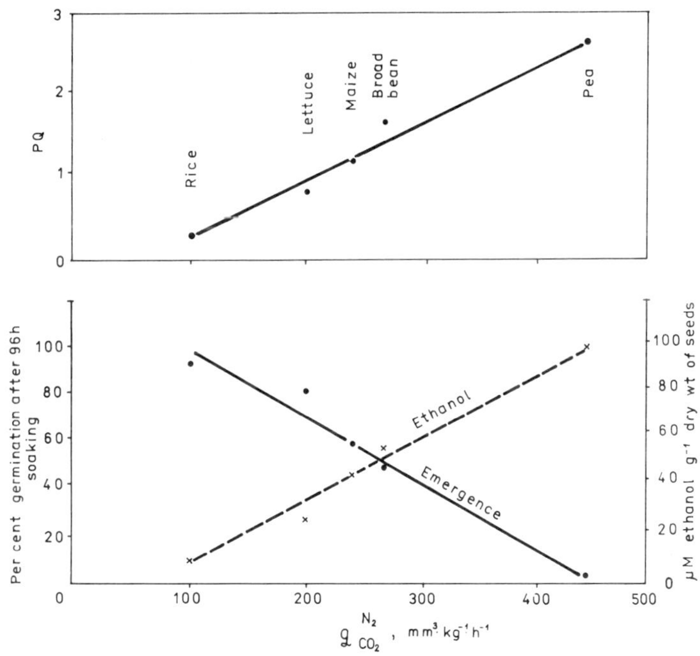


FIGURE 5. Dependence of Pasteur effect as measured by Pasteur quotient $PQ = (3q^{n2}_{CO_2} - q^{air}_{CO_2})/q^{air}_{CO_2}$, seed viability of five crop plants after 96 hr soaking and ethanol accumulation in seeds on anoxic $CO_2$ production measured in $N_2$ atmosphere ($q^{N2}_{CO_2}$). (Modified from Crawford, R. M. M., in *Nitrogen in the Environment,* Vol. 1, Nielsen, D. R. and McDonald, E. G., Eds., Academic Press, New York, 1978, 427; *Izv. Akad. Nauk. SSSR, Ser. Biol.,* 1, 77, 1981.)

Examining the ethanol balance in the tomato plant, Bolton and Erickson[66] and Fulton and Erickson[217] showed that its concentration in the plant increases with increase in soil temperature, atmospheric temperature, and light intensity. It must be emphasized that ethanol accumulation appears quickly (in the course of a few hours) and extends to the whole plant. The maximum ethanol concentration was found in the bottom of the stem and in the top roots. Its highest concentrations in the plant were 300 to 350 $\mu g\ g^{-1}$.

The decrease of ethanol concentration in the lower part of the root is explained by its secretion to the environment to the extent of about 20% of the total ethanol formed. Its decrease in the upper parts of plant is primarily due to the increased oxygen concentration in those parts, which accelerates ethanol remetabolization. Ethanol losses caused by transpiration are not high (about 5% of the amount formed).

The possibility of toxic effects of ethanol on plant tissues depends on its concentration, which is the result of the ethanol formation processes (the magnitude of the Pasteur effect in a given plant), and of the processes of its removal by roots and shoots and also on the extent of its remetabolization within the plant tissues.

Crawford[130] gives as an example rice and swamp tupelo tree (*Nyssa sylvatica*), which easily remove ethanol by adventitious roots and tolerate flooding well despite the fact that they show Pasteur effect and that ethanol is a main product of anoxic respiration.

All the factors are enumerated and ultimately the particular plants tolerance to high ethanol concentrations decide the plant's ability to survive under anoxic conditions.

Injury caused by ethanol consists of a deleterious effect on the cell membranes[370] resulting in leakage of electrolite, organic acids, amino acids, and sugars from the roots.[284,465] This cell injury by ethanol explains the Pasteur effect, i.e., acceleration of the rate of glycolysis

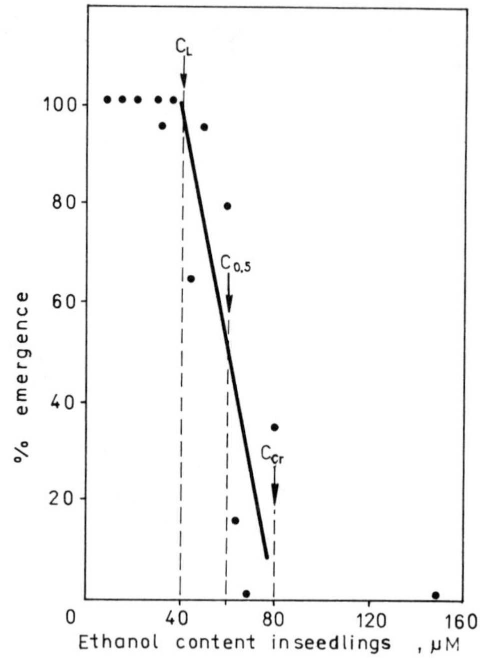


FIGURE 6. Relationship between pea seedling emergence after anoxic incubation for various periods and the ethanol content in seedlings at the time of their removal from the anoxic jars; $C_l$, limiting concentration; $C_{0.5}$, half emergence concentration; $C_{cr}$, critical concentration. (Modified from Barclay, A. M. and Crawford, R. M. M., *J. Exp. Bot.*, 32, 943, 1981.)

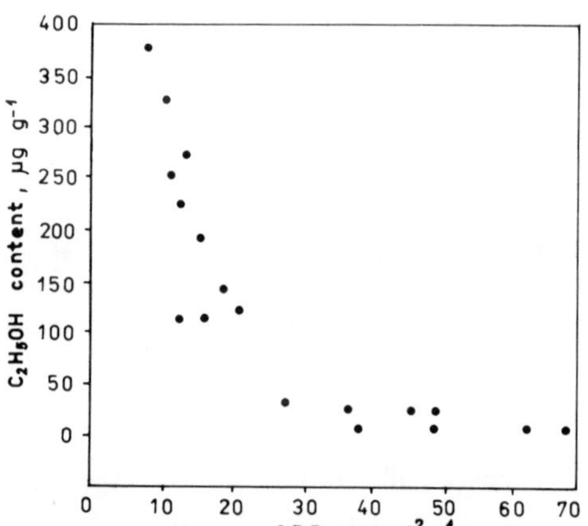


FIGURE 7. Ethanol content in xylem exudate of tomato plants vs. soil ODR. (Modified from Fulton, J. M. and Erickson, A. E., *Soil Sci. Soc. Am. Proc.*, 28, 610, 1964.)

which is observed in flood-intolerant plants. This, in turn, results in a decreasing ATP level in the cells and, as a consequence, increasing ethanol production. This autocatalitic process, according to Crawford,[130] explains the susceptibility of many plants even to oxygen stresses of short duration.[191,324]

### *c. Ethylene in Plants*

Ethylene is always present in plants and an increased concentration is caused by various stresses: thermal, water, mechanical, chemical, and disease.[2,288] Moreover, root oxygenation is a very important factor. As already mentioned (cf. Chapter 3, Section II.D), ethylene formation in the root requires the presence of some oxygen. In shoots, however, its production is stimulated by anoxic conditions in the root. The mechanism of this stimulation has been recently explained. It consists of increased synthesis of ACC (being the precursor of ethylene) in the root under anoxic conditions.[8,72] The ACC is transported through the xylem to the shoots, where (in the presence of oxygen) it is transformed to ethylene. The appearance of greater amounts of ACC in tomato shoots (sometimes reaching 3 $\mu M$) occurs after the soil containing the roots had been flooded for 12 hr and only 4 hr after the nutrient solution containing other roots had been flushed with nitrogen.[72]

An increase of ethylene production in tomato shoots has also been observed within 24 hr of flooding the roots.[71,337]

There is much further evidence that oxygen deficiency in the roots results in a significant increase of ethylene production in the shoots. Thus, in the tomato caps, normally producing ethylene at the rate of 0.7 to 1.7 $mm^3kg^{-1}hr^{-1}$ with oxygen present in the root environment the rate of ethylene production increased to 3.2 to 9.6 $mm^3kg^{-1}hr^{-1}$ with the roots in flooded conditions.[337] In other parts of the shoots ethylene production is less intensive and with a sufficient oxygen supply to roots is only 0.04 to 0.4 $mm^3kg^{-1}hr^{-1}$. However, when the roots are flooded with water, ethylene production in the shoots increases to 0.4 to 1.5 $mm^3$ $kg^{-1}hr^{-1}$.[72,335,337]

As a result of increased ethylene production under flooding conditions its content in tomato shoots increases to about 4 $cm^3m^{-3}$, whereas with good oxygenation of the roots, this content is less than 0.4 $cm^3m^{-3}$.[334,362] An increase of ethylene concentration in plants after only 24 hr of flooding has been noted.[334] It should be mentioned that this increase can also be caused, at least partly, by stomata closure (cf. Chapter 5, Section I.A.3) and the subsequent decrease in ethylene evolution to the atmosphere.[337]

It has been stated that ethylene appearance in shoots causes petiole epinasty, i.e., a change in the angle between the petiole and the stem. This phenomenon occurs as early as 8 to 24 hr after flooding the root system,[72,334,335,337,703] or after its treatment with nitrogen.[71]

With the presence of ethylene, adventitious roots containing aerenchyma are connected.[171]

### *d. Other Anoxic Metabolites in Plant Tissues*

Besides carbon dioxide, which is discussed in another section, and metabolites such as ethanol, ethylene, and its precursor ACC which were discussed above, there are other substances which may appear in seeds, roots, and plant tissues when the supply of oxygen is insufficient. Their formation and role are not yet fully explained.

Crawford[128,131] and Stolzy and Flühler[706] report that in anoxic conditions lactic and succinic acids are formed in seeds, while in roots pyruvic, fumaric, acetic, aspartic, glutamic, propionic, butyric, malic, shikimic, glycolic, and α-aminobutyric acids, as well as glycerol, alanine, serine and proline, have been found.

In turn Labanauskas et al.[398,399] found that in the leaves of lemon and orange cultivated in soils at a reduced oxygen concentration there was a decrease in the sum of protein amino acids and an increase in the sum of free amino acids. For individual free amino acids an increase of free tryptophan, histidine, arginine, and proline and a decrease in threonine was observed. A decrease of protein amino acids such as lysine, histidine, aspartic acid, threonine, serine, glutamic acid, glycine, alanine, valine, methionine, isoleucine, leucine, tyronine, and phenylalanine was also found. An increase in amino acid production under anoxic conditions has also been reported by other authors.[130,220,409]

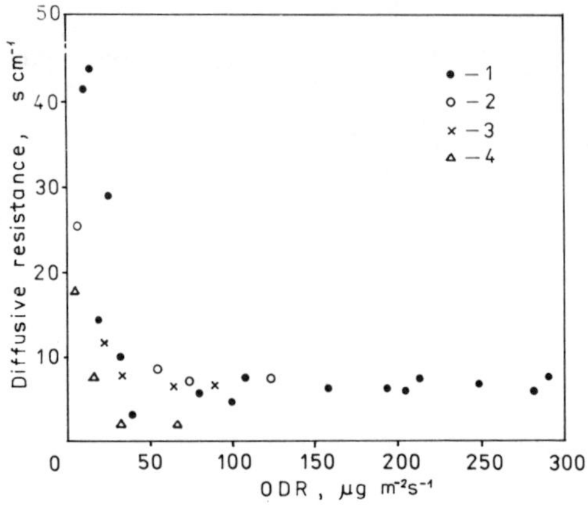


FIGURE 8. Diffusive resistance of (1) wheat, (2) jojoba, (3) sunflower, and (4) cotton leaves as affected by ODR. (Data from Sojka, R. E. and Stolzy, L. H., *Soil Sci.*, 130, 350, 1980.)

On the other hand, in wheat, no significant changes in the content of the 18 amino acids mentioned above either in the free or protein forms, were noted, under conditions of flooding the soil or forcing nitrogen into it.[396]

However Heinrichs[307] reported a decrease of raw protein in nine leguminous plants after flooding their roots.

The carbohydrate content in plants may also depend on the oxygen status of the soil. Visser et al.[770] observed an increase in the carbohydrate content of apple leaves both under flooded soil conditions and when growing in a nutrient solution without aeration. In the roots, however, the amount of càrbohydrates decreased in the case of the nonaerated nutrient solution but for soil grown plants increased after the soil was flooded. Similar phenomenon of increasing carbohydrates content in barley after flooding the soil was observed by Savickaya.[636]

In tobacco, under anoxic conditions a decrease of the nicotine concentration in roots and leaves and an increase in the stems was noted.[301] The total content of the nicotine in all parts of plants decreased under such conditions.

Murch and Paxton[500] found a decrease of as much as 50% in glyceollin accumulation in soybean, when its roots were subjected to anoxic conditions by flooding or by treating them with pure nitrogen or helium. The decrease of glyceollin accumulation probably diminishes the physiological resistance of the soybean to diseases.

The results obtained by Blevis et al.[61] show an increase of particulate viologen utilizing nitrate reductase (EC 1.9.6.1) in barley roots situated in sterile, anoxic conditions.

### 3. *Stomata Closure*

Adjustment of the stomata aperture is one of the most important regulatory mechanisms in the plant. It determines the transpiration and heat balance of the plant and probably also its photosynthesis and respiration. It is well known that the main factors regulating stomatal aperture are partial pressure of carbon dioxide, light, water stress, and temperature.[631]

Recently it was found that stomata closure is also caused by insufficient oxygen supply to the roots. This phenomenon was observed indirectly by measuring the diffusive resistance of leaves of tomato,[337] sunflower and jojoba,[682] cotton,[473] wheat,[683] poplar and willow,[554] and bean.[678,679]

The dependence of diffusive resistance $R_D$ on ODR is shown in Figure 8. A sharp decrease in stomatal diffusive resistance is observed at an ODR below 35 $\mu g\ m^{-2}s^{-1}$. This relationship is well described by the exponential equation:[682]

$$R_D = a \cdot ODR^b \quad (1)$$

where a and b are parameters characterizing the plant with regard to its age as well as its variety.

The phenomenon of stomata closure under the influence of oxygen stresses in the roots was well reviewed recently by Sojka and Stolzy.[682] According to these authors, and also Jackson et al.[337] it is not possible to explain this phenomenon by the assumption of a passive mechanism causing a fall in the water potential in the plant owing to decreased water permeability of the root in anoxic conditions (see further), because stomata closure was observed even at a high water potential in the plants. In many cases[337] the water potential in plants with flooded roots is higher than that in plants with a good oxygen supply to their roots, simply because of the stomata closure and the resultant decrease in transpiration.

Sojka and Stolzy[682] suggest the existence of at least two mechanisms which can be involved in the regulation of stomata closure under conditions of oxygen deficiency in the roots.

One mechanism may be connected with the possible production of a physiologically active chemical such as ethylene. This is supported by the fact that ethylene production increases during drought and is accompanied by stomata closure. As the other possible mechanism they consider an alternation of potassium ion flux into guard cells, resulting in a change in their turgidity. This hypothesis is supported by the decrease of potassium content in shoots when their roots are oxygen deficient (see further) and by the relation between increase of potassium concentration and stomata closure.[125,551] Moreover, Sojka and Stolzy[682] suggest that stomata closure when the roots are short of oxygen is a cause of the observed increase in plant resistance to ozone toxicity in such conditions.[712,713]

Because the decrease of transpiration rate in conditions of root anoxia is at least in part, the result of stomata closure, this phenomenon may be partly responsible for the observed reduction in photosynthesis under these conditions.

*4. Water Uptake*

It has been known that flooding the roots with water causes partial or total inhibition of transpiration, and in some cases[382,386] the paradoxical phenomenon of the plant wilting after soil submersion. A review of earlier papers on this subject has been made by Bergman.[50]

The occurrence of this phenomenon has also been confirmed in later papers.[5,6,97,318,335,337,422,738,795,802] It was shown that the first symptom of the disturbance of the water conditions in the plant as a result of flooding the roots is the disappearance of guttation.[738] An example of the dependence of transpiration on oxygen concentration is shown in Figure 9. A suggested explanation is that the decrease in the water permeability of the roots in conditions of anoxia,[339,382-386,475] leads to a decrease in the water potential of the leaves as a result of transpiration and, in consequence, the closure of the stomata.

It is true that some authors[180,233,382,386] have observed the intensification of the water stress after flooding the roots. In many cases, however, this phenomenon was not confirmed, and they even observed the opposite effect, i.e., an increase of water potential in the leaves in conditions of root anoxia caused either by flooding[337,554,738] or by introducing nitrogen to the soil.[683] This means that a decrease in the water permeability of the roots is not the only factor decreasing transpiration. The phenomenon of stomata closure plays a significant role here but in these cases the signal causing it is not connected with any decrease of water potential in plants.

We are unaware of any papers containing data from simultaneous investigations of the diffusive resistance of the leaves (stomata closure) and the water permeability of the roots. Accordingly it is not easy to evaluate the relative contributions of the decrease in the water permeability of the roots and of stomata closure, to the decrease in plant transpiration when the roots are flooded with water or treated with transpiration inhibitors.[611,762]

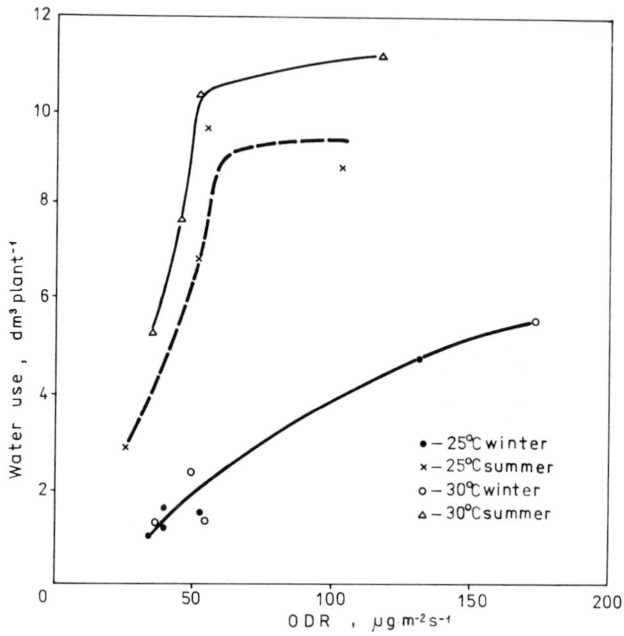


FIGURE 9. Water use by orange seedlings (*Citrus sinensis*) as related to soil ODR. (Based on the data of Stolzy, L. H., Van Gundy, S. D., Labanauskas, C. K., and Szuszkiewicz, T. E., *Soil Sci.*, 96, 292, 1963.)

Holder and Brown[318] found a positive linear correlation between oxygen uptake by roots and water use by plants with undamaged root systems. This correlation disappeared when the root system had been injured by sudden oxygen stress.

Willey[795] found that with tobacco plants a decreasing rate of transpiration appears within a few minutes of nitrogen being forcing into the root environment and that the rate becomes stabilized after about half an hour. This is shown in Figure 10. The decrease of transpiration was accompanied by an increase in the water stress of the plant, as measured by stem diameter. After the roots had been in the atmosphere of nitrogen for 6 hr, access of air to root was restored. This was followed by the return of the transpiration rate to the initial level within a few minutes and by an increase in the stem diameter. Unfortunately the author did not measure the degree of stomata closure.

Irrespective of whether stomata closure took place or not, the decrease in stem diameter during the period of oxygen stress seems to prove that there was a concurrent decrease in the water permeability of the root system.

Next, Grineva[283] stated that maize and sunflower subjected to an oxygen stress lasting 18 hr reduced water uptake and transpiration during the stress. After this period the water conditions in plant returned to their initial state. Oxygen stress lasting 24 hr permanently destroyed the normal functioning of the plants and hence the water uptake.

Very little is known concerning the transpiration response of plants tolerant to prolonged or permanent flooding.

It should be added that, in addition to the direct influence of root anoxia on plant transpiration caused by stomata closure or decreased water permeability of the roots (as shown in Figure 10), during prolonged root anoxia indirect effects connected with changes in the size of root system and/or of the leaf surface are involved.

Trought and Drew,[739] examining young seedlings of wheat, have established that their water use (per plant) did not change during 2 weeks of flooding, but transpiration (in terms

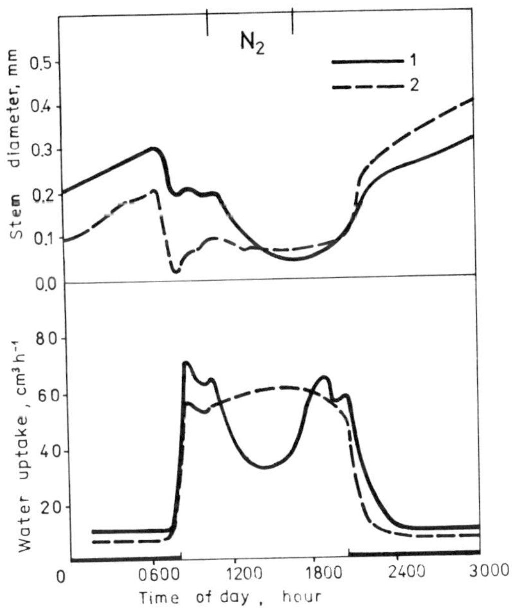


FIGURE 10. Diurnal water uptake and changes of stem diameter of tobacco seedlings grown in nutrient solution treated with (1) air and (2) $N_2$. (Data from Willey, C. R., *Agron. J.*, 62, 224, 1970.)

of leaf surface) decreased. In the control plants water use (per plant) increased by a factor of 8, due to increase of the transpiration surface.

It was also determined, in the case of wheat[16,97] and tomato[601] that during root anoxia the efficiency of water use by plants decreases, i.e., the water required for the production of unit mass of crop increases.

### 5. *Plant Growth and Yield*

#### *a. Root Response*

The reaction of roots to anoxic conditions manifests itself in various ways.

Completely anoxic conditions stop root growth entirely within 2 to 3 min. This has been observed with cotton and soybean.[324] When the stress did not last longer than 30 min the growth rate subsequently returned to normal. Anoxic stress lasting 3 hr for cotton and 5 hr for soybean killed the tap roots.

Roots of plants which are very sensitive to anoxic conditions (e.g., tobacco, legumes) die after a few days of flooding,[308,382,800,802] but the first symptoms of damage by waterlogging may appear after 1 to 2 hr and permanent damage may appear after 24 hr, especially on warm sunny days.[386]

As already mentioned, plants which can survive an anoxic stress nevertheless decrease the rate of growth of their roots. This has been observed with tomato,[335] maize,[228,229,235,747] barley,[227-229] wheat,[654] pea,[118,227-229] oak,[653] snapdragon,[711] tulips,[792] sugar beets,[789] and mustard.[281] As a result of these root growth disturbances (due to an insufficient supply of oxygen to the roots) the mass of the root system decreases. This was observed with sorgo,[321] barley,[172] oats,[778] wheat,[616,738,767,778,790] maize,[170,442] apple tree,[770] sugar beets,[789] lemon,[398,400,708] orange,[397,708,774] jojoba,[597] avocado,[404] grasses,[423,773] cotton,[542] legumes,[118] sugar cane,[548] broad bean, cabbage, and potato.[790]

Under anoxic conditions the root mass of flood intolerant plants decreases earlier and to a higher degree than the mass of the shoots. Hence the ratio of shoots to roots is higher in

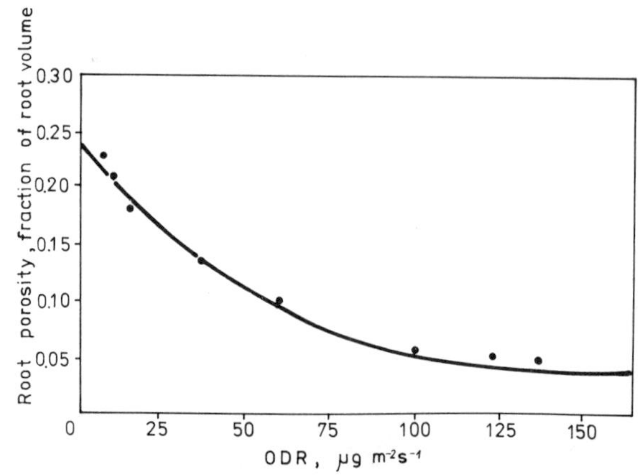


FIGURE 11. Rice root porosity in active tillering phase as affected by soil O.D.R. (Based on the data of Ghildyal, B. P., in *Vertisols and Rice Soils of the Tropics*, Shri S. N. Mehta, New Delhi, India.)

anoxic conditions than in the presence of oxygen. This was found in the cases of wheat,[6,738] lemon,[395] jojoba,[597] orange,[774] and apple tree.[770]

Roots growing in an environment with restricted oxygen supply are thicker, shorter, less branched, of light color,[118,773] have larger cells,[558] and, in the case of legumes, possess fewer nodules.[118] Increased porosity of roots grown under anoxic conditions was found with wheat,[6,767] barley,[558] rice,[230,442] maize,[442,487] and fescue.[487] This is the result of stimulation of the growth of adventitious roots having a greater porosity because of the formation of aerenchyma, e.g., tomato,[335] barley,[172] maize.[170] This is shown in Figure 11.

Roots which are near to the anoxic zone in the soil show an ability to avoid this zone,[654] which results in formation of a shallow root system,[171,439,723] and in some cases the roots become aerotropic (negatively geotropic) and emerge from the soil.[335,773,791] The formation of the shallow root system results from stimulation of the growth of lateral roots following the death of the tap root tips.[324]

Soil compaction is a factor affecting root extension both by increased mechanical resistance and by deterioration of oxygen conditions in the soil. Scott and Erickson,[645] improving the oxygen status of a compacted soil (to 1.9 mg $m^{-3}$) by the addition of $CaO_2$ observed an increase in the number of roots in the soil in the case of lucerne, tomato, and sugar beet.

It is widely agreed that the best indicator of soil oxygen availability for plant roots is the ODR. The reaction of roots of different plants to ODR is shown in Figure 12. We see that the number of roots in the soil (Figure 12A) already begins to be reduced at an ODR of 50 μg $m^{-2}s^{-1}$ and drops rapidly to less than one fifth of the original number over the ODR interval 50 to 40 μg $m^{-2}s^{-1}$. The ODR value of 40 μg $m^{-2}s^{-1}$ may, accordingly, be accepted as critical, and 50 μg $m^{-2}s^{-1}$ as limiting for wheat root growth. Similarly, the root growth rate of desert shrubs is restricted at an ODR of 50 μg $m^{-2}s^{-1}$ and is completely suppressed at an ODR of 30 to 35 μg $m^{-2}s^{-1}$ (Figure 12B).

The critical ODR values for other plants are differentiated and (in μg $m^{-2}s^{-1}$) are equal to 25 for barley,[427] 42 and 20 for maize,[53,757] 33 for sunflower,[425,427] 33 for snapdragon,[711] and 33 for cotton.[425] Growth of grasses was inhibited at an ODR of 25 μg $m^{-2}s^{-1}$ for turfgrass and 17 μg $m^{-2}s^{-1}$ for ryegrass,[266,423] 8 to 15 μg $m^{-2}s^{-1}$ for Kentucky bluegrass and 33 μg $m^{-2}s^{-1}$ for Newport Kentucky bluegrass.[426] Desert shrubs such as *Artemisia tridentata*, *Larrea tridentata*, and *Franseria dumosa*[440] have critical values of ODR in the range 30 to 35 μg $m^{-2}s^{-1}$.

Root growth of plants adapted to prolonged flooding (e.g., rice) is better in such conditions than when not flooded. This causes a decrease of shoot to root ration due to flooding the

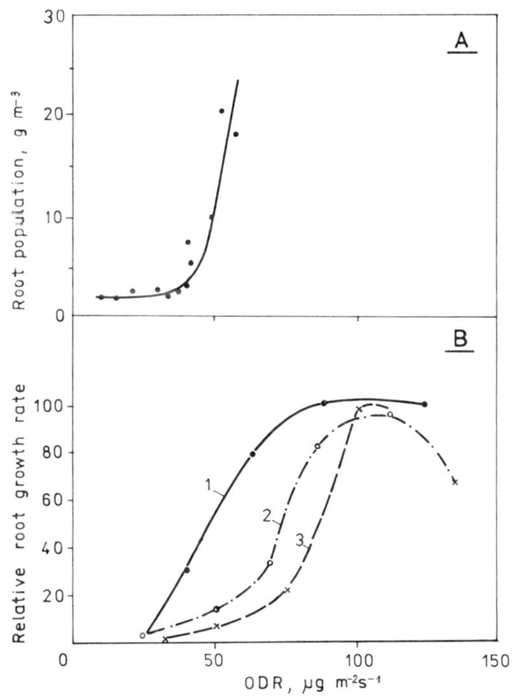


FIGURE 12. Root response to ODR in soils; (A) population of wheat roots in soil under field conditions,[654] (B) relative root growth rate of three desert shrubs: *Franseria dumosa* [1], *Larrea tridentata* [2], *Artemisia tridentata* [3].[440]

soil.[355,442] These plants generally have more porous roots and sometimes, e.g., black mangroves (*Avicennia nitida*), form special aerial roots (pneumatophores) containing aerenchyma, which supply oxygen to the normal roots.

***b. Shoot Response***

The growth response of shoots to oxygen conditions in the soil is closely connected with that of the roots; the former being later in relation to the latter, which results in an increased shoot to root ratio in anoxic conditions (cf. previous section). The visual symptoms in the shoots, which indicate an insufficient oxygen supply to the roots are chlorosis, epinasty, leaf senescence, and abscission,[172,173,382,386,738,778,802] decreasing number of tillers,[172,173,683,778] and partial or complete suppression of the shoot growth rate which was observed for tomato,[335,382] wheat,[6] barley,[172,228,229] oats,[629] tobacco,[801] apple trees,[770] maize,[227-229] broad bean,[228] *Narthecium osgifragum*,[721] peas, and potato[229] which results in a decrease in the mass and yield of the plants. This was found with oats,[778] wheat,[16,97,98,403,738,782,790] barley,[173,227,428,636,778] maize,[235,747,799] legumes,[307,471] broad bean,[790] tomato,[86] jojoba,[597] sorghum,[799] potato,[790] rape,[102] tobacco,[299] snapdragon,[422] avocado,[404] white clover,[267] cabbage,[790] peas,[117] cotton,[473] lemon, and orange.[395,399,400,710,774]

An example of the relation between yield and ODR is shown for several plants in Figure 13. In some cases, with short lasting anoxia stresses, there is only a delay in plant growth without a negative influence on the final yield.

Grasses[773] and young seedlings of lemon and oranges[399,708] did not react to oxygen depression in soil.

In contrast to the plants listed above rice does not exhibit a negative growth response to anoxic conditions in the soil and may even show an increase of yield in such conditions.[355,543,748]

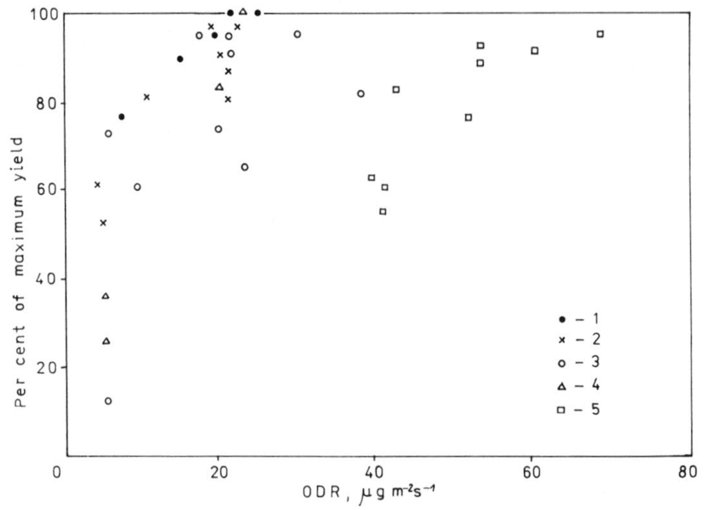


FIGURE 13. Crop yield as related to soil ODR; (1) sorghum, (2) cabbage, (3) soybean, (4) maize,[799] (5) potato.[630]

Campbell and Ferguson[97] and Campbell et al.[98] found that a decrease in the yield of wheat grain under the influence of deficient aeration of the roots was due to pollen sterilization, which commenced at the top of the ear. Artificial pollination of the plants subjected to oxygen stress greatly reduced its detrimental effects. A similar phenomenon with barley was reported by Aleksandrova and Skazkin.[13]

*6. Nutrient Uptake and Mineral Composition of Plants*

Oxygen stress in the soil causes the disturbances in the functioning of the whole plant, especially in nutrient uptake and in the composition of the plant. Generally, the changes in nutrient uptake are seen first because they are connected with a distinct decrease in the mass of the plant exposed to oxygen deficiency, while changes in the mineral composition of the plant appear only in conditions of more drastic or prolonged oxygen stress.

***a. Nitrogen***

A review of papers on the changes of nitrogen content in plants under the influence of oxygen stress is presented in Table 2. As can be seen, the direction of the changes of nitrogen content in plants in anoxic conditions is different in different plants and even in the same plant, as was found in the case of lemon, wheat, and avocado. The latter effect is probably connected with the stage in the plant's development at which the oxygen stress was applied.

In wheat[739] with flooded roots nitrogen uptake was stopped completely while the shoot growth was only partly inhibited. This resulted in a decrease in nitrogen content in the plant tissue (per mass unit) and also in the xylem sap due to the ''dilution'' effect. Simultaneously, nitrogen was displaced from old to young leaves, the effect being significant after only 4 days of flooding. After 2 weeks of submergence about 80% of the nitrogen had been displaced from the old leaves.

Ohyama and Kumazawa[519] found that the activity of nitrogen fixation in nodules of soybean was proportional to the oxygen concentration. In anoxic conditions nitrogen fixation dropped to 1.5% of its initial value (at atmospheric concentration).

The nitrate form of nitrogen is especially sensitive to root anoxia. Flooding the roots reduced the nitrate content by a factor of 8 in wheat[739] and by a factor of 30 in barley[172] in comparison with the control.

## Table 2
## CHANGES IN NITROGEN CONTENT AND UPTAKE BY VARIOUS PLANTS IN CONDITIONS OF INSUFFICIENT SUPPLY OF OXYGEN TO THEIR ROOTS

| | Content | | | | | Uptake | | | |
|---|---|---|---|---|---|---|---|---|---|
| **Plant** | **Roots** | **Tops** | **Stem** | **Leaves** | **Grain** | **Total plant** | **Roots** | **Tops** | **Ref.** |
| Lemon | + | − | | | | − | | | 402 |
| | + | − | | | | − | − | − | 395 |
| | + | | + | + | | − | | | 400 |
| Lemon and orange | − | | | − | | − | | | 708 |
| | | | + | − | | ○ | | | 398 |
| Apple tree | − | − | | | | | | | 770 |
| Jojoba | + | | + | + | | − | | | 597 |
| Avocado | + | − | | | | − | | | 404 |
| | − | | ○ | − | | − | | | 661 |
| Potato | | | | | | | − | − | 790 |
| Wheat | | | | | | | | − | 790 |
| | | | | | ○ | | | − | 403 |
| | | − | | | | | | | 767 |
| | | − | | | | − | | | 739 |
| Barley | − | − | | | | | − | − | 173 |
| | | − | | | | | | − | 172 |
| | | − | | | | | | | 428 |
| Broad bean | | | | | | | | − | 790 |
| Cabbage | | | | | | | | − | 790 |
| Tobacco | | | | − | | | | − | 790 |
| Peas | − | − | | | | | − | − | 118 |
| Maize | | − | | | | | | | 170 |
| Rice | | ○ | | | | | | | 748 |
| Cotton | | − | | | | | | | 473 |

Explanations: − decrease, + increase, ○ insignificant changes.

In all the plants investigated (Table 2) nitrogen uptake was always less under oxygen stress than with a sufficient oxygen supply to the roots. This was accompanied by a general decrease in the mass and yield of the plants in anoxic conditions.

### *b. Phosphorus*

Data on the changes due to oxygen stress in the content and uptake of phosphorus by plants are compiled in Table 3. They show unambiguously a decrease of phosphorus uptake by all plants. Its content decreased or remained unchanged.

From Figure 14 it can be seen that a decrease of phosphorus content in plants appears after the oxygen concentration drops below 0.05 $m^3m^{-3}$.

When roots of wheat are flooded[739] the phosphorus uptake (like that of nitrogen) is almost completely stopped; in spite of this plant growth continues, although, of course, more slowly. As a result of this a decrease in the phosphorus content of the older leaves becomes evident after only 2 days of flooding. Younger leaves, however, showed an increase in phosphorus content. After 2 weeks of flooding more than 80% of the phosphorus was displaced from old to young leaves.

Harris and van Bavel,[300] who used isotope methods with tobacco plants found that in these plants phosphorus uptake was inhibited after 1 day of oxygen deficiency at the roots.

## Table 3
## CHANGES IN PHOSPHORUS CONTENT AND UPTAKE BY VARIOUS PLANTS IN CONDITIONS OF INSUFFICIENT SUPPLY OF OXYGEN TO THEIR ROOTS

| | Content | | | | | | Uptake | | |
|---|---|---|---|---|---|---|---|---|---|
| **Plant** | **Roots** | **Tops** | **Stem** | **Leaves** | **Grain** | **Total plant** | **Roots** | **Tops** | **Ref.** |
| Lemon | ○ | – | | | | – | | | 402 |
| | ○ | – | | | | – | | | 395 |
| | | | | | | – | – | – | 400 |
| Lemon and orange | – | – | | | | – | | | 708 |
| | | | ○ | – | | – | | | 398 |
| Jojoba | – | | ○ | ○ | | – | | | 597 |
| Avocado | ○ | – | | | | – | | | 404 |
| | ○ | | ○ | – | | – | | | 661 |
| Potato | | | | | | | – | – | 790 |
| Wheat | | | | | | | | – | 790 |
| | | | | | | | | – | 403 |
| | | – | | | | – | | – | 739 |
| Barley | | | | | | – | | | 173 |
| | | – | | | | – | | | 172 |
| | | – | | | | | | | 428 |
| Broad bean | | | | | | – | | | 790 |
| Cabbage | | | | | | – | | | 790 |
| Tobacco | | | | – | | | | – | 299 |
| Peas | – | – | | | | | – | – | 118 |
| Rice | | – | | | | – | | | 355 |
| | | – | | | | | | | 748 |
| | | + | | | | | | + | 543 |
| Maize | | – | | | | | | | 170 |
| Snapdragon | | – | | | | | | | 422 |
| Cotton | | + | | | | – | | | 473 |

Explanations: – decrease, + increase, ○ insignificant changes.

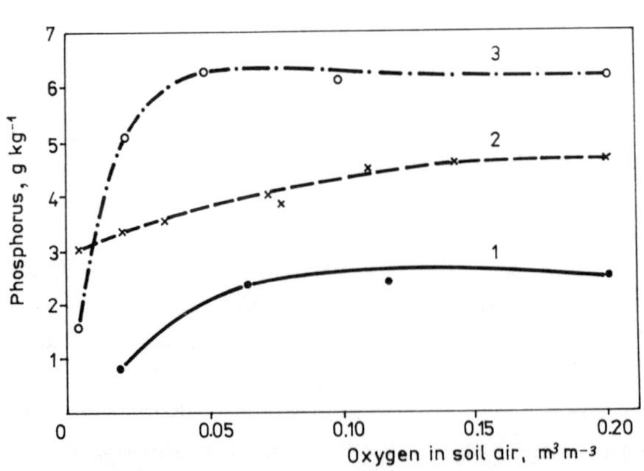


FIGURE 14. Phosphorus content in plants as related to oxygen concentration in soil air; (1) jojoba roots,[597] (2) snapdragon,[422] (3) barley leaves.[427]

**Table 4**
**CHANGES IN POTASSIUM CONTENT AND UPTAKE BY VARIOUS PLANTS IN CONDITIONS OF INSUFFICIENT SUPPLY OF OXYGEN TO THEIR ROOTS**

| | Content | | | | | Uptake | | | |
|---|---|---|---|---|---|---|---|---|---|
| Plant | Roots | Tops | Stem | Leaves | Grain | Total plant | Roots | Tops | Ref. |
| Lemon | ○ | – | | | | – | | | 402 |
| | – | – | | | | | | | 395 |
| | | | | | | – | – | – | 400 |
| Orange | | | | ○ | | | | | 774 |
| Lemon and orange | – | – | | | | – | | | 708 |
| | | | – | – | | – | | | 398 |
| Jojoba | – | | – | – | | | | | 597 |
| Avocado | – | – | | | | | | | 404 |
| | – | | – | – | | | | | 661 |
| Potato | | | | | | | – | | 790 |
| Broad bean | | | | | | | | – | 790 |
| Cabbage | | | | | | | | – | 790 |
| Wheat | | | | | ○ | | | | 790 |
| | | – | | | | – | | | 403 |
| | | – | | | | – | | | 739 |
| Tobacco | | | | – | | – | | | 299 |
| Barley | | | | | | – | | | 173 |
| | – | | | | | – | | | 172 |
| | – | | | | | | | | 428 |
| Peas | – | – | | | | | – | – | 118 |
| Maize | | – | | | | | | | 170 |
| Snapdragon | | – | | | | | | | 422 |
| Apple tree | – | – | | | | | | | 770 |
| Rye | | – | | | | | | | 182 |
| Ryegrass | | – | | | | | | | 182 |
| Clover | | – | | | | | | | 182 |
| Cotton | | – | | | | – | | | 473 |

Explanations: – decrease, ○ insignificant changes.

### *c. Potassium*

The changes in the potassium status of plants, under anoxic conditions in the soil also have an unequivocal character. As shown in Table 4 and in Figure 15, anoxic conditions in the root environment cause a decrease in potassium content in all parts of the plant. This of course implies a decrease in potassium uptake by the plants.

Flooding the roots, e.g., of wheat,[739] completely inhibits potassium uptake, as in the case of nitrogen and phosphorus. In this case also a displacement of potassium (like that of nitrogen and phosphorus) from old to young leaves is noticeable after 2 days of flooding and reaches 80% after 2 weeks.

Some authors conclude that potassium is the plant nutrient most sensitive to root anoxia. This means that a decrease of its content in the plant is seen earlier and more distinctly than in the case of other ions. One result is that a decrease of the K/(Ca + Mg) ratio is observed (e.g., by Elkins and Hoveland[182] for rye).

### *d. Calcium and Magnesium*

The content of calcium and magnesium in plants seems to be less dependent on root anoxia than that of nitrogen, potassium, and phosphorus (Tables 5 and 6). This was confirmed by Harris and van Bavel[300] and Letey et al.[422] According to Trought and Drew[739] the uptake

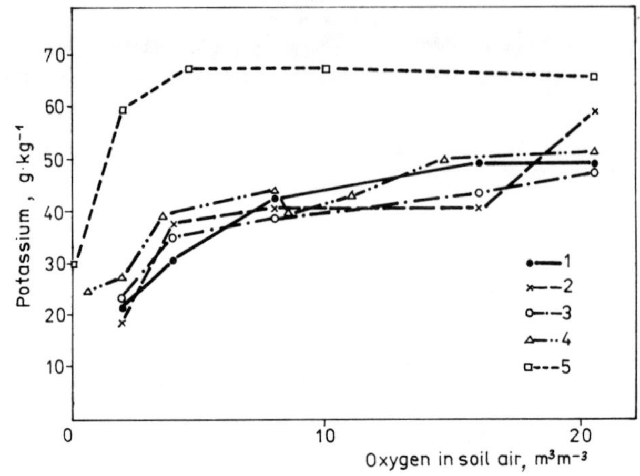


FIGURE 15. Potassium content in plant shoots as related to oxygen concentration in soil air; (1) rye, (2) ryegrass, (3) clover,[182] (4) snapdragon,[422] and (5) barley.[427]

## Table 5
## CHANGES IN CALCIUM CONTENT AND UPTAKE BY VARIOUS PLANTS IN CONDITIONS OF INSUFFICIENT SUPPLY OF OXYGEN TO THEIR ROOTS

| | Content | | | | | Uptake | | | |
|---|---|---|---|---|---|---|---|---|---|
| **Plant** | **Roots** | **Tops** | **Stem** | **Leaves** | **Grain** | **Total plant** | **Roots** | **Tops** | **Ref.** |
| Lemon | ○ | – | | | | – | | | 402 |
| | – | ○ | | | | – | | | 395 |
| | | | | | | – | – | – | 400 |
| Lemon and orange | – | | – | – | | – | | | 708 |
| | | | – | – | | – | | | 398 |
| Jojoba | ○ | | + | – | | – | | | 597 |
| Avocado | + | – | | | | – | | | 404 |
| | ○ | | ○ | – | | – | | | 661 |
| Potato | | | | | | – | – | | 790 |
| Broad bean | | | | | | – | – | | 790 |
| Cabbage | | | | | | – | – | | 790 |
| Wheat | | | | | | – | – | | 790 |
| | | – | | | | – | | | 739 |
| | | | | | ○ | | | – | 403 |
| Tobacco | | | | – | | | | – | 299 |
| Peas | | – | + | | | | – | – | 118 |
| Maize | | – | | | | | | | 170 |
| Snapdragon | | ○ | | | | | | | 422 |
| Rice | | – | | | | | | | 748 |
| Apple tree | – | | | – | | | | | 770 |
| Rye | | ○ | | | | | | | 182 |
| Ryegrass | | – | | | | | | | 182 |
| Clover | | ○ | | | | | | | 182 |
| Cotton | | – | | | | – | | | 473 |

Explanations: – decrease, + increase, ○ insignificant changes.

**Table 6**
**CHANGES IN MAGNESIUM CONTENT AND UPTAKE BY VARIOUS PLANTS IN CONDITIONS OF INSUFFICIENT SUPPLY OF OXYGEN TO THEIR ROOTS**

| | Content | | | | | Uptake | | | |
|---|---|---|---|---|---|---|---|---|---|
| Plant | Roots | Tops | Stem | Leaves | Grain | Total plant | Roots | Tops | Ref. |
| Lemon | – | – | | | | – | | | 402 |
| | ○ | + | | | | – | | | 395 |
| | | | | | | – | – | – | 400 |
| Orange | | | | ○ | | | | | 774 |
| Orange and lemon | ○ | – | | | | – | | | 708 |
| | | – | – | | | – | | | 398 |
| Jojoba | – | | + | – | | – | | | 597 |
| Avocado | – | – | | | | – | | | 404 |
| | – | | ○ | – | | – | | | 661 |
| Tobacco | | – | | | | – | | | 299 |
| Peas | ○ | + | | | | | – | – | 118 |
| Wheat | | | | | ○ | | | – | 403 |
| Rye | | ○ | | | | | | | 182 |
| Ryegrass | | – | | | | | | | 182 |
| Clover | | – | | | | | | | 182 |
| Cotton | | + | | | | – | | | 473 |

Explanations: – decrease, + increase, ○ insignificant changes.

of these elements by wheat in conditions of root flooding is less impeded than that of nitrogen, phosphorus, and potassium and the plants did not manifest any tendency to displacement of calcium and magnesium from old to young leaves.

For tobacco, Harris and van Bavel[300] ranked five macroelements in descending order of their susceptibility to a restricted oxygen supply at the roots, as follows: K > N > P > Mg = Ca. The order reported by Letey et al.[422] for snapdragon is similar: K > P > (Ca + Mg).

Elkins and Hoveland[182] carried out investigations under identical experimental conditions and found that various plants show different reactions to root anoxia in terms of the calcium and/or magnesium content. Thus the calcium content in rye and clover does not depend on the level of oxygen supply to the roots but that in ryegrass is reduced by root anoxia to one third of its normal value. Magnesium content does not depend on the root oxygen supply for rye but shows a tendency to decrease under root anoxia in ryegrass and clover.

### *e. Sodium*

Oxygen deficiency in the root environment usually results in an increased sodium content in the shoots but in other cases has no significant effect on sodium content and distribution (Table 7). In some cases a reduction in the content of this element was observed in roots. An increase of sodium uptake was found in the orange while the general tendency to decrease results from the reduction of plant mass under conditions of insufficient oxygen supply to the roots that outweighs the increased sodium concentration in the plants.

### *f. Chlorine*

Anoxic conditions in soils bring about a decrease of chlorine content in roots and its increase in shoots (Table 8). The uptake of this element by plants under these conditions is generally decreased.

### Table 7
### CHANGES IN SODIUM CONTENT AND UPTAKE BY VARIOUS PLANTS IN CONDITIONS OF INSUFFICIENT SUPPLY OF OXYGEN TO THEIR ROOTS

| | Content | | | | | Uptake | | | |
|---|---|---|---|---|---|---|---|---|---|
| **Plant** | **Roots** | **Tops** | **Stem** | **Leaves** | **Grain** | **Total plant** | **Roots** | **Tops** | **Ref.** |
| Lemon | ○ | + | | | | ○ | | | 402 |
| | + | + | | | | + | | | 395 |
| | | | | | | − | − | + | 400 |
| Lemon and orange | | | + | ○ | | ○ | | | 708 |
| | | | + | ○ | | ○ | | | 398 |
| Jojoba | − | | + | + | | − | | | 597 |
| Avocado | − | + | | | | − | | | 404 |
| | + | | + | ○ | | − | | | 661 |
| Snapdragon | | + | | | | | | | 422 |
| Wheat | | | | | ○ | | | | 403 |
| Rice | | ○ | | | | | | | 748 |
| Cotton | | + | | | | − | | | 473 |

Explanations: − decrease, + increase, ○ insignificant changes.

### Table 8
### CHANGES IN CHLORINE CONTENT AND UPTAKE BY VARIOUS PLANTS IN CONDITIONS OF INSUFFICIENT SUPPLY OF OXYGEN TO THEIR ROOTS

| | Content | | | | | Uptake | | | |
|---|---|---|---|---|---|---|---|---|---|
| **Plant** | **Roots** | **Tops** | **Stem** | **Leaves** | **Grain** | **Total plant** | **Roots** | **Tops** | **Ref.** |
| Lemon | − | + | | | | − | | | 402 |
| | − | ○ | | | | − | | | 395 |
| | | | | | | − | − | + | 400 |
| Lemon and orange | − | | + | ○ | | − | | | 708 |
| | | | + | ○ | | − | | | 398 |
| Avocado | − | ○ | | | | − | | | 404 |
| | − | | ○ | ○ | | − | | | 661 |
| Rice | | + | | | | | | | 748 |
| Barley | − | | | | | | | | 340 |
| Cotton | | + | | | | | | | 473 |

Explanations: − decrease, + increase, ○ insignificant changes.

#### *g. Micronutrients*

Investigations on the influence of oxygen deficiency in the soil on the content and uptake of micronutrients by plants have been carried out in relation to boron, zinc, copper, manganese, and iron (Tables 9 to 13).

An example of this relationship is shown, for jojoba, in Figure 16. As can be seen oxygen deficiency in the roots influences micronutrient content in plants, but (cf. Tables 9 to 13) the direction of the changes is not the same for all plants. Only for manganese is there a general tendency to increase in roots and decrease in leaves.

Micronutrient uptake by plants, like the uptake of most macronutrients, is decreased by oxygen deficiency in the soil, and only in some special conditions are there no significant changes. Rice, which is a plant well adapted to flooded conditions, shows an increase in iron and manganese uptake in such conditions.

Table 9
**CHANGES IN BORON CONTENT AND UPTAKE BY VARIOUS PLANTS IN CONDITIONS OF INSUFFICIENT SUPPLY OF OXYGEN TO THEIR ROOTS**

| | Content | | | | | Uptake | | | |
|---|---|---|---|---|---|---|---|---|---|
| **Plant** | **Roots** | **Tops** | **Stem** | **Leaves** | **Grain** | **Total plant** | **Roots** | **Tops** | **Ref.** |
| Lemon | ○ | | ○ | − | | − | | | 398 |
| | ○ | + | | | | ○ | | | 402 |
| | + | − | | | | − | | | 395 |
| | | | | | | − | − | − | 400 |
| Lemon and orange | ○ | − | | | | − | | | 708 |
| Jojoba | + | | + | − | | − | | | 597 |
| Avocado | | − | | | | − | | | 404 |
| Cotton | | + | | | | − | | | 473 |

Explanations: − decrease, + increase, ○ insignificant changes.

Table 10
**CHANGES IN ZINC CONTENT AND UPTAKE BY VARIOUS PLANTS IN CONDITIONS OF INSUFFICIENT SUPPLY OF OXYGEN TO THEIR ROOTS**

| | Content | | | | | Uptake | | | |
|---|---|---|---|---|---|---|---|---|---|
| **Plant** | **Roots** | **Tops** | **Stem** | **Leaves** | **Grain** | **Total plant** | **Roots** | **Tops** | **Ref.** |
| Lemon | + | | ○ | + | | − | | | 398 |
| | ○ | − | | | | − | | | 402 |
| | + | ○ | | | | − | | | 395 |
| | | | | | | − | − | ○ | 400 |
| Orange | | | | ○ | | | | | 774 |
| Orange and lemon | ○ | ○ | | | | ○ | | | 708 |
| Wheat | | | | | − | − | | | 403 |
| Jojoba | ○ | | + | − | | − | | | 597 |
| Avocado | | ○ | | | | − | | | 404 |
| | + | | ○ | − | | − | | | 661 |
| Rice | | | | | | − | | | 355 |

Explanations: − decrease, + increase, ○ insignificant changes.

## B. Plant Response to Oxygen Surplus

Besides the well-known detrimental effect of oxygen deficiency in roots a deleterious effect of excessive (supraoptimum) oxygenation of the roots may also appear.

This phenomenon is not unexpected because the effects of nearly all factors of the environment on living organisms have the form of a curve with a maximum. The harmful influence of oxygen surplus on animals and man is also well known. In the case of plants it is important to know if there are plants for which the concentration that causes excessive root oxygenation lies below the atmospheric concentration of oxygen, i.e., within the range which is found in soil.

This phenomenon has not been intensively investigated. The possibility of its occurrence was suggested by Anderson and Kemper[18] and Eliasson[181] and this was later supported by Anaya and Stolzy[16] who found a decrease in winter wheat yield when the oxygen concentration in the soil air was above 0.1 $m^3m^{-3}$. The latter authors[16] explain this phenomenon by luxury consumption of oxygen and carbohydrates in the roots.

### Table 11
### CHANGES IN COPPER CONTENT AND UPTAKE BY VARIOUS PLANTS IN CONDITIONS OF INSUFFICIENT SUPPLY OF OXYGEN BY THEIR ROOTS

| | Content | | | | | Uptake | | | |
|---|---|---|---|---|---|---|---|---|---|
| **Plant** | **Roots** | **Tops** | **Stem** | **Leaves** | **Grain** | **Total plant** | **Roots** | **Tops** | **Ref.** |
| Lemon | + | | ○ | − | | − | | | 398 |
| | ○ | + | | | | ○ | | | 402 |
| | − | − | | | | − | | | 395 |
| | | | | | | − | − | ○ | 400 |
| Orange | | | | − | | | | | 774 |
| Orange and lemon | + | ○ | | | | − | | | 708 |
| Jojoba | ○ | | − | ○ | | − | | | 597 |
| Avocado | | ○ | | | | − | | | 404 |
| | ○ | | − | − | | − | | | 661 |
| Wheat | | | | | ○ | | | − | 403 |
| Cotton | | − | | | | − | | | 473 |

Explanations: − decrease, + increase, ○ insignificant changes.

### Table 12
### CHANGES IN MANGANESE CONTENT AND UPTAKE BY VARIOUS PLANTS IN CONDITIONS OF INSUFFICIENT SUPPLY OF OXYGEN TO THEIR ROOTS

| | Content | | | | | Uptake | | | |
|---|---|---|---|---|---|---|---|---|---|
| **Plant** | **Roots** | **Tops** | **Stem** | **Leaves** | **Grain** | **Total plant** | **Roots** | **Tops** | **Ref.** |
| Lemon | + | | − | − | | − | | | 398 |
| | ○ | ○ | | | | ○ | | | 402 |
| | + | + | | | | − | | | 395 |
| | | | | | | − | − | ○ | 400 |
| Orange | | | | − | | | | | 774 |
| Orange and lemon | + | ○ | | | | ○ | | | 708 |
| Jojoba | + | | + | − | | − | | | 597 |
| Avocado | | ○ | | | | − | | | 404 |
| | ○ | | − | − | | − | | | 661 |
| Grasses | | + | | | | | | | 773 |
| Rice | | | | | | + | | | 355 |
| Wheat | | + | | | | | | | 739 |
| | | | | | + | − | | | 403 |
| Barley | | ○ | | | | | | | 428 |

Explanations: − decrease, + increase, ○ insignificant changes.

The relationship between the length of maize roots and oxygen concentration over a wide range of concentrations is shown in Figure 17. In this case the optimum oxygen concentration is close to the atmospheric value and the marked decrease in root length takes place at much higher concentrations. The many investigations made hitherto which show a positive reaction of plants to increasing oxygen concentration up to the atmospheric value prove that the problem of excessive oxygenation of the soil is marginal. It may be significant but only in the case of plants with a low oxygen requirement.

**Table 13**
**CHANGES IN IRON CONTENT AND UPTAKE BY VARIOUS PLANTS IN CONDITIONS OF INSUFFICIENT SUPPLY OF OXYGEN TO THEIR ROOTS**

| | Content | | | | | Uptake | | | |
|---|---|---|---|---|---|---|---|---|---|
| Plant | Roots | Tops | Stem | Leaves | Grain | Total plant | Roots | Tops | Ref. |
| Lemon | + | | ○ | − | | ○ | | | 398 |
| | ○ | − | | | | − | | | 402 |
| | + | + | | | | − | | | 395 |
| | | | | | | − | − | − | 400 |
| Orange | | − | | | | − | | | 774 |
| Orange and lemon | + | − | | | | | | | 708 |
| Jojoba | − | | + | − | | − | | | 597 |
| Avocado | | + | | | | − | | | 404 |
| | ○ | | ○ | + | | − | | | 661 |
| Rice | | − | | | | + | | | 355 |
| | | − | | | | | | | 543 |
| Wheat | | − | | | | | | | 739 |
| | | | | | ○ | − | | | 403 |

Explanations: − decrease, + increase, ○ insignificant changes.

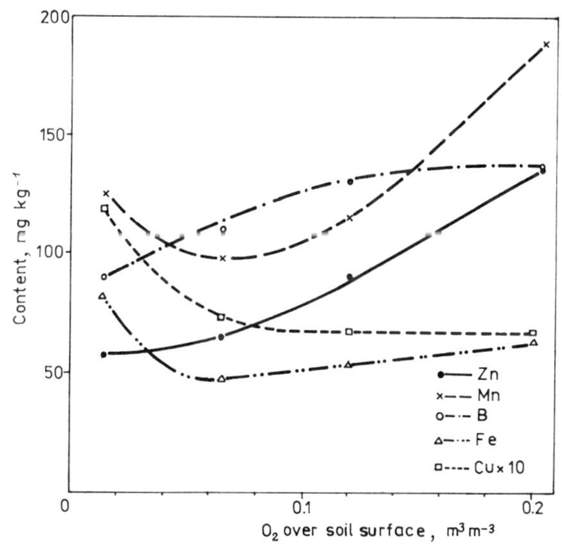


FIGURE 16. Concentration of Zn, Mn, B, Fe, and Cu in leaves of 2-year-old jojoba (*Simmondsia chinensis*) grown for 6 months at different oxygen concentration over the soil surface. (Based on the data of Reyes, D. M., Stolzy, L. H., and Labanauskas, C. K., *Agron. J.*, 69, 647, 1977.)

## C. Plant Response to Carbon Dioxide

The direct effect which carbon dioxide in the soil has on plants can be either positive or negative depending on the plant.

Many authors observed that a low concentration of $CO_2$ in the soil had a stimulating effect on plants. Thus Geisler[228,229] found a stimulating effect of $CO_2$ in concentrations of 0.01 to 0.02 $m^3m^{-3}$ in the soil air on the growth of both roots and shoots of winter barley, maize,

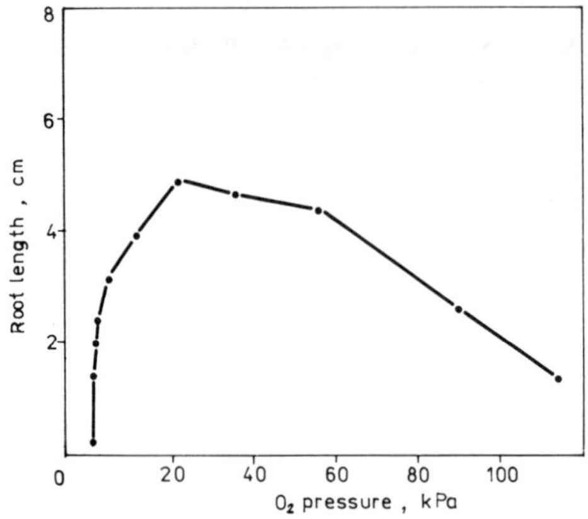


FIGURE 17. Radicle length of 65-hr-old maize seedlings as affected by $O_2$ pressure during 17-hr imbibition period in pure $O_2$ at the indicated pressure. (Data from Unger, P. W. and Danielson, R. E., *Agron. J.*, 57, 56, 1965.)

peas, and broad bean. This influence was especially distinct in the range of low concentrations of oxygen from 0 to 0.07 $m^3m^{-3}$ and disappeared at higher oxygen concentrations.

The stimulating effect of $CO_2$ was also observed in peas grown in a water culture to which air containing 0.02 to 0.1 $m^3m^{-3}$ $CO_2$ was introduced.[226] A similar stimulating effect, of 0.01 $m^3m^{-3}$ $CO_2$ in the presence of ethylene, on the lateral roots of radish was reported by Radin and Loomis.[578] Carbon dioxide stimulated germination of maize and soybean[261,262] even in concentrations up to 0.2 $m^3m^{-3}$. Carbon dioxide concentrations in the range up to 0.13 $m^3m^{-3}$ stimulated on the growth of radicals of maize; the optimum concentration of growth was 0.02 $m^3m^{-3}$.[747]

A stimulatory effect of $CO_2$ concentrations (up to 0.15 $m^3m^{-3}$) on the germination and the initial growth of four grasses and wheat was observed by Dasberg et al.[150] Carbon dioxide at a concentration of 0.08 $m^3m^{-3}$ may stimulate stolone tuberization of potato.[488]

Labanauskas et al.[401] found that a $CO_2$ concentration of 0.12 $m^3m^{-3}$ caused an increase in the height and mass of shoots and a decrease of root mass in the case of lemon seedlings. In these conditions a decrease of N, P, Ca, Mg, and Mn content in the shoots took place. In the roots a drop in N, P, K, and B and an increase in Mg and Mn content was observed. A significant increase of uptake occurred only in the case of Mg and K.

There is also much information concerning the deleterious effect of carbon dioxide on plants. The values of the $CO_2$ concentrations at which this effect exists are very differentiated. Thus Stolwijk and Thimann[704] observed a negative influence of $CO_2$ at a concentration of 0.01 $m^3m^{-3}$ on the root growth of peas, bean, sunflower, and broad bean in gravel cultures. At a concentration of 0.065 $m^3m^{-3}$ the growth of roots of these plants was completely suppressed. The growth of oat and barley roots was limited to a small extent. Radin and Loomis[578] also found an inhibitory effect of 0.01 $m^3m^{-3}$ $CO_2$ on the elongation and the number of lateral roots in radish.

Similarly Goss[251] lists soybean, bean, and peas as sensitive to carbon dioxide in the presence of ammonium ions and maize and wheat as plants which are less sensitive.

In the papers by Geisler[228,229] mentioned above he reported a statistically significant decrease in the mass of the shoots and roots of barley, peas, maize, and broad bean at $CO_2$ concentrations in soil air, above 0.04 to 0.06 $m^3m^{-3}$.

In turn, Harris and van Bavel[301] found complete inhibition of root respiration in tobacco, maize, and cotton at a $CO_2$ concentration of 0.05 $m^3m^{-3}$.

There are also plants which are resistant to high $CO_2$ concentrations in the soil. Thus, in soybean and maize the toxic effect of $CO_2$ appeared only at concentrations higher than 0.4 $m^3m^{-3}$, and even at such a high concentration as 0.8 $m^3m^{-3}$ plant mortality was not observed.[263] The root system of soybean was denser, roots were thinner and longer at $CO_2$ concentrations below 0.14 $m^3m^{-3}$ in comparison to higher concentrations at which the roots were thick, short, less branched, and dark colored. Both in soybean and in maize, at $CO_2$ concentrations higher than 0.2 $m^3m^{-3}$, leaf chlorosis appeared and, moreover, in maize a decrease in iron uptake occurred.

According to Tackett and Pearson[724] $CO_2$ concentrations in soil air of less than 0.1 $m^3m^{-3}$ had little influence on the root growth of soybean, and even at 0.25 $m^3m^{-3}$ $CO_2$ root growth was slowed down only by 30%.

Tobacco, which is one of the most sensitive plants to flooding, as mentioned earlier, does not respond to $CO_2$ concentrations in the soil air below 0.1 $m^3m^{-3}$.[299,300] Some negative effects of $CO_2$ appear at 0.2 $m^3m^{-3}$ $CO_2$,[801,802] but these authors, and also others[299,300,795,800] conclude that the sensitivity of tobacco to flooding does not result from $CO_2$ accumulation in the soil, but from oxygen deficiency. Also the mineral composition of tobacco did not change at $CO_2$ concentrations below 0.1 $m^3m^{-3}$, although at higher concentrations a decrease in the content of the main nutrients was observed, the sensitivity of which decreased as follows: K > N > P > Mg > Ca.[300] A similar order has been given by Chang and Loomis.[111]

Both the stimulating and deleterious effects of soil carbon dioxide on plants have not been fully explained. However, the stimulating effect of $CO_2$ seems to be connected with its transport from root to shoot[691] where it may be utilized as a source for photosynthesis, and also with suppression of ethanol accumulation in plant.[477] The latter phenomenon may partly explain the fact observed by Geisler,[227] that a stimulating effect of low $CO_2$ concentration was connected with a lack of oxygen or with its low concentration, for ethanol accumulation is one of the basic causes of plant injury at insufficient oxygen supply to the roots (cf. Section I.A.2.b).

As is known, $CO_2$ concentration in soil air hardly ever exceeds 0.1 $m^3m^{-3}$. Therefore, a view is sometimes expressed that $CO_2$ in the soil in such concentration does not influence plants. From this review, however, it is seen that the above opinion is not always justified and that it is necessary to extend the investigations, because at the lower $CO_2$ concentrations both negative and positive effects may appear.

### D. Plant Response to Ethylene

Ethylene has been known as a root growth inhibitor ever since 1915.[304] Recently, however, it was found that it can be a stimulator at lower concentrations (below 1 $cm^3m^{-3}$). Such stimulation was noted in broad bean,[675] tomato and mustard,[376] rye,[578] and some varieties of rice.[376,675]

Higher ethylene concentrations reduce the growth of roots of these plants, and also that of others, such as maize,[336] radish,[578] barley,[136,675,676] oats and wheat,[675] pea and white clover,[250] grapevine,[331] and some varieties of rice.[675] An example of the dependence of root growth on ethylene concentration is shown in Figure 18.

Other effects of ethylene on roots are decreasing the number of nodules in bean at ethylene concentrations below 0.4 $cm^3m^{-3}$ [285] and in pea and clover at concentration below 10 $cm^3m^{-3}$.[250] In these plants the decrease in the number of nodules was followed by a decrease of nitrogenase activity. In turn, Mingo-Castel et al.[488] noted the inhibition of tuberization and accumulation of starch in potato.

Exposure of grapevine roots to 5 $cm^3m^{-3}$ ethylene stunted them; the roots were brownish in color, had swollen tips, the number of lateral roots decreased, and those with broken lenticel increased.[331] Moreover, ethylene reduced P and K content in the leaves and that of P and Mg in the roots.

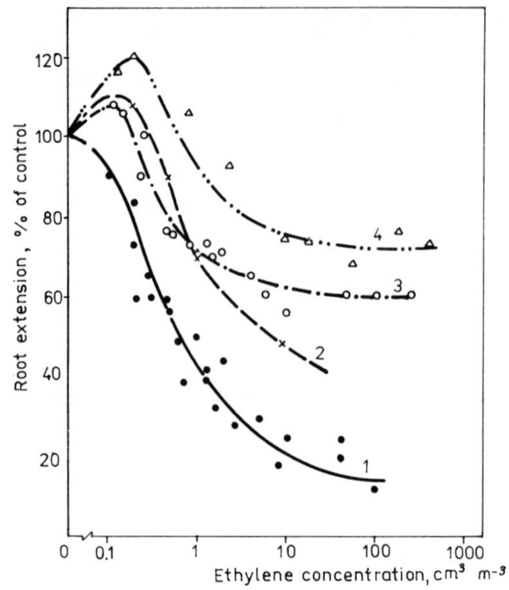


FIGURE 18. Elongation rate of roots exposed to atmospheres containing different concentrations of ethylene; (1) barley, (2) tomato, (3) rye,[675] and (4) rice.[376]

An important effect of ethylene on roots is stimulation of adventitious roots and aerenchyma formation.[171]

The influence of ethylene in the soil environment on shoots has been investigated less than its influence on roots. Ethylene at a concentration of 10 $cm^3m^{-3}$ significantly decreased the stem length and fresh mass of clover (after 6 weeks of ethylene treatment) and the dry mass of pea (after 7 weeks of ethylene treatment).[250] In barley[136] a decrease of shoot mass at 1 to 10 $cm^3m^{-3}$ of ethylene in the soil air after 7 to 21 days was observed, whereas the uptake of K, Ca, and P and their content in the plants were not affected. On the other hand Ishii et al.[331] reported that 5 $cm^3m^{-3}$ ethylene in the soil air reduced P and K content in the leaves and that of P and Mg in the roots of grapevines.

A negative influence of ethylene on the shoot mass of tomato[335] and grapevine[331] was also reported.

Some authors also reported a favorable effect of ethylene in the soil on shoots. It stimulated the growth and mass of cotton and sorghum.[215] Stimulation of soybean growth has been found by Nakayama et al.[503] at 1 to 10 $cm^3m^{-3}$ of ethylene in the soil air and only higher ethylene concentration impeded growth.[434]

## II. INDIRECT EFFECTS OF DEFICIENT AERATION ON PLANTS

As mentioned earlier, the indirect effects of soil air on plants arise from all those changes in the soil environment which are induced by changes in the amount and composition of the soil air. These are redox potential, soil reaction, nutrient availability, the status of pathogens and pests, and also the appearance of toxic compounds in the soil.

### A. Redox Potential

To the best of our knowledge, redox potential per se does not influence plants directly, but it is connected with changes in the chemical composition of the soil solution which itself is determined by the oxygen status of the soil. It can, therefore, be used as an index for

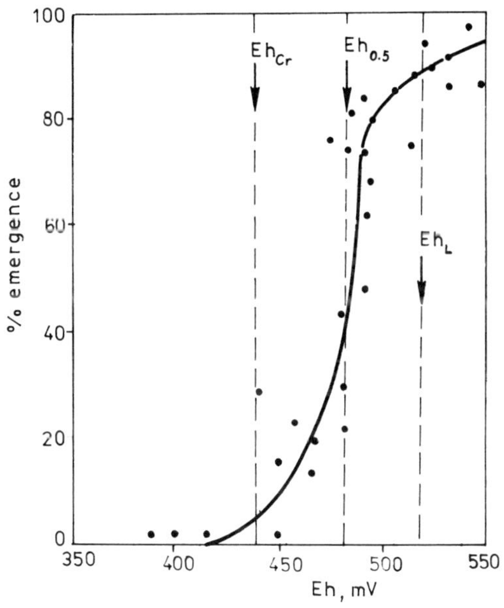


FIGURE 19. Final emergence of oats as related to soil redox potential (for explanations see Figure 3). (Modified from Glínski, J., Stępniewski, W., Łabuda, S., and Przywara, G., *Roczn. Glebozn.*, 35, 3, 1984.)

evaluation and prediction of plant reaction to soil conditions. The redox potential is particularly useful in the range of very low oxygen concentrations for which ODR index (which gives a direct evaluation of oxygen availability to the plant roots) becomes of little use.

Redox potential as an index for evaluation of plant reaction for soil oxygen conditions has been used less frequently than ODR. It was used for evaluation of soil oxygen status during emergence of barley,[244] oats,[244] bean,[245] maize,[245] tomato,[236,245] wheat,[238,245] and rye.[244,245] An example of this relationship is shown in Figure 19.

Similarly, as in the case of ODR (cf. Table 1), the values — limiting, half emergence, and critical — were distinguished (Table 14). For flax[238,245] and sugar beets[239,245] only critical values could be obtained.

Carter[106] demonstrated a very interesting approach, which makes it possible to combine the changes in redox potential with the crop yield. In microplot experiments he found that the yield of sugar cane was negatively correlated with the number of days in which the redox potential was below 332 mV. The decrease in crop yield for each day of such reduced conditions was 200 to 300 kg $ha^{-1}$.

## B. Soil Reaction

From Chapter 4 we know that the changes of soil reaction under the influence of oxygen conditions may amount to more than two units of pH and are caused both by changes in $CO_2$ concentration in the soil air and the involvement of hydrogen ions in the redox reactions.

Plant behavior connected with the changes of soil reaction is commonly known and it is discussed in many books.[56,624] Hence, it will not be treated here in detail. It seems that pH changes in soils under the influence of oxygenation may be significant to plants, though we cannot give an experimental proof of this statement.

## C. Nutrient Availability

It has been shown previously (Chapter 4) that the soil aeration status affects the solubility and thus the availability of many nutrients.

**Table 14**
**THRESHOLD Eh VALUES FOR EMERGENCE OF SOME CROP PLANTS**

| Plant | $Eh_l$ | $Eh_{0.5}$ | $Eh_{cr}$ |
|---|---|---|---|
| Barley | 400 | 380 | 370 |
| Oats | 520 | 480 | 440 |
| Beans | 510 | 450 | 370 |
| Wheat | 460 | 390 | 370 |
| Flax | — | — | 430 |
| Maize | 400 | 370 | 340 |
| Tomato | 520 | 470 | 450 |
| Sugar beet | — | — | 400[a] |
| Rye | 540 | 480 | 440 |

*Note:* $Eh_l$, limiting value; $Eh_{0.5}$, half emergence value corresponding to 50% of emergence; $Eh_{cr}$, critical value.

[a] Critical value above which emergence is possible but further survival of plant is not assured.

Data from Glínski, J., Stępniewski, W., Łabuda, S., and Przywara, G., *Roczn. Glebozn.*, 35, 3, 1984.

According to Sillanpää,[655] solubility represents availability, but in many cases there is no correlation between extractable fractions of nutrients and their uptake by plants.

Oxygen may influence nutrient uptake both by changes in nutrient availability occurring in the soil and by changes in the functioning of the roots themselves. It is not easy to distinguish between the effects of these two factors and their relative contribution to plant response is rather obscure. The changes in nutrient uptake ascribed to root functioning have been discussed previously.

There is, however, some data indicating that changes in nutrient uptake are related to their availability. This is true first of all with respect to rice, the functioning of its root being little affected by the oxygen status in the soil. This data indicates that the availability of phosphorus, silicon, iron, manganese, and aluminum increases when the soil is submerged.[562] In extreme cases it may result in toxicity. Toxicity problems in flooded soil have been reported with respect to iron,[825] manganese, and aluminum.[546] A range of 50 to 100 mg $kg^{-1}$ of water soluble $Fe^{++}$ is tentatively considered as a critical limit for rice growth.[825]

In the case of zinc, the problem of its deficiency may occur in flooded soils.[546] A well-known example of redox related availability problems is copper in organic soils. Its availability after draining these soils sharply decreases and severely affects grain formation.[580]

### D. Toxic Substances in a Reduced Soil Environment

Reduced soil may contain some phytotoxic products of anoxic metabolism. Beside the toxicity of iron mentioned in the previous section (at concentrations of more than 50 to 100 mg $kg^{-1}$ of water soluble forms[825]) and of manganese and aluminum,[546] connected with their excessive availability to toxic metabolites belong sulfides and organic acids.

Sulfide ions are formed under strongly reduced conditions (cf. Chapter 4, Section VIII) and their solubility is decreased by the precipitation of FeS and also of sulfides of heavy metals.

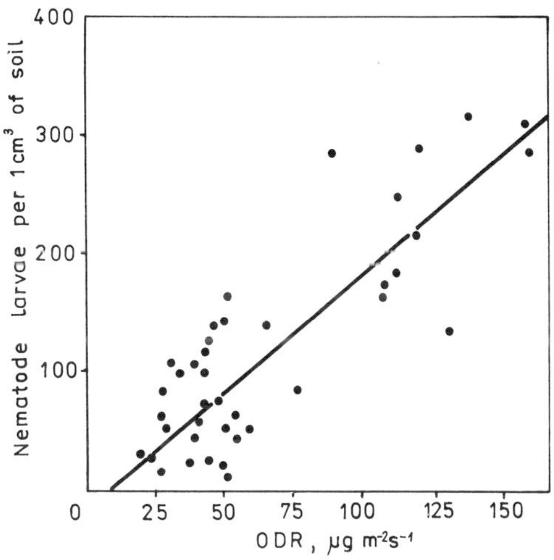


FIGURE 20. Relationship between soil ODR and counts of *Tylenchulus semipenetrans* larvae during two 19-day periods of different soil-oxygen treatment. (Modified from Stolzy, L. H., Van Gundy, S. D., Labanauskas, C. K., and Szuszkiewicz, T. E., *Soil Sci.*, 96, 292, 1963.)

Among the organic acids occurring in soil under anoxic conditions (cf. Chapter 4, Section XII.B) acetic, propionic, and butyric acids at concentrations 1 to 10 m*M* have been reported[512] to reduce rice weight, height, and root elongation, to inhibit initiation of new roots, and to decrease P and K uptake and P concentration in the shoots. At 1 m*M* the rice seedlings withered, at 5 m*M* bronzing of leaf tips occurred, while at 10 m*M* the seedlings died after 24 hr.

No toxic effect of $CH_4$ on rice seedlings have been found.[511,772]

### E. Pathogens and Pests

A review of earlier papers concerning the influence of soil oxygen conditions on plant diseases was made by Bergman.[50] From this review we know that in anoxic conditions an increase in root diseases (rotting) of many plants may occur. It was also shown that pathogenic fungi (e.g., *Phytomatotrichum omnivorum, Fusarium eumartii,* and *F. oxysporum*) develop in conditions of oxygen deficiency or of excessive carbon dioxide. It was not, however, unequivocally stated whether insufficient soil oxygenation primarily affects the host organism, decreasing its resistance to diseases or pathogens attacking it.

Plant diseases in the conditions of soil oxygen deficiency may have two reasons: a decrease of the resistance of plants to infestation or an increase in the population of pathogens or of their virulence. As to the susceptibility to infestation, there are data suggesting that in soybean an oxygen shortage decreases the tissue content of glyceoline being phytoalexine, i.e., the compound increasing the resistance of the plant to disease.[500]

Concerning the pathogen population several authors studied the influence of soil oxygen status on fungi attacking roots of citrus trees[372,397,399,710] and Rishbach cocoa.[603] The results obtained, however, do not establish a relationship between soil oxygen and the population of the fungi.

The results concerning the influence of soil oxygen conditions on nematodes are clearer than those concerning pathogens. The number of larvae of *Tylenchulus semipenetrans* Cobb which attack citrus roots increases in soil with increase of ODR[714] which is shown in Figure 20.

Other soil nematodes, *Hemicycliophora arenaria* Raski, which attack citrus and tomato are also closely related to soil oxygen conditions and show a linear growth in egg production when ODR increases from 0 to 70 $\mu g\ m^{-2}s^{-1}$ and a distinct mortality at ODR $< 20\ \mu g\ m^{-2}s^{-1}$.[761] This is the cause of a considerable decrease in nematode population in field conditions with frequent soil irrigation.[760]

## III. FLOODING TOLERANCE AND ADAPTATION OF PLANT SPECIES

The relative tolerance of particular plant species and varieties to flooding the soil is not always constant as it depends on such factors as the properties of the soil, its temperature, and the development stage of the plant. Nevertheless, under established experimental conditions some comparisons of the plants can be made and such experiments are reported in the literature. For example Kramer[382] reported that sunflower was more flood tolerant than tomato which in turn was more tolerant than tobacco. The latter plant as well as peas belong to the flood intolerant species.[338]

There is a high differentiation among legumes. Their tolerance varies from 20 to 5 days of flooding and is in the following order of decreasing tolerance: birdsfoot trefoil (*Lotus corniculatus* L. — cultivar Leo) = white clover (*Trifolium repens* L. — cultivar Pilgrim) = strawberry clover (*T. fragiferum* L.) > red clover (*T. pratense* L. — cultivar Altaswede) = aslike clover (*T. hybridum* L. — cultivar Aurora) = alfalfa (*Medicago media* Pers. — cultivar Rambler) > sweet clover (*Melilotus alba* Desr. — cultivar Arctic) = cicer milk vetch. (*Astargalus cicer* L.) > sainfoin (*Onobrychis viciaefolia* Scop. — cultivar Krasnodar).[307,308]

Grasses also differ widely in flooding tolerance. For example, "Toronto" creeping bentgrass (*Agrostis palustris* Huds.) is highly tolerant, "Merion" Kentucky bluegrass (*Poa pratensis* L.) and annual bluegrass (*P. annua* L.) are intermediate, while red fescue (*Festuca rubra* L.) is quite intolerant.[45]

A comparative study by Tadano et al.[725] showed that the order of plant tolerance to flooding is not necessarily the same as that for tolerance to high $CO_2$ concentrations or lack of oxygen in the soil air. On the basis of all these types of stresses, the overall tolerance to "deficient aeration" of 18 plant species from the most tolerant (rice) to the least tolerant (Chinese cabbage) is presented in Table 15.

The differentiation among the plants with respect to survival in media deprived of oxygen (predominantly by temporary or prolonged submergence of the soil) is due to several mechanisms of plant adaptation to such conditions. This adaptation is directed to an alternative means of oxygen supply to the plant roots rather than to an alternative metabolism in the roots.

Numerous authors identified three different kinds of plant adaptation to anoxia in the soil viz. morphological adaptation to internal transport, shallow rooting, and metabolic adaptation.

Internal transport of gases due to the development of continuous air space system within the plant (cf. Chapter 2) is the most important form of adaptation. This ability may be a permanent feature (as in rice or other aquatic plants, the roots of which transport enough oxygen to supply their own needs and to oxidize the surrounding soil) or may become apparent or amplified under unfavorable conditions (cf. Section I.A.5.a).

The basic method of this adaptation is the formation of adventitious roots containing aerenchyma. It is a reaction of plants to oxygen shortage in the root environment which is frequently met. Some plant species develop specialized knee roots and pneumatophores, constituting a diffusion path for gases.[233,735] Plants whose roots do not readily form aerenchyma (peas[169]) belong to most sensitive to flooding.

Shallow rooting is a second way of plant adaptation to oxygen deficient media. In this way the roots remain in the surface layer of the soil which is more abundant in oxygen than

### Table 15
### COMPARISON OF TOLERANCE TO HIGH MOISTURE, CARBON DIOXIDE, AND LACK OF OXYGEN IN THE SOIL AMONG CROP PLANTS

| Plant | Crop yield under stress conditions (percent of control) | | | | |
|---|---|---|---|---|---|
| | **Flooding 12 days** | **$N_2$ 7 days** | **5 kPa $CO_2$ 3 days** | **20 kPa $CO_2$ 13 days** | **Mean value** |
| Rice (*Oryzae sativa* L.) | 106 | 95 | 95 | 90 | 96.5 |
| Bean (*Phaseolus radiatus* L. var. aurea Prain) | 105 | 66 | 105 | 33 | 77 |
| Onion (*Allium cepa* L.) | 102 | 80 | 75 | 36 | 73 |
| Maize (*Zea mays* L.) | 93 | 30 | 95 | 60 | 70 |
| Wheat (*Triticum aestivum* L.) | 77 | 60 | 85 | 55 | 69 |
| Carrot (*Daucus carrota* L.) | 55 | 41 | 102 | 35 | 66 |
| Soybean (*Glycine max.* L. Merill) | 92 | 37 | 90 | 37 | 64 |
| Bean (*Phaseolus vulgaris* L.) | 88 | 70 | 85 | 13 | 64 |
| Cucumber (*Cucumis sativus* L.) | 91 | 51 | 93 | 15 | 62.5 |
| Red pepper (*Capsicum annum* L.) | 101 | 65 | 65 | 12 | 61 |
| Chrysanthemum (*Chrysanthemum coronarium* L.) | 55 | 49 | 100 | 28 | 58 |
| Eggplant(*Solanum melongena* L.) | 75 | 37 | 90 | 27 | 57 |
| Radish (*Raphanus sativus* L. var. hortensis Backer) | 107 | 37 | 55 | 12 | 53 |
| Cabbage (*Brassica cleracea* L. var. capitata L.) | 75 | 51 | 52 | 27 | 51.2 |
| Sugar beet (*Beta vulgaris* L. var. saccharifera Allfeld) | 57 | 30 | 85 | 30 | 50.5 |
| Peas (*Pisum sativum* D.C.) | 87 | 40 | 60 | 10 | 50 |
| Tomato (*Lycopersicum esculentum* Mill.) | 64 | 25 | 50 | 12 | 37.7 |
| Chinese cabbage (*Brassica peleneusis* Rupr.) | 45 | 30 | 44 | 10 | 32.5 |

Data from Tadano, T., Kirimoto, K., Aoyama, J., and Tanaka, A., *J. Sci. Soc. Manure*, 50, 261, 1979.

deeper horizons. Moreover, the path for internal diffusion in such roots, which are shorter than in plants grown in well-oxygenated media, is usually also shorter. According to Huck[324] shallow horizontal rooting is explained by the death of the top roots upon reaching the anoxic layer followed by stimulation of lateral root growth. Another mechanism responsible for shallow rooting is aerotropic root growth,[773,791] i.e., upward growth to better oxygenated soil layers. Due to this sometimes roots grow just at or even above the soil surface.[336]

Metabolic adaptation to withstand anoxia have been discussed recently by Crawford,[130] Drew,[169] and Wiedenroth.[787] The mechanism of such adaptation, which seems to be important, at least during short term oxygen deficiency, is rather obscure. In such adaptation a decrease of Pasteur effect, ethanol removal from plant roots, and alternative metabolic paths leading not to ethanol but to less toxic end products such as malate, lactate, or glycerol, are involved.

Another problem relates to differentiation of plant susceptibility to toxic metabolites formed in the soil environment. It was suggested[271] that roots of some adapted plants, due to internal oxygen transport and by thus oxidizing the surrounding soil, may detoxify it, thus contributing to the plant's tolerance.

## IV. SUMMARY

Plant response to deficient soil aeration depends on the plant and on certain environmental factors, chiefly temperature. Regarding plant tolerance to deficient soil aeration, there is a wide spectrum of plants from entirely adapted hygrophytes (aquatic plants) to very sensitive species severely damaged after short periods of the order of 1 or 2 days. A plant's response to deficient aeration also depends on the stage of its development.

The differentiation of plant species with respect to deficient aeration results from several adaptive mechanisms such as shallow rooting, morphological adaptation of roots to internal gas transport, capability of performing anoxic respiration, and resistance to toxic metabolites in the soil environment.

Visual symptoms connected with deficient aeration may appear on the roots as well on the shoots. The symptoms concerning roots are poor root growth or in extreme case root death, negative geotropism in some cases, accelerated adventitious rooting connected with aerenchyma formation and increased root porosity. Characteristic symptoms likely to appear in shoots are low rates of shoot growth, wilting, leaf epinasty and senescence, abortion of young flowers, and premature termination of growth at the shoot apex.

The physiological disturbances connected with deficient aeration are decreased transpiration and accumulation of ethylene, ethanol and other metabolites, and changes in nutrient uptake.

The effects of deficient aeration on plants may be direct and indirect; the direct being related to the interaction of the main components of soil air (i.e., oxygen and carbon dioxide) with the plant roots, while indirect effects result from air-related changes occurring in the soil environment.

It is not easy, and in some cases it is impossible, to separate the direct and indirect effects of deficient aeration. Most of the deleterious effects observed after short term stress conditions seem to be directly related to oxygen deficiency, the role of carbon dioxide being of minor importance.

The indirect effects comprise changes in redox processes, nutrient availability, soil reaction, appearance of toxic metabolites in the soil, and the viability of pathogens and pests.

The complexity of soil air-plant relations is such that they cannot be evaluated by a single aeration indicator and due to this several indicators concerning particular aeration aspects are in use.

The best measure of indirect aeration effects connected with soil reduction processes is the redox potential while oxygen availability to plant roots is most comprehensively reflected by the ODR index.

Future work is needed to provide a fuller understanding of the complex interrelations described above, with particular reference to the changes occurring in root functioning and to finding oxygen threshold values for particular plant species at different stages of growth.

Chapter 6

# MEASURING SOIL AERATION

The complexity of the interactions of soil air with the soil environment and with plant roots is such that although there are many indicators of soil aeration status, each reflecting particular aspects of it, none of them adequately describes the whole of the phenomenon. The indicators currently in use are air-filled porosity, air permeability, gas diffusion coefficient, respiration rate, soil air composition, oxygen diffusion rate (ODR), redox potential, and some further indicators employing chemical and biological procedures.

All these methods are briefly described below. Because work on methods never ceases and none of the methods is as yet ultimate and ideal, no recommendation of, or preference for, particular methods will be given here but only a presentation of the measurement principle and the related literature.

## I. AIR-FILLED POROSITY

Determining the air-filled porosity is the oldest and the simplest method of assessing soil aeration status. It is reasonable to assume that an air content above 0.25 $m^3m^{-3}$ is sufficient for good aeration. In the range of 0.10 to 0.25 $m^3m^{-3}$ aeration may be deficient under some conditions while values below 0.10 $m^3m^{-3}$ characterize decidedly deficient aeration. Air-filled porosity is usually calculated from knowledge of the actual moisture content, the bulk density of the soil and the density of its solid phase; these values being determined by well-known methods which will not be described here.

A more rapid determination can be made by the use of air pycnometers; numerous models being described in the literature.[222,393,436,478,513,528,622,771] Some of these are portable and can be used directly in field, their readings being unaffected by variations in atmospheric pressure.[222]

## II. AIR PERMEABILITY

Air permeability appears as the constant included in the Darcy equation (cf. Chapter 2, Section I) for the mass flow of gases in porous solids. This parameter indicates the ability of the soil to transmit gases when a difference in pressure exists. The Darcy equation is the basis for the measurement of air permeability in soil.

The air permeability measurement is more frequently used in soil structure studies than to assess gas transport in the soil in relation to its aeration, as the main mechanism involved in the exchange of soil gases is diffusion and not mass flow.

The methods can be steady-state[38,78,287,346,759] or nonsteady-state,[195,369] and can be used in laboratory[38,41,346] or under field conditions.[31,195,272,369,728,759]

The construction of the apparatus varies and may be based on applying, e.g., a descending float chamber,[194,287,346,728] a diaphragm type pressure regulator,[602,689] a Cartesian monstat,[759] or on the "falling head" principle[369] where the volume flux is calculated from the pressure reduction in a known time interval. Barden and Pavlakis[41] described a triaxial cell permeameter which maintains separate continuity of both the air and water phases during permeation.

In air permeability measurements, high pressure gradients must be avoided in order to eliminate turbulent gas flow and thus to preserve the validity of the Darcy equation. The pressure gradients applied, e.g., by Van Groenewoud[759] was 8 hPa $cm^{-1}$, by Green and Fordham,[272] 4 hPa $cm^{-1}$, by Turski et al.[744] 2 hPa $cm^{-1}$, and by Bakker et al.[35] 1 hPa $cm^{-1}$.

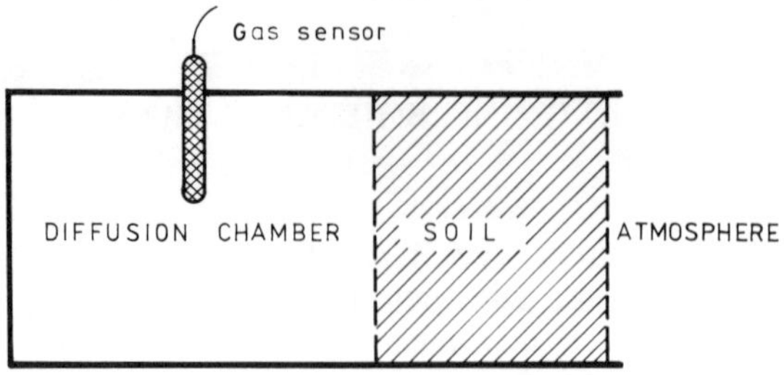


FIGURE 1. A scheme of the device for measurement of gas diffusion coefficient through the soil sample attached to the diffusion chamber. The measurement is based on recording the change in concentration of the diffusing substance inside the diffusion chamber (see text).

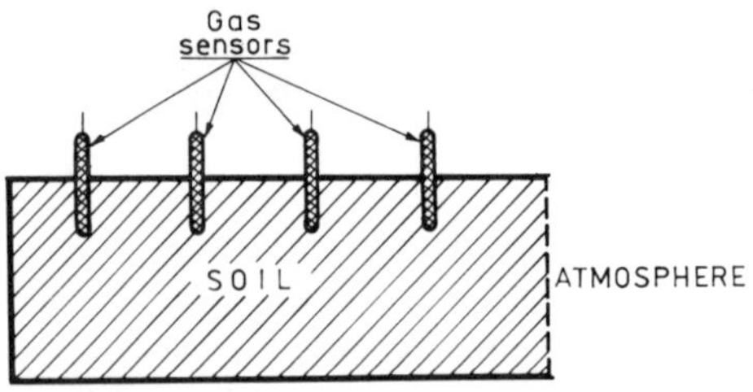


FIGURE 2. Principle of gas diffusion coefficient measurement in a soil column. The gas diffusion coefficient is calculated from the change in concentration distribution with time (see text).

## III. DIFFUSION COEFFICIENT OF GASES

The coefficient of gas diffusion in soil or in any porous body can be measured by steady-state or nonsteady-state methods.

In the steady-state methods, the diffusion flow of a diffusing substance (gas or vapor) is measured at a constant gradient of its concentration or partial pressure. This method was first applied to soil studies in 1904 by Buckingham[87] who used carbon dioxide as the diffusing substance. Penman[553] measured the rate of diffusion of carbon dioxide, carbon disulfide and acetone vapors[552] in soils. Other authors applying the steady-state method used methanol,[751] ethylene dibromide,[95] and oxygen[797] as diffusing substances. All these methods have been used under laboratory conditions. They are rather slow, several hours often being required for one measurement.

In nonsteady-state methods (where the gradient of concentration or partial pressure varies with time) used under laboratory conditions, the rate of the diffusing substance concentration increase or decrease in a closed volume connected through the soil sample to the atmosphere is usually measured. Two general models of laboratory devices which may be used are shown in Figures 1 and 2.

In the first model the concentration of the diffusing gas is measured in a chamber connected via the soil sample to the atmosphere and in the second one the chamber is omitted and the measurement is performed directly in the soil sample or column.

For the first model, when a gas diffuses from the chamber to the atmosphere its concentration in the chamber is described by the formula:[36,139]

$$\frac{\Delta C_t}{\Delta C_o} = \sum_{n=1}^{\infty} \frac{2E_g}{l_c} + \frac{\exp\left(-D\alpha_n^2 \frac{t}{E_g}\right)}{\left[l_s\left(\alpha_n^2 + \frac{E_g^2}{l_c^2}\right) + \frac{E_g}{l_c}\right]} \quad (1)$$

where $\Delta C_t$ and $\Delta C_o$ denote concentration difference between both ends of the sample at time t and 0, respectively, D = gas diffusion coefficient in the soil, $E_g$ = air-filled porosity of the sample, t = time, $l_s$ and $l_c$ = length of the sample, and of the diffusion chamber, respectively, $\alpha_n$ = nth root of $\alpha l_s \tan \alpha l_s = E_g \cdot l_s / l_c$. For $t \geqslant 0$ only the first term is sufficient.

When the gas storage within soil pores is negligible this equation is simplified to:[36]

$$D = \frac{l_s\, l_c}{t} \ln \frac{\Delta C_o}{\Delta C_t} \quad (2)$$

For the second model presented in Figure 2, Equations (106) and (109) from Chapter 2, dependent on boundary conditions, are applicable.

The nonsteady-state methods are quicker and because of this are more frequently employed. The following diffusing substances have been used: acetone, carbon disulfide,[552] carbon dioxide,[87,553,561,625] nitrogen,[731] hydrogen,[139] oxygen,[36,192,264,532,649,696] krypton,[38,781] and xenon.[781]

The choice of the diffusing gas for the measurement is of importance. Most of the techniques employing $O_2$ and $CO_2$ are subjected to interference from soil respiration. Due to this, other diffusing gases are preferred, such as hydrogen, krypton,[38,781] and xenon.[781] The choice of the diffusing gas and the analytical method are closely connected as the measurement of a rapidly changing concentration without interfering with the process constitutes one of the main difficulties in nonsteady-state methods. For example, for oxygen concentration measurement polarographic membrane covered sensors can be used.[36,532,696,797] Currie[139] used a katharometer to measure the diffusion of hydrogen in soil samples. Radioactive isotope methods were used in measuring the diffusion of carbon dioxide[625] (with $^{14}CO_2$ isotope), krypton[38,781] (with $^{85}Kr$ isotope), and xenon[781] (with $^{133}Xe$ and $^{131}Xe$ isotopes).

All the above methods were used under laboratory conditions and most of them permit determination of the diffusion coefficient of gases in samples which can be removed from the soil with the minimum possible disturbance. However, a reliable field method is needed for measurements of soil aeration under crop plants.

Melhuish et al.[476] measured the oxygen diffusion coefficient, under nonsteady state conditions, in a rhizotron. The procedure consisted of (1) sterilizing the soil in the rhizotron by treating it with methyl bromide for 48 hr, (2) passing nitrogen, or a nitrogen/air mixture containing about 0.03 to 0.04 $m^3m^{-3}$ of oxygen, through the rhizotron (which was covered with a polyethylene sheet) until equilibrium of $O_2$ content was attained throughout the whole soil volume, (3) measuring the oxygen concentration increase with time during subsequent exposure of the rhizotron to the atmosphere. The diffusion coefficient was calculated from Equation (106) from Chapter 2, putting the respiration term $q^* = 0$. By performing procedures 2 and 3 before sterilizing the soil, it was possible to calculate respiratory activity from Equation (106) from Chapter 2, the diffusion coefficient being known from the subsequent measurements, i.e., after sterilizing the soil and repeating 2 and 3. This procedure is applicable to any soil column or core (provided the diffusion coefficient and respiration rate are uniform throughout it) in laboratory conditions (cf. Figure 2) and it was used by Papendick and Runkles.[532] The method described cannot, however, be used in an open field.

The first field method was proposed by Raney[581] who placed a cylindrical diffusion chamber in the soil with the diffusion ports at a desired depth. After flushing the chamber with nitrogen and opening a special valve allowing diffusion from the soil to the chamber to take place, the rate of increase of the oxygen concentration inside the chamber was measured. The drawback of this method is its sensitivity to the geometry of the placement of the diffusion chamber and disruption of the site of the measurement.

The method proposed later by Lai et al.[407] is based on the theory of radial diffusion of a finite quantity of gas in a semi-infinite medium. The experimental procedure consists of injecting 50 $cm^3$ of oxygen of concentration $C_o$ into the soil with a gastight syringe and hypodermic access needle and thereafter withdrawing small samples (0.25 $cm^3$) at intervals of a few minutes which are analyzed for oxygen concentration (C) by a portable gas chromatograph. Before the injection of oxygen its initial concentration in the soil ($C_f$) is determined.

The solution of the radial diffusion equation (Equation [110] from Chapter 2) for the boundary conditions:

$$\begin{array}{lll} C = C_o & 0 < r \leq a & t = 0 \\ C = C_f & r > a & t = 0 \\ C = C_f & r = \infty & t > 0 \end{array} \quad (3)$$

where C is the $O_2$ concentration in the soil air at time t at a distance r from the center of the spherical $O_2$ source of radius a, as follows (neglecting soil respiration):

$$\frac{C - C_f}{C_o - C_f} = \frac{1}{2}\,\mathrm{erf}\left(\frac{a + r}{2\sqrt{D^*t}}\right) + \frac{1}{2}\,\mathrm{erf}\left(\frac{a - r}{2\sqrt{D^*t}}\right) - \frac{1}{r}\sqrt{\frac{D^*t}{\pi}}\left\{\exp\left[-\frac{(a + r)^2}{4D^*t}\right] - \exp\left[-\frac{(a + r)^2}{4D^*t}\right]\right\} \quad (4)$$

where D* is apparent diffusivity as defined in Equation (10) from Chapter 2.

The value of D* can not be calculated explicitly from Equation (4). However, measuring experimentally $(C - C_f)/(C_o - C_f)$ with respect to time it is then possible to find the best value of D* (by least square fit between the two sides of Equation [4]) on the right hand side of this equation to fit the experimental values $(C - C_f)/(C_o - C_f)$. Similarly the r value which is initially unknown (the pressure of injection causes a shift of the center of the $O_2$ pocket away from the needle point) can be determined by the least square fit.

This method cannot, however, be used in dry powdered soils, heavy clay soils, saturated soils, near the soil surface, or near cracks.

## IV. SOIL RESPIRATION

The soil respiration rate can be determined on the basis of measurement of oxygen consumption or on the basis of carbon dioxide evolution in the soil. The measurement is performed in the laboratory or *in situ* under field conditions.

The laboratory techniques permit studying the effects of different factors, e.g., temperature, moisture content, $O_2$, $CO_2$ and other gases, organic amendments, etc. but their main drawback is that the conditions are modified in comparison to the field situation.

The field techniques are limited to total soil respiration, measured usually by $CO_2$ evolution from the soil surface. However there is no reliable nondestructive method of measuring the distribution of soil respiratory activity within the soil profile, under field conditions.

### A. Laboratory Techniques

A large variety of laboratory techniques have been developed for soil respiration studies.[715] All these techniques involve the measurement of $O_2$ consumed by, or $CO_2$ evolved from, known quantities of soil incubated under controlled environmental conditions. As the incubation vessels, Erlenmayer flasks, Warburg vessels, dessicators, different sizes and types of bottles and jars have been employed.

The differences in laboratory methods are with respect to the incubation methods and the methods of gas ($O_2$ or $CO_2$) analysis.

The methods of incubation besides differentiation in the size and shape of the soil containing vessels can be divided into three groups with respect to the methods of air exchange: (1) no air flow, (2) continuous air flow (through the soil sample or over it), and (3) intermittent air flow.[715]

In the first type of incubation, the soil sample is incubated in a closed system containing a known volume of air. The respiration rate is determined from the $O_2$ uptake (a decrease in gas phase concentration) or from the $CO_2$ evolution (accumulation in the gas phase or its absorption in an alkali solution).

In the second type of incubation, i.e., incubation with a continuous air flow, the incubation vessels are connected to an aspiration manifold and are flushed with a stream of air (usually $CO_2$ free). Carbon dioxide in the outgoing air is usually absorbed in an alkali solution, but can be determined by, e.g., infrared analysis,[327] provided a constant flow rate is assured. Oxygen uptake can be determined, with a precision $O_2$ analyzer, from the difference in concentration between the incoming and outgoing air streams and the flow rate.

This type of incubation is particularly suitable for rapid and routine analyses and a device for 250 simultaneous determinations has been described.[779] The air can flow through the sample or over its surface. In the latter method a diffusion within the soil sample may be involved causing appearance of gas concentration gradients, but in the case of layered soil columns it is an approximation to the field situation.

The third incubation method is based on intermittent air flow and respiration is usually determined from the amount of $CO_2$ absorbed in an alkali solution.

Numerous analytical methods have been applied in respirometric measurements. Early volumetric methods based on absorption of oxygen (in sodium antraquinone-β-sulfonate) and carbon dioxide (in alkali solution) have been replaced by more modern techniques (see next section).

Carbon dioxide, in techniques with continuous or intermittent air-flow, is as a rule absorbed in an alkali solution[486,755] but it may be determined by such methods as infrared spectrofotometry or gas chromatography.[91,454,755]

Oxygen in incubation without air flow can be determined by gas chromatography or by a membrane electrode, while in incubation with a continuous air flow it can be determined with a precision paramagnetic analyzer. Analytical methods for $O_2$ and $CO_2$ estimations are described in the next section.

The size of the soil samples used in laboratory techniques varies from below 1 g in manometric respirometers[113,755] to many kilograms, as in the case of lysimeters with trees.[485,486]

## B. Field Methods

Several methods have been developed for measuring the total respiration rate (per unit area of soil surface) in the field. They are all based on measurement of carbon dioxide evolution and comprise chamber methods, micrometeorological methods, and a method of carbon dioxide profile in the soil.[153]

The chamber method proposed by Lundegårdh[438,439] has static and dynamic versions. In the static version, the soil is covered with a chamber and carbon dioxide accumulation inside it is followed by its absorption in an alkali solution,[153] or by measuring the increase in $CO_2$ concentration with time.[458,501,535] This method was also used *in situ* for a comparative study of soil respiration at different depths.[99]

In the dynamic version, a stream of air flows through the chamber and is then analyzed for $CO_2$ content.[153,359] This method, applied to soil lysimeters[79,145] is intermediate between field and laboratory methods of incubation with a continuous air flow. The drawback of the chamber methods is the disturbance of the ecosystem's microclimate. In the dynamic method, mass flow effects are also involved, causing differences as large as one order of magnitude between air blowing and air drawing procedures.[359]

The micrometeorological or "aerodynamic" method exists in two forms, the "in canopy" and the "above canopy" method. Both forms consist in the measurement of the $CO_2$ concentration gradient in the atmosphere within and above the vegetation and estimation of the appropriate transfer coefficients.[153]

The last procedure called the "$CO_2$ profile method" is based on calculation of the $CO_2$ flux from the $CO_2$ gradients and the experimentally determined diffusion coefficients of $CO_2$ in the soil.[153,154]

As was mentioned, all field methods are unsatisfactory and none of them provides a reliable determination of respiration rate distribution within the soil profile although attempts at this have been undertaken.[99,154,598] A comparative study showed that the highest respiration values were obtained from the $CO_2$ profile method, while the dynamic chamber method (with continuous air circulation) usually gave the lowest results.[153]

## V. SOIL AIR COMPOSITION

Determination of soil air composition comprises a sampling procedure and subsequent analysis of the sample. As the sampling technique must be suited to the analytical method, these methods will be described first.

### A. Analytical Methods

Soil air analysis comprises determination of such components as $O_2$, $CO_2$, $C_2H_4$, $CH_4$, $N_2O$, and $H_2$ and sometimes $NH_3$, NO, and $NO_2$. Obviously not all these gases are determined in every situation. Sometimes determination of only one component, e.g., $O_2$ or $CO_2$, is sufficient.

#### *1. Oxygen*

The oldest analytical methods are volumetric techniques based on absorption in sodium anthraquinone-β-sulfonate. Modification of these methods made possible microanalysis of samples as small as 0.5 to 0.001 $cm^3$ [289,640,641] for $O_2$ and $CO_2$. These laborious methods have been replaced by more modern and rapid techniques.

Paramagnetic oxygen analyzers are based on the principle that oxygen is the only gas attracted by a magnetic field. They require a rather large air sample usually more than 50 $cm^3$ to flush the internal chamber containing the magnetic balance. Portable paramagnetic analyzers can be used directly in field.[752] More complicated high precision $O_2$ analyzers can only be used in the laboratory for measuring respiration by continuous air flow procedures or for analysis of air samples brought from field.

Polarographic membrane covered sensors[796] are very convenient in use as they require relatively small air samples (up to 10 $cm^3$) which can be analyzed in the field with the portable oxygen meters.[457,746] The water condensation effects encountered when the $O_2$ sensors were buried in the soil for continuous $O_2$ monitoring have been eliminated by a double membrane system.[188] The drawback of the polarographic sensors is the variability of the calibration and frequent recalibration is required. A miniaturized version of the membrane covered electrode has been applied to measure $O_2$ gradients over distances below 1 mm.[275,280]

Gas chromatography is a modern technique allowing small air samples of the order of several cubic millimeters to be analyzed accurately in a matter of minutes. For chromatographic separation of oxygen, the common practice is to use molecular sieve 5 A column at room temperature.[46,58,59,89,293,516,671]

Katharometers providing a detection limit of 75 $cm^3m^{-3}$ [293,671] and an ultrasonic detector with a detection limit of 1 $cm^3m^{-3}$ [59] have been used for oxygen detection in gas chromatography.

Oxygen dissolved in water can be detected in 10 $mm^3$ samples on a column with Porapak® T (for $H_2O$ absorption) and molecular sieve 5 A column with the use of an electron capture detector.[292]

A determination of oxygen in soil air by mass spectrometry has also been described.[607] It should be mentioned that this method can be applied both in the laboratory and in field conditions.[504]

*2. Carbon Dioxide*

Older analytical methods for carbon dioxide determinations in soil air samples were based on absorption in alkali solutions.[289,508,640,641]

The microanalytical techniques can be used with samples as small as 0.5 to 0.001 $cm^3$ [289,640,641] while in macrotechniques samples of several cubic centimeters are used.

Gas chromatography is at present commonly used for carbon dioxide determination in small samples of air. The columns are filled with Porapak® Q.[46,59,91,671] Katharometers provide a detection limit of 20 $cm^3m^{-3}$ [671] while ultrasonic detectors[59] and helium ionization detectors[91,246,479] have at least a tenfold sensitivity advantage over katharometers with detection limits of the order of 1 $cm^3m^{-3}$.[59]

Other methods which can be used for $CO_2$ analysis in soil air employ membrane covered glass electrodes[351] or katharometers,[560] mass spectrometers,[607] and infrared analyzers.[588,752]

*3. Ethylene*

Ethylene is most conveniently analyzed by gas chromatography with flame ionization detectors providing a limit of detection of 0.02 $cm^3m^{-3}$ [671] and even 0.005 $cm^3m^{-3}$.[806] Other detectors, e.g., helium ionization,[91] are less sensitive.

Ethylene can be separated from atmospheric gases on Porapak® Q columns.[91]

Separation of ethylene from other light hydrocarbons can be accomplished by utilizing a combination of columns with alumina F1, Porapak® P, and Molecular Sieve® 13X[806] or a column of alumina deactivated with sodium iodide.[671]

*4. Methane*

Methane is determined by gas chromatography with the use of Porapak® Q,[58] alumina deactivated with sodium iodide,[671] or Molecular Sieve® 5 A[46] columns. Flame ionization detectors with a detection limit of 0.02 $cm^3m^{-3}$,[671] ultrasonic detectors with a detection limit of about 1 $cm^3m^{-3}$,[59] or helium ionization detectors with similar sensitivity[91] can be used. The detection limit using a katharometer is about 20 $cm^3m^{-3}$.[671]

*5. Nitrogenous Gases*

A review of methods developed for analysis of nitrogenous gases ($N_2$, $N_2O$, NO)[114] described diverse techniques. From this, it appears that NO and $NO_2$ can be readily estimated by absorption in an acid oxidizing reagent and determined as nitrate, while $N_2$ (and also $N_2O$) can be estimated by gas chromatography.[46,81,91,155,164,293,494,664,671]

Gaseous nitrogen is separated from oxygen on a column with Molecular Sieve® 5 A at room temperature[46,67,89,293] or with Porapak® Q at the temperature of a dry ice-acetone bath.[58,59,516]

Nitrous oxide is readily determined by gas chromatography with the use of columns with Porapak® Q,[46,91,164,293,494,671] Porapak® R,[155] or Molecular Sieve® 5 A.[664]

With the use of a katharometer, a detection of 1 $cm^3m^{-3}$ of $N_2O$ in air is possible,[164] while with a helium ionization detector[155,673] or an electron capture detector,[494] concentrations of $N_2O$ as low as 0.1 $cm^3m^{-3}$ can be detected. Atmospheric air usually contains about 0.3 $cm^3m^{-3}$ of nitrous oxide.

Methods of gas chromatography for NO determination with the use of Molecular Sieve® 5 A[664] and Porapak Q[59] have also been reported. They are, however, only suitable for gas analysis in closed systems free of oxygen, and have not been applied to soil atmosphere analysis.

*6. Hydrogen*

Hydrogen can be determined by gas chromatography on a column with Porapak® Q[59,91] or Molecular Sieve® 5 A[46] with the use of a helium ionization detector[91] or an ultrasonic detector.[59] The detection limit is of about 1 $cm^3m^{-3}$ with either detector.

*7. Hydrogen Sulfide*

Hydrogen sulfide in soil air can be determined by gas chromatography on Porapak® Q[48,91] with the use of a helium ionization detector.[91]

*8. Simultaneous Determination of Many Soil Gases*

In order to obtain a gas chromatographic analysis of soil air components using a single sample injection, various techniques have been applied. If separation of $O_2$, $N_2$, and Ar is not required all other gases, viz. $H_2$, $CH_4$, Kr, $CO_2$, $N_2O$, $C_2H_4$, and $H_2S$, can be determined using a single Porapak® Q column.[33,48,91,105] A column with Molecular Sieve® 5 A at a programmed temperature separates $O_2$, $N_2$, NO, CO, $N_2O$, and $CO_2$.[664]

Burford[89] used three separate columns with Molecular Sieve® 5 A at different temperatures in series and a katharometer to determine $N_2$, $O_2$, $CO_2$, $N_2O$, and Ar. One of the columns was maintained at −98°C to separate Ar from $O_2$.

Obermiller and Charlier[516] separated Ar, $O_2$, $N_2$, $H_2S$, $SO_2$, $CO_2$, and CO on two Porapak® Q columns in series, one of which was cooled to −70°C. A similar system with one column immersed in a dry ice-acetone bath at a temperature of −78°C was used by Blackmer et al.[58] and Blackmer and Bremner.[59] Porter[567] used Porapak® Q and Molecular Sieve® 5 A columns in series and a katharometer to separate NO, $O_2$, $N_2$, $CO_2$, and $N_2O$ at 40°C. The same set of columns but with two detectors, i.e., a katharometer and an electron capture detector was used for analysis of $N_2$, $O_2$, $N_2O$, and $CO_2$ by Hall and Dowdell.[293] Similarly two parallel columns and a katharometer were used by Beard and Guenzi[46] for $H_2$, $N_2O$, $O_2$ + Ar, $CH_4$, and NO determinations.

Smith and Dowdell[671] applied four parallel columns (one with Porapak® Q, one with Molecular Sieve® 5 A and two with deactivated alumina) with three separate detectors (a flame-ionization detector and two katharometers) for separation of $O_2$, $N_2$, Ar, $CO_2$, $N_2O$ and $C_1$-$C_4$ hydrocarbons.

Tackett[722] used two parallel columns with Molecular Sieve® 5 A and silica gel and one katharometer for determination of $O_2$, $N_2$, and $CO_2$ in soil air.

## B. Sampling the Soil Air

As was mentioned earlier, the sampling method must be suited to the method of analysis. It was found by Hack[289] that the size of the air sample extracted from the soil highly influences its composition. Thus small samples (0.1 $cm^3$) contained less oxygen and more carbon dioxide than larger ones. Moreover, the variability in the composition of the microsamples was much higher than in the case of the macrosamples, and the chance of obtaining low $O_2$ or high $CO_2$ contents is much less in the macrosamples. This finding is often quoted as confirming the heterogeneity of soil air composition and the microsite concept that anoxic conditions may occur within crumbs although the "average" concentration of oxygen in interaggregate pores may be rather high (cf. Chapter 2).

Despite some differences in the size and the construction of the probes for sampling the soil air the principle is always the same. The air is extracted by means of tubing inserted in the soil. The tubing may be of plastic or metal, sometimes it is a hypodermic needle. It may be installed permanently in the soil or reinserted before each sampling.

Macrosampling devices have been described by Russell and Appleyard,[621] Boynton and Reuther,[70] Jastrebov,[347] and Dowdell et al.[168]

Microsampling devices for samples of several cubic centimeters have been described by Robertson and Bracewell,[607] Maidl and Fischbeck,[457] and Patrick et al.[548]

The permanently installed probes are often equipped with a porous or perforated reservoir.[168,548,746]

Another method of soil air sampling was adopted by Boggie[63] who buried reservoirs of special construction in the soil enabling the air inside them to attain equilibrium with the atmosphere in their vicinity. After excavating the reservoirs the composition of their contents could be determined.

In another procedure Boyce and McCalla[69] extracted air under vacuum from soil samples brought to laboratory in an oil.

If the samples are not analyzed in the field with a portable oxygen meter or gas chromatograph they are brought to laboratory in gas pipets or in syringes. In some measuring techniques such as placing an oxygen probe directly in the soil[188,275,280,797] no air sampling is needed.

## VI. OXYGEN DIFFUSION RATE (ODR)

### A. Principle of the Method

The method of ODR measurement consists of amperometric measurement of electric current intensity corresponding to oxygen reduction on a platinum cathode placed in the soil and negatively polarized with respect to a saturated calomel electrode or silver chloride electrode. This was first proposed by Lemon and Erickson[419] in 1952 and since that date it has been treated in numerous publications.[55,420,424,468-470]

When a negative voltage within the range 0.2 to 0.8 V with respect to the saturated calomel electrode is applied to the platinum electrode placed in the soil the current initially decreases with time but stabilizes after a few minutes (cf. Figure 3).

By plotting this ''limiting'' current, corresponding to the attainment of a quasistationary state (after 4 or 5 min) vs. gradually increasing applied voltage two essentially different types of current voltage characteristics may be obtained (cf. Figure 4).

In a saturated soil or its suspension an increase in current intensity is observed, then a ''plateau'' range corresponding to oxygen reduction is reached and thereafter a steep increase in the current intensity, due to commencement of hydrogen reduction, is observed (see Figure 4 and also Figure 5). The plateau range usually lies within $-0.4$ to $-0.7$ V vs. the saturated calomel electrode,[57,468,582,599,816] although there may be some changes depending on the soil pH. In particular, under acid conditions the plateau values are shifted to the left.[27,57,470]

In unsaturated soils, however, there is no plateau at all (cf. curve 3 in Figure 4 and curves 2, 5, 6, 9, 10 in Figure 5). This is a serious deficiency of the method and its principle must be a matter of question.

The process of oxygen reduction on the electrode surface according to the McIntyre[470] is as follows, in the pH range 5 to 12:

$$O_2 + 2H_2O + 4e \rightarrow 4\ OH^- \tag{5}$$

and

$$O_2 + 4H^+ + 4e \rightarrow 2\ H_2O \tag{6}$$

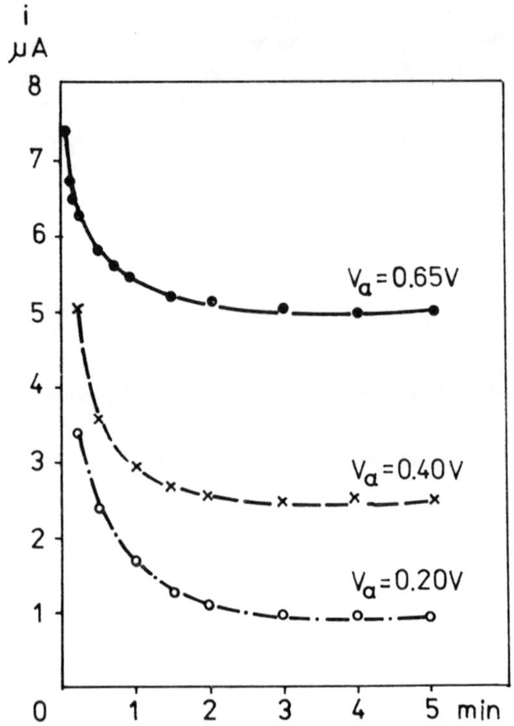


FIGURE 3. Current-time curves obtained by polarizing 0.5 × 4 mm platinum electrode with a constant applied voltage $V_a$ vs. saturated calomel electrode. The electrode placed in unsaturated sandy loam soil.

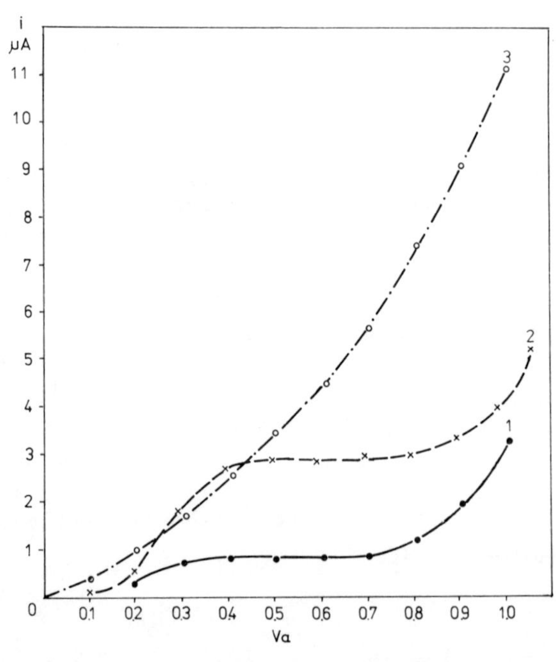


FIGURE 4. Current-applied voltage characteristics of a platinum electrode in (1) saturated sandy loam soil, (2) stirred soil suspension, and (3) unsaturated sandy loam soil.

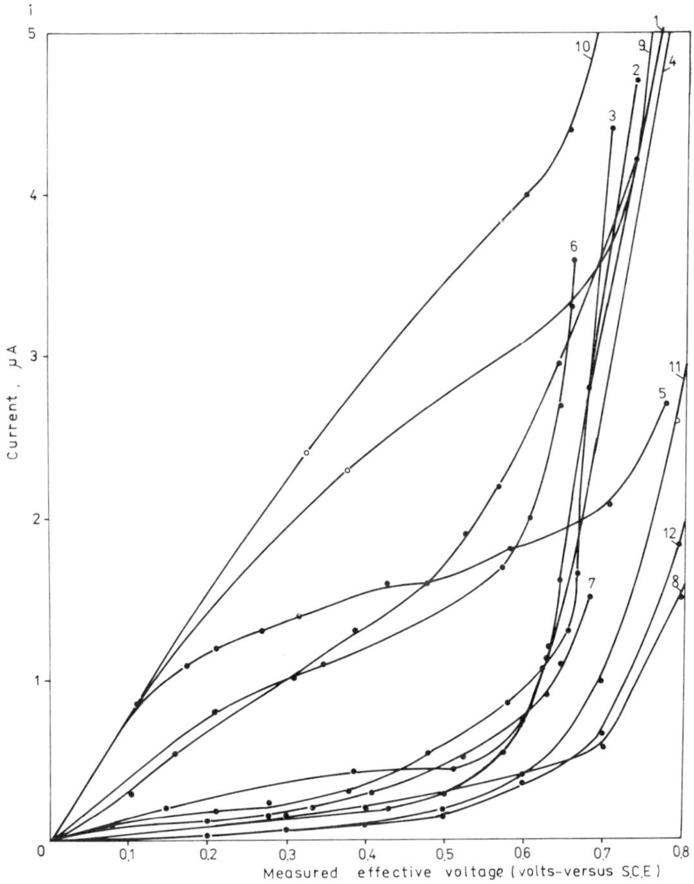


FIGURE 5. Current-effective voltage characteristics for 12 different soils (curves 1, 3, 4, 7, 8, 11, and 12 — soils of moisture tension within 0.25 to 2.5 kPa; curves 2, 5, and 6 — soils of moisture tension from 3 to 10 kPa; curves 9 and 10 — soils of moisture tension 12 to 60 kPa).

at pH <5.

Thus, independently of pH, reduction of one oxygen molecule is connected with uptake of four electrons. Current intensity is described as follows:

$$i = n\,F\,A\,f_r \tag{7}$$

where i = current intensity in amperes, n = 4 = number of equivalents per mole of oxygen, F = the Faraday (=96.500 C per equivalent) A = electrode area in square meters, $f_r$ = oxygen flux in moles $m^{-2}s^{-1}$.

Oxygen flux $f_r$ can be calculated as follows:

$$f_r = \frac{i}{n\,F\,A} \tag{8}$$

Expressing $f_r$ in SI units ($\mu g\ m^{-2}s^{-1}$) and denoting it as ODR we obtain:

$$\mathrm{ODR}[\mu g\ m^{-2}s^{-1}] = \frac{M\,i}{n\,F\,A} = 8.29 \cdot 10^{-5}\,\frac{i\ [\mu A]}{A\ [m^2]} \tag{9}$$

where M is the molecular weight of $O_2$ (32 g $mol^{-1}$).

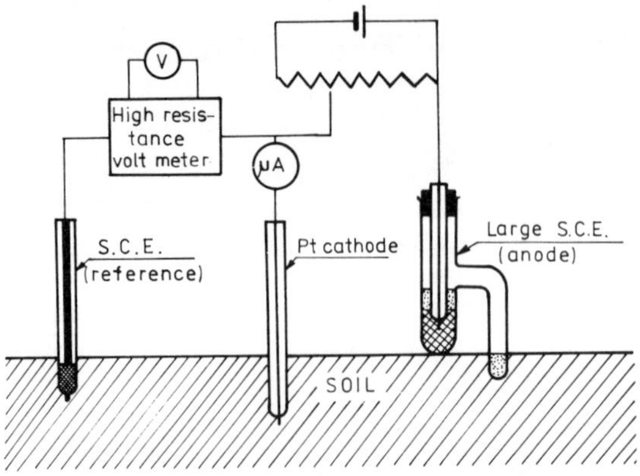


FIGURE 6. A scheme of the system allowing for measurement of applied and effective voltage during ODR measurement.

A comprehensive review on the ODR method has been given by McIntyre.[470] He lists several types of electrode polarization that are likely to occur during ODR measurement viz. resistance polarization, electrolyte transport polarization, activation polarization, concentration polarization, and reaction polarization.

The resistance polarization arises from the voltage drop ($V_R$) across the resistance of the cathode-soil-anode system (R) due to the flow of current i, and, according to Ohm's law:

$$V_R = i\,R \tag{10}$$

Thus the effective voltage of reduction ($V_e$) is lower than the applied voltage ($V_a$) by the value of $V_R$, i.e.,

$$V_e = V_a - i\,R \tag{11}$$

The resistance R of the path in the soil is of order of one to several tens of kilohms[388,469] giving, at current intensities of several microamperes, a voltage drop of as much as 0.2 to 0.4 V, i.e., more than half of the applied voltage.

Several methods have been proposed in order to eliminate resistance polarization. The most convenient method is the application of an independent (noncurrent) reference electrode to monitor the effective reduction voltage using a voltmeter having a high input resistance ($>10^{10}\Omega$).[225] The scheme of such a system is shown in Figure 6. The original method used by Lemon and Erickson[420] did not contain the "noncurrent" reference electrode and high resistance voltmeter. The current intensity and thus the resistance polarization are negligible in the second (monitor) circuit.

Electrolyte transport polarization $V_{el}$ arises from accumulation of hydroxide ions on the cathode surface due to inadequate removal of these ions from the electrode. This polarization is a cause of some overvoltage $V_{el}$ in addition to that necessary to perform the reaction of oxygen reduction.

Activation polarization is the result of an insufficient reaction rate at the electrode surface. This causes the current to be controlled by the rate of reaction and not by the flux of oxygen. The presence of activation polarization is manifested by a nonlinear increase of current with increase in oxygen concentration.

Concentration polarization is caused by low reactant concentration at high current densities. The degree of polarization increases with voltage until the concentration at the electrode reaches zero. In this situation the reaction rate is completely diffusion-controlled and the diffusion current is obtained.

Reaction polarization may be involved when a chemical reaction precedes or succeeds the electrochemical one. This is likely to occur during catalytic decomposition of $H_2O_2$ which is an intermediate in $O_2$ reduction to $H_2O$.

In the ideal case only concentration polarization and resistance polarization should occur while the other types should be eliminated. The resistance polarization is unavoidable but its contribution can be assessed either by calculation or by direct measurements as shown in Figure 6. This permits the measurement to be performed with a constant known effective voltage.

However, the absence of the polarographic plateau, which is never seen in tests with unsaturated soils, implies that besides concentration polarization, activation polarization and electrolyte transport polarization are probably present.[469,470] The presence of activation polarization is indicated by the nonlinear current-oxygen concentration dependence found by many authors.[468,470,695]

The possibility of eliminating electrolyte transport polarization has been considered and discussed by Stępniewski.[695] No satisfactory way of eliminating undesired polarizations and thus obtaining a plateau in unsaturated soils has been found as yet.

Despite these numerous limitations and uncertainties, ODR is at present still the best index of potential oxygen availability for plant roots in the soil (as shown in Chapter 5).

## B. Equipment

The first forms of apparatus used for ODR measurements in soil did not allow for the effective voltage to be controlled. However, they enabled a constant voltage to be applied to many platinum electrodes and measurement of the individual currents.[344,419,420]

Some subsequent arrangements permitting the measurement of soil resistance and correction of the applied voltage according to Equation (11) of this chapter have been described.[67,96,582] Also some devices for automated measurements have been reported.[27,96,208,556]

The method providing automatic control of the effective electrode voltage was described by Malicki and Walczak.[459]

Platinum cathodes, usually several simultaneously, are used for ODR measurement. They are constructed of 10 mm Pt wire in length with diameters in the range of 0.4 to 1.2 mm but most commonly about 0.5 mm. This wire is connected to a copper cable and mounted in a sheath which may be glass, plastic, or metal allowing some 4 to 5 mm of bare Pt wire to project. In the case of a glass sheath the attachment may be made by fusion or with special adhesives, e.g., epoxy resins, which are also suitable for plastic materials and metals. To minimize the physical effects of electrode insertion in the soil it is desirable to make the electrode sheath as thin as possible. However, it is not easy to construct electrodes thinner than 3 mm. The measurement time is usually 4 to 5 min, permitting quasistationary conditions to be achieved.

It is recommended that the electrodes are inserted into the soil immediately before the measurement; leaving the electrodes in the soil may cause a decrease of the readings obtained with them. This phenomenon, sometimes called "poisoning" of the electrodes has been observed by several authors but its mechanism has not been explained satisfactorily.[55,600,758] It was suggested that precipitation of some compounds such as carbonates or hydroxides may occur due to $OH^-$ formation at the electrode surface.[377,599,758] Electrode cleaning by washing and abrasion removes the "poisoning" effect. Between the measurements electrodes can be kept in air.

As reference electrodes, saturated calomel electrodes or silver chloride electrodes have been used. In the original Lemon and Erickson[420] system the reference electrode is also the anode through which a current of up to some tens of microamperes flows. Because of this, large electrodes of special construction with a surface of mercury (in the case of calomel electrodes) or of silver covered with silver chloride, which have an area of several square centimeters, are used to preserve a constant potential despite the flow of current.[424] In the system described by Gawlik et al.[225] and shown in Figure 6, a copper or brass rod can be used as the current carrying anode while the second "noncurrent" reference electrode can be a standard calomel one as commonly used for pH measurements.

It should be pointed out that ODR measurement is reliable only when the entire electrode surface is wetted. The ODR value, normally increasing with soil moisture, decrease sharply below a certain moisture content corresponding to break up of the water film covering the platinum surface. Thus, for proper interpretation of ODR changes in a particular soil in the field, the relationship between ODR and moisture content should be checked. The apparent values of ODR corresponding to the moisture interval at which continuity of moisture films on the platinum cathode disappears should be discarded as artificial. Usually the moisture range for reliable ODR measurements corresponds to a moisture tension up to 50 to 100 kPa.[697,763]

## VII. REDOX POTENTIAL

Redox potential in soil is measured with bright platinum electrodes of different shapes and sizes, their potential being measured with respect to a saturated calomel electrode. As the measuring device, a high input resistance voltmeter is used. This requirement is fulfilled by all modern portable pH-meters which have an input resistance of $10^{10}$ to $10^{13}$ $\Omega$.

Some authors[32] reported a phenomenon of "poisoning" the Pt electrodes during redox measurement but this phenomenon is not so evident as in ODR measurement. Some authors obtained good records of the variation in redox potential in the soil over a long period by leaving the electrodes in the soil.[27a]

It should be added that use of the same electrodes immediately after ODR measurements causes the redox potential values indicated to be less than the true values. This is due to negative polarization of the electrodes which should not be used for redox measurements. The duration of this polarization after-effect, in the case of an electrode left in the soil, is more than 24 hr. Reinserting the electrode decreases the magnitude and duration of the effect. The exact conditions of the disappearance of this effect have not been described. It should be mentioned that some authors recommend short polarization of the electrode (1 min with a voltage about 600 mV) before the measurement in order to shorten the time of reaching equilibrium potential.[435]

The range of moisture contents in which reliable redox potential measurements are possible is wider than for ODR as partial electrode moistening is sufficient for the electrode to acquire the potential of the soil.

The redox potential values with respect to a standard hydrogen electrode are calculated by adding the reference electrode potential. The potential of a saturated calomel electrode is 254, 251, 247, and 244 mV at temperatures 10, 15, 20, and 25°C, respectively.

Redox potential is the best measure of soil reduction processes in waterlogged soil and in conditions of oxygen exhaustion in the soil when other indicators such as ODR or $O_2$ content are insensitive and of little value.

Devices and facilities for continuous multiple measurements of redox potential have been described.[96,208,433] Some of them are coupled with simultaneous ODR measurements.[96,208]

The problem of measuring of the redox potential and its maintenance on a desired level in soil suspensions has been discussed by Patrick et al.[549] and in soil cores by Patrick and Henderson.[544]

# VIII. OTHER INDICATORS

Apart from the indicators of soil oxygen status described above, some other methods have been used or suggested. These are reduced iron content in the soil, observation of color, and detection of a characteristic odor.[278]

Bartlett[43] suggested determining the transformations of nitrogen occurring in a standard soil placed in polyethylene bags (impermeable to water and ions but permeable to carbon dioxide and oxygen) buried in the soil profile under study for an appropriate period.

Some plant symptoms characteristic of poor oxygen supply to the roots can also be utilized here. These are visual symptoms such as epinasty, yellowing of leaves, adventitious rooting near the soil surface, as well as changes in the chemical composition of the plants. Thus Schlichting et al.[639] proposed the ratio $N_pMn_s/N_sMn_p$ (subscripts s and p denoting N and Mn concentrations in soil and plant, respectively) as an integral soil aeration index which is high when oxygen is abundant and low when oxygen is scarce. Other chemical methods include determination of anoxic metabolites such as ethanol in plant tissue.[29,66,217]

Chapter 7

# A LOOK AHEAD

A comprehensive review of the extensive literature of the subject which has been presented shows, on the one hand, the progress in understanding the numerous determinants of soil aeration status and its complex multifactorial influence on soil processes and plant response and, on the other hand, discovers the gaps in our knowledge and the need for their rectification in future research.

Considerable progress has been made in mathematical modelling of gas transport in soil, in learning about the chemical transformations occurring in soils and in improving methods of measurement.

There is insufficient knowledge concerning the oxygen requirements of particular plant species as well as insufficient understanding of plant behavior under oxygen stress conditions. Also soil and root respiration *in situ* under field conditions have not been satisfactorily determined.

Moreover, the role of anoxic metabolites appearing in soil has not been sufficiently estimated.

The multifactorial character of soil air-plant relationships accounts for the existence of several indicators of soil aeration reflecting particular aspects of the problem but there is a lack of a single comprehensive indicator involving all these relations together. At present, the best index of soil oxygen availability to plant roots is the ODR index, while for the chemical and biochemical oxygen related processes in soil it is the redox potential.

Almost all the literature presented aimed at

1. Determination of the oxygen requirements of particular plant species
2. Description of the oxygen status in the soil on the basis of its physical parameters
3. Amelioration of the soil environment so that it would suit the plant

Recently, Crawford[130] questioned both the feasibility and the desirability of the approach. He proposed to increase the crop plants tolerance of flooding by selecting appropriate varieties. In his opinion: "After centuries of manipulating the environment to suit the plant it is time to spend some efforts in understanding the plants. It is to be hoped that future developments in this subject will teach us how to adapt the plant to suit the environment instead of relying on the brute force of technological power to make the environment suit the plant."

# REFERENCES


1. **Abdel-Ghaffar, A. S., El-Shakweer, M. H. A., and Bakakat, M. A.,** Effect of organic matter and salts on the activity of some soil enzymes, in *Proc. Symp. Organized by IAEA, FAO and Agrochimica. Soil Organic Matter Studies,* Braunschweig, International Atomic Energy Agency, Vienna, Austria, 1977, II, 319.
2. **Abeles, F. B.,** *Ethylene in Plant Biology,* Academic Press, New York, 1973.
3. **Abrasimova, L. N.,** Use of manometric method for studying gas exchange processes in soil (Russian, English summary), *Pochvovedenie,* 11, 133, 1973.
4. **Abrasimova, L. N.,** Hysteresis and temperature dependences of $O_2$ and $CO_2$ gas exchange processes in soils (Russian, English summary), *Pochvovedenie,* 6, 86, 1979.
5. **Aceves-Novarro, E., Stolzy, L. H., and Mehuys, G. R.,** Combined effect of low oxygen and salinity on germination of a semidwarf Mexican wheat, *Agron. J.,* 67, 530, 1975.
6. **Aceves-Novarro, E., Stolzy, L. H., and Mehuys, G. R.,** Response of three semidwarf Mexican wheats to different aeration conditions in the rooting medium at a constant salinity level, *Soil Sci. Soc. Am. Proc.,* 39, 515, 1975.
7. **Adamczyk-Winiarska, Z., Król, M., and Kobus, J.,** Microbiol oxidation of elemental sulphur in brown soil, *Plant Soil.,* 43, 95, 1975.
8. **Adams, D. O. and Yang, S. F.,** Ethylene biosynthesis: identification of 1-aminocyclopropane-1-carboxylic acid as an intermediate in the conversion of methionine to ethylene, *Proc. Natl. Acad. Sci. U.S.A.,* 76, 170, 1979.
9. **Adamse, A. D., Hoeks, J., de Bont, J. A. M., and van Kessel, J. F.,** Microbial activities in the soil near a leak of natural gas (Dutch, English summary), *SIAB-Rep.* 6, The Hague, The Netherlands, 1972.
10. **Adamson, J. A., Francis, A. J., Duxbury, J. M., and Alexander, M.,** Formation of volatile organic products in soils under anaerobiosis. I. Metabolism of glucose, *Soil Biol. Biochem.,* 7, 45, 1975.
11. **Aimi, R.,** Cell-physiological study on the function root. IV. Active oxygen supply into the root from leaves in rice plant, *Proc. Crop Sci. Soc. Jpn.,* 29, 51, 1960.
12. **Albertsen, M.,** Carbon dioxide balance in the gas-filled part of the unsaturated zone, demonstrated at a podzol (German, English summary), *Z. Pflanzenernaehr. Bodenkd.,* 142, 39, 1979.
13. **Aleksandrova, A. Z. and Skazkin, F. D.,** Influence of surplus soil moisture to barley at different development stages (Russian), *Dokl. Akad. Nauk. SSSR,* 159, 205, 1964.
14. **Amer, F. M. and Bartholomew, W. V.,** Influence of oxygen concentration in soil air on nitrification, *J. Paper No. J-1833 of the Iowa Agric. Exp. Stn.,* Ames, Project 1070, 215, 1950.
15. **Amoore, J. E.,** Dependence of mitosis and respiration in roots upon oxygen tension, *Proc. R. Soc. London,* 154, 109, 1961.
16. **Anaya, M. G. and Stolzy, L. H.,** Wheat response to different soil water-aeration conditions, *Soil Sci. Soc. Am. Proc.,* 36, 485, 1972.
17. **Anderson, J. M.,** Carbon dioxide evolution from two temperate, deciduous woods and soils, *J. Appl. Ecol.,* 10, 361, 1973.
18. **Anderson, W. E. and Kemper, W. D.,** Corn growth as affected by aggregate stability, soil temperature, and soil moisture, *Agron. J.,* 56, 453, 1964.
19. **André, M., Massimino, D., and Daguenet, A.,** Daily patterns under the life cycle of a maize crop. I. Photosynthesis, transpiration, respiration, *Physiol. Plant.,* 43, 397, 1978.
20. **Arashi, K. and Nitta, H.,** Studies on the lysigenous intercellular space as the ventilating system in the culm of rice and some other graminaceous plants, *Proc. Crop Sci. Soc. Jpn.,* 24, 78, 1955.
21. **Arikado, H.,** Supplementary studies on the development of the ventilating system in various plants growing on lowland and on upland, *Bull. Fac. Agric., Mie Univ.,* 20, 1, 1959.
22. **Arima, Y.,** Respiration and efficiency of $N_2$ fixation by nodules formed with a $H_2$-uptake positive strain of *Rhizobium japonicum, Soil Sci. Plant Nutr.,* 27, 115, 1981.
23. **Armstrong, W.,** Oxygen diffusion from the roots of some British bog plants, *Nature (London),* 204, 801, 1964.
24. **Armstrong, W.,** The relationship between oxidation-reduction potentials and oxygen-diffusion levels in some waterlogged organic soils, *J. Soil Sci.,* 18, 27, 1967.
25. **Armstrong, W.,** Aeration in higher plants, *Adv. Bot. Res.,* 7, 325, 1979.
26. **Armstrong, W. and Gaynard, T. J.,** The critical oxygen pressures for respiration in intact plants, *Physiol. Plant.,* 37, 200, 1976.
27. **Armstrong, W. and Wright, E. J.,** A polarographic assembly for multiple sampling of soil oxygen flux in the field, *J. Appl. Ecol.,* 13, 849, 1976.

27a. **Armstrong, W.,** private communication.

28. **Artyushenko, O. F.,** A contribution to biological soil activity (Russian, English summary), *Pochvovedenie,* 4, 134, 1969.

29. **Aubertin, G. M., Rickman, R. W., and Letey, J.,** Plant ethanol content as an index of the soil-oxygen status, *Agron. J.*, 58, 305, 1966.
30. **Ayres, K. W., Button, R. G., and de Jong, E.,** Soil morphology and soil physical properties. I. Soil aeration, *Can. J. Soil Sci.*, 52, 311, 1972.
31. **Babel, U.,** Field measurement of air permeability of soil with the use of Janert's device (German), *Z. Pflanzenernaehr. Dueng. Bodenkd.*, 91, 14, 1960.
32. **Bailey, L. D. and Beauchamp, E. G.,** Nitrate reduction, and redox potentials measured with permanently and temporarily placed platinum electrodes in saturated soils, *Can. J. Soil Sci.*, 51, 51, 1971.
33. **Bailey, L. D. and Beauchamp, E. G.,** Gas chromatography of gases emanating from a saturated soil system, *Can. J. Soil Sci.*, 53, 122, 1973.
34. **Bailey, L. D. and Beauchamp, E. G.,** Effects of temperature on $NO_3^-$ and $NO_2^-$ reduction, nitrogenous gas production, and redox potential in a saturated soil, *Can. J. Soil Sci.*, 53, 213, 1973.
35. **Bakker, J. W., Dasberg, S., and Verhaegh, W. B.,** Effect of soil structure on diffusion coefficient and air permeability of soils (manuscript unpublished).
36. **Bakker, J. W. and Hidding, A. P.,** The influence of soil structure and air content on gas diffusion in soils, *Neth. J. Agric. Sci.*, 18, 37, 1970.
37. **Balderston, W. L., Sher, B., and Payne, W. J.,** Blockage by acetylene of nitrous oxide reduction in *Pseudomonas perfectomarinus*, *Appl. Environ. Microbiol.*, 31, 504, 1976.
38. **Ball, B. C., Harris, W., and Burford, J. R.,** A laboratory method to measure gas diffusion and flow in soil and other porous materials, *J. Soil Sci.*, 32, 323, 1981.
39. **Barber, D. A., Ebert, M., and Evans, N. T. S.,** The movement of $^{15}O_2$ through barley and rice plants, *J. Exp. Bot.*, 13, 397, 1962.
40. **Barclay, A. M. and Crawford, R. M. M.,** Temperature and anoxic injury in pea seedlings, *J. Exp. Bot.*, 32, 943, 1981.
41. **Barden, L. and Pavlakis, G.,** Air and water permeability of compacted unsaturated cohesive soils, *J. Soil Sci.*, 22, 302, 1971.
42. **Bartha, R., Lanzilotta, R. P., and Pramer, D.,** Stability and effects of some pesticides in soil, *Appl. Microbiol.*, 15, 67, 1967.
43. **Bartlett, R. J.,** A biological method for studying aeration status of soil in situ, *Soil Sci.*, 100, 403, 1964.
44. **Baver, L. D.,** *Soil Physics*, 3rd ed., John Wiley & Sons, New York, 1956.
45. **Beard, J. B. and Martin, D. P.,** Influence of water temperature on submersion tolerance of four grasses, *Agron. J.*, 62, 257, 1970.
46. **Beard, W. E. and Guenzi, W. D.,** Separation of soil atmospheric gases by gas chromatography with parallel columns, *Soil Sci. Soc. Am. J.*, 40, 319, 1976.
47. **Belkovsky, V. I. and Reshetniak, A. P.,** Dynamics of $CO_2$ evolution from a peaty soil under various crop, (Russian, English summary), *Pochvovedenie*, 6, 57, 1981.
48. **Bell, R. G.,** Separation of gases likely to be evolved from flooded soils by gas chromatography, *Soil Sci.*, 105, 78, 1968.
49. **Bergersen, F. J.,** The effects of partial pressure of oxygen upon respiration and nitrogen fixation by soybean root nodules, *J. Gen. Microbiol.*, 29, 113, 1962.
50. **Bergman, H. F.,** Oxygen deficiency as a cause of disease in plants, *Bot. Rev.*, 25, 417, 1959.
51. **Berry, L. J.,** The influence of oxygen tension on the respiration rate in different segments of onion roots, *J. Cell. Comp. Physiol.*, 33, 41, 1949.
52. **Berry, L. J. and Norris, W. E.,** Studies of onion root respiration. Velocity of oxygen consumption in different segments of root of different temperatures as a function of partial pressure of oxygen, *Biochim. Biophys. Acta*, 3, 593, 1949.
53. **Bertrand, A. R. and Kohnke, H.,** Subsoil conditions and their effects on oxygen supply and growth of corn roots, *Soil Sci. Soc. Am. Proc.*, 21, 135, 1957.
54. **Bird, R. B., Stewart, W. E., and Lightfoot, E. N.,** *Transport Phenomena*, John Wiley & Sons, New York, 1960.
55. **Birkle, D. E., Letey, J., Stolzy, L. H., and Szuszkiewicz, T. E.,** Measurement of oxygen diffusion rates with the plantinum microelectrode. II. Factors influencing the measurement, *Hilgardia*, 35, 555, 1964.
56. **Black, C. A.,** *Soil-Plant Relationships*, 2nd ed., John Wiley & Sons, New York, 1968.
57. **Black, J. D. F. and Buchanan, A. S.,** Polarographic reduction of oxygen at a platinum surface in relation to the measurement of oxygen flux in soils, *Aust. J. Chem.*, 19, 2169, 1966.
58. **Blackmer, A. M., Baker, J. H., and Weeks, M. E.,** A simple gas chromatographic method for separation of gases in soil atmosphere, *Soil Sci. Soc. Am. Proc.*, 38, 689, 1974.
59. **Blackmer, A. M. and Bremner, J. M.,** Gas chromatographic analysis of soil atmosphere, *Soil Sci. Soc. Am. J.*, 41, 908, 1977.
60. **Blake, G. R. and Page, J. B.,** Direct measurement of gaseous diffusion in soils, *Soil Sci. Soc. Am. Proc.*, 13, 37, 1948.

61. **Blevis, D. G., Lowe, R. H., and Staples, L.,** Nitrate reductase in barley root under sterile, low oxygen conditions, *Plant Physiol.*, 57, 458, 1976.
62. **Blijev, Yu. K. and Martynov, A. N.,** Effect of chemical maintenance of forest on the biological activity and agrochemical properties of the soil (Russian, English summary), *Pochvovedenie*, 12, 89, 1978.
63. **Boggie, R.,** Water-table depth and oxygen content of deep peat in relation to root growth of *Pinus contorta*, *Plant Soil*, 48, 447, 1977.
64. **Bohn, H. L.,** Electromotive force of inert electrodes in soil suspensions, *Soil Sci. Soc. Am. Proc.*, 32, 211, 1968.
65. **Bollag, J. M. and Tung, G.,** Nitrous oxide release by soil fungi, *Soil Biol. Biochem.*, 4, 271, 1972.
66. **Bolton, E. F. and Erickson, A. E.,** Ethanol concentration in tomato plants during soil flooding, *Agron. J.*, 62, 220, 1970.
67. **Bornstein, J. and McGuirk, M.,** Modifications to a soil oxygen diffusion ratemeter, *Soil Sci.*, 126, 280, 1978.
68. **Boussignault, J. and Léwy, A.,** The composition of air in the cultivated soil (French), *Ann. Chim. Phys.*, Serie 3, 37, 5, 1853.
69. **Boyce, J. S. and McCalla, T. M.,** Aeration status of subtilled and plowed soils as determined by the partial vacuum method, *Soil Sci.*, 108, 241, 1969.
70. **Boynton, D. and Reuther, W.,** A way of sampling soil gases in dense subsoils and some of its advantages and limitations, *Soil Sci. Soc. Am. Proc.*, 3, 37, 1938.
71. **Bradford, K. J. and Dilley, D. R.,** Effects of root anaerobiosis on ethylene production, epinasty, and growth of tomato plants, *Plant Physiol.*, 61, 506, 1978.
72. **Bradford, K. J. and Yang, S. F.,** Xylem transport of 1-aminocyclopropane-1-carboxylic acid, an ethylene precursor, in waterlogged tomato plants, *Plant Physiol.*, 65, 322, 1980.
73. **Brandt, G. H., Wolcott, A. R., and Erickson, A. E.,** Nitrogen transformations in soil as related to structure, moisture, and oxygen diffusion rate, *Soil Sci. Soc. Am. Proc.*, 28, 71, 1964.
74. **Bremner, J. M. and Blackmer, A. M.,** Nitrous oxide: emission from soils during nitrification of fertilizer nitrogen, *Science*, 199, 295, 1978.
75. **Bremner, J. M. and Blackmer, A. M.,** Effects of acetylene and soil water content on emission of nitrous oxide from soils, *Nature (London)*, 280, 380, 1979.
76. **Bridge, B. J. and Rixon, A. J.,** Oxygen uptake and respiratory quotient of field soil cores in relation to their air-filled pore space, *J. Soil Sci.*, 27, 279, 1976.
77. **Brinkman, R.,** Ferrolysis, a hydromorphic soil forming process, *Geoderma*, 3, 199, 1970.
78. **Brooks, R. M. and Reeve, R. C.,** Measurement of air and water permeability of soils, *Trans. ASAE*, 2, 125, and 128, 1959.
79. **Brown, N. J., Fountaine, E. R., and Holden, M. R.,** The oxygen requirement of crop roots and soils under near field conditions, *J. Agric. Sci.*, 64, 195, 1965.
80. **Brown, K. W. and Rosenberg, N. J.,** Energy and $CO_2$ balance of an irrigated sugar beet (*Beta vulgaris*) field on the Great Plains, *Agron. J.*, 63, 207, 1971.
81. **Bruening, M. L. and Wullstein, L. H.,** Controlled atmosphere technique for measurement of molecular nitrogen, nitric oxide, nitrous oxide and oxygen by gas chromatography, *Environ. Sci. Technol.*, 8, 72, 1974.
82. **Brümmer, G.,** Redox reactions as processes characterizing hydromorphic soils (German, English summary), in *Trans. Comm. V and VI of the Int. Soc. Soil Sci. Pseudogley and Gley*, Schlichting, E. and Schwertmann, U., Eds., Verlag Chemie GmbH, Weinheim, W. Germany, 1973, 17.
83. **Brümmer, G.,** Redox potentials and redox processes of manganese, iron and sulphur compounds in hydromorphic soils and sediments (German, English summary), *Geoderma*, 12, 207, 1974.
84. **Brümmer, G. and Herms, U.,** Influence of soil reaction and organic matter on the solubility of heavy metals in soils, in *Effects of Accumulation of Air Pollutants in Forest Ecosystems*, Ulrich, B. and Pankrath, J., Eds., D. Reidel, Dodrecht, The Netherlands, 1983, 233.
85. **Bryant, R. D., Gordy, E. A., and Laishley, E. J.,** Effect of soil acidification on the soil microflora, in *Water, Air, and Soil Pollution*, D. Reidel, Dodrecht, The Netherlands, 1979, 437.
86. **Bryce, J. H., Focht, D. D., and Stolzy, L. H.,** Soil aeration and plant growth response to urea peroxide fertilization, *Soil Sci.*, 134, 111, 1982.
87. **Buckingham, E.,** Contributions to our knowledge of the aeration of soils, *U.S. Dep. Agric. Bur. Soils Bull.*, 25, 1, 1904.
88. **Buresh, R. J. and Patrick, W. H.,** Nitrate reduction to ammonium in anaerobic soil, *Soil Sci. Soc. Am. J.*, 42, 913, 1978.
89. **Burford, J. R.,** Single sample analysis of $N_2$-$N_2O$-$CO_2$-A-$O_2$ mixtures by gas chromatography, *J. Chromatogr. Sci.*, 7, 760, 1969.
90. **Burford, J. R.,** Ethylene in grassland soil treated with animal excreta, *J. Environ. Qual.*, 4, 55, 1975.
91. **Burford, J. R. and Bremner, J. M.,** Gas chromatographic determination of carbon dioxide evolved from soils in closed systems, *Soil Biol. Biochem.*, 4, 191, 1972.

92. **Burford, J. R., Dowdell, R. J., and Crees, R.,** Emission of nitrous oxide to the atmosphere from direct drilled and ploughed clay soils, *J. Sci. Food Agric.*, 32, 219, 1981.
93. **Burford, J. R. and Millington, R. J.,** Nitrous oxide in the atmosphere of a red-brown earth, *Trans. 9th Int. Congr. Soil Sci.*, Adelaide, Australia, 2, 505, 1968.
94. **Burns, R. G.,** Microbial activity at soil colloid surfaces, *Bull. ISSS*, 60, 79, 1981/82.
95. **Call, F.,** Soil fumigation. V. Diffusion of ethylene dibromide through soils, *J. Sci. Food Agric.*, 8, 143, 1957.
96. **Callebaut, F., Balcaen, M., Gabriels, D., and De Boodt, M.,** Data acquisition system for field determination of redox potential, oxygen diffusion rate and soil electrical resistance, *Med. Fac. Landbouw., Rijksuniv. Gent*, 45, 15, 1980.
97. **Campbell, C. A. and Ferguson, W. S.,** Influence of air temperature, light intensity, soil moisture stress and soil aeration on moisture use by wheat, *Can. J. Plant Sci.*, 49, 129, 1969.
98. **Campbell, C. A., McBean, D. S., and Green, D. G.,** Influence of moisture stress, relative humidity and oxygen diffusion rate on seed set and yield of wheat, *Can. J. Plant Sci.*, 49, 29, 1969.
99. **Campbell, J. A. and Frascarelli, L.,** Measurement of $CO_2$ evolved from organic soil at different depths in situ, *Can. J. Soil Sci.*, 61, 137, 1981.
100. **Campbell, R. B. and Moreau, R. A.,** Ethylene in a compacted field soil and its effect on growth, tuber quality, and yield of potatoes, *Am. Pot. J.*, 56, 199, 1979.
101. **Cannell, R. Q.,** Soil aeration and compaction in relation to root growth and soil management, *Appl. Biol.*, 2, 1, 1977.
102. **Cannell, R. Q. and Belford, R. K.,** Effects of waterlogging at different stages of development on the growth and yield of winter oil seed rape *(Brassica napus* L.*)*, *J. Sci. Food Agric.*, 31, 963, 1980.
103. **Cannon, W. A. and Free, E. E.,** Physiological features of roots with especial reference to the relation of roots to aeration of the soil, *Carnegie Inst. Washington Publ.*, 368, 1, 1925.
104. **Carey, R. W. and Berry, J. A.,** Effects of low temperature on respiration and uptake of rubidium ions by excised barley and corn roots, *Plant Physiol.*, 61, 858, 1978.
105. **Carle, G. C.,** Gas chromatographic determination of hydrogen, nitrogen, oxygen, methane, krypton and carbon dioxide at room temperature, *J. Chromatogr. Sci.*, 8, 550, 1970.
106. **Carter, C. E.,** Redox potential and sugarcane yield relationship, *Trans. ASAE*, 23, 924, 1980.
107. **Černa, S.,** Relation between respiratory and dehydrogenase activity of soil, *Acta Univ. Carol. Biol.*, 1/2, 1, 1978.
108. **Chancellor, W. J.,** Compaction of soil by agricultural equipment, *Div. Agric. Sci. Univ. Calif., Bull.*, 1881, 1977.
109. **Chandra, P., Furtick, W. R., and Bollen, W. B.,** The effect of four herbicides on microorganisms in nine Oregon Soils, *Weeds*, 8, 589, 1960.
110. **Chandrasekaran, S. and Yoshida, T.,** Effect of organic acids on growth of rice plant, *Soil Sci. Plant Nutr.*, 19, 39, 1973.
111. **Chang, H. T. and Loomis, W. E.,** Effect of $CO_2$ on absorption of water and nutrients by roots, *Plant Physiol.*, 20, 221, 1945.
112. **Charyulu, P. B. B. N. and Rajaramamohan Rao, V.,** Influence of ammonium nitrogen on nitrogen fixation in paddy soils, *Soil Sci.*, 131, 140, 1981.
113. **Chase, F. E. and Gray, P. H. H.,** Use of the Warburg respirometer to study metabolic activity in soil, *Nature (London)*, 171, 481, 1953.
114. **Cheng, H. H. and Bremner, J. M.,** Gaseous forms of nitrogen, in *Methods of Soil Analysis*, Part 2, Black, C. A., Ed., American Society of Agronomy, Madison, Wisc., 1965, 1287.
115. **Ćirić, M. and Škorić, A.,** Diagnostic characteristics of iron-manganese concretions in some pseudogleys in Yugoslavia, in *Trans. Comm. V and VI Int. Soc. Soil Sci. Pseudogley and Gley*, Schlichting, E. and Schwertmann, U., Eds., Verlag Chemie GmbH, Weinheim, W. Germany, 1973, 63.
116. **Clark, W. M.,** *Oxidation-Reduction Potentials of Organic Systems*, Williams & Wilkins, Baltimore, 1960.
117. **Clements, F. E.,** Aeration and air content. The role of oxygen in root activity, *Carnegie Inst. Washington Publ.*, 315, 1, 1921.
118. **Cline, R. A. and Erickson, A. E.,** The effect of oxygen diffusion rate and applied fertilizer on the growth, yield, and chemical composition of peas, *Soil Sci. Soc. Am. Proc.*, 23, 333, 1959.
119. **Collins, J. F. and Buol, S. W.,** Patterns of iron and manganese precipitation under specified Eh-pH conditions, *Soil Sci.*, 110, 157, 1970.
120. **Connell, W. F. and Patrick, W. H.,** Sulfate reduction in soil: effect of redox potential and pH, *Science*, 159, 86, 1968.
121. **Conrad, R. and Seiler, W.,** Field measurement of hydrogen evolution by nitrogen-fixing legumes, *Soil Biol. Biochem.*, 11, 689, 1979.
122. **Conrad, R. and Seiler, W.,** Contribution of hydrogen production by biological nitrogen fixation to the global hydrogen budget, *J. Geophys. Res.*, 85, 5493, 1980.

123. **Conrad, R. and Seiler, W.,** Decomposition of atmospheric hydrogen by soil microorganisms and soil enzymes, *Soil Biol. Biochem.*, 13, 43, 1981.

124. **Considine, D. J., Flynn, N., and Patching, J. W.,** Ethylene production by soil microorganisms, *Appl. Environ. Microbiol.*, 33, 977, 1977.

125. **Cooper, R. B., Blaser, R. E., and Brown, R. H.,** Potassium nutrition effects on net photosynthesis and morphology of alfalfa, *Soil Sci. Soc. Am. Proc.*, 31, 231, 1967.

126. **Cornforth, I. S.,** The persistence of ethylene in aerobic soils, *Plant Soil*, 42, 85, 1975.

127. **Coult, D. A. and Vallance, K. B.,** Observations on the gaseous exchanges which take place between *Manyanthes trifoliata*, L. and its environment, *J. Exp. Bot.*, 9, 384, 1958.

128. **Crawford, R. M. M.,** Physiological ecology: a comparision of adaptation to oxygen-poor environments in plants and animals (German, English summary), *Flora*, 161, 209, 1972.

129. **Crawford, R. M. M.,** Tolerance of anoxia and the regulation of glycolysis in tree roots, in *Tree Physiology and Yield Improvement*, Cannell, M. G. R. and Last, F. T., Eds., Academic Press, London, 1976.

130. **Crawford, R. M. M.,** Metabolic indicators in the prediction of soil anaerobiosis, in *Nitrogen in the Environment*, Vol. 1, Nielsen, D. R. and McDonald, J. G., Eds., Academic Press, New York, 1978, 427.

131. **Crawford, R. M. M.,** Biochemical and ecological similarities in marsh plants and diving animals (Russian, English summary), *Izv. Akad. Nauk. SSSR, Ser. Biol.*, 1, 77, 1981.

132. **Crawford, R. M. M. and Baines, M. A.,** Tolerance of anoxia and the metabolism of ethanol in intact roots and seeds, *New Phytol.*, 79, 519, 1977.

133. **Crawford, R. M. M. and Palin, M. A.,** Root respiration and temperature limits to the North-South distribution of four perennial maritime plants, *Flora*, 171, 338, 1981.

134. **Criswell, J. G., Havelka, U. D., Quebedeaux, B., and Hardy, R. W. F.,** Adaptation of nitrogen fixation by intact soybean nodules to altered rhizosphere $pO_2$, *Plant Physiol.*, 58, 622, 1976.

135. **Criswell, J. G., Havelka, U. D., Quebedeaux, B., and Hardy, R. W. F.,** Effect of rhizosphere $pO_2$ on nitrogen fixation by excised and intact nodulated roots, *Crop. Sci.*, 17, 39, 1977.

136. **Crossett, R. N. and Campbell, D. J.,** The effects of ethylene in the root environment upon the development of barley, *Plant Physiol.*, 42, 453, 1975.

137. **Croswell, E. T. and Waring, S. A.,** Effect of grinding on the decomposition of soil matter. II. Oxygen uptake and nitrogen mineralization in virgin and cultivated cracking clay soils, *Soil Biol. Biochem.*, 4, 435, 1972.

138. **Culler, R. C., Hanson, R. L., Myrick, R. M., Turner, R. M., and Kipple, F. P.,** Evapotranspiration before and after clearing phreatophytes, Gila River flood plain, Graham County, Arizona, U.S. Geological Survey Professional Paper 655-P, 1982.

139. **Currie, J. A.,** Gaseous diffusion in porous media. I. A nonsteady state method, *Br. J. Appl. Phys.*, 11, 314, 1960.

140. **Currie, J. A.,** Gaseous diffusion in porous media. II. Dry granular materials, *Br. J. Appl. Phys.*, 11, 318, 1960.

141. **Currie, J. A.,** Gaseous diffusion in porous media. III. Wet granular materials, *Br. J. Appl. Phys.*, 12, 275, 1961.

142. **Currie, J. A.,** Gaseous diffusion in the aeration of aggregated soils, *Soil Sci.*, 92, 40, 1961.

143. **Currie, J. A.,** Diffusion within soil microstructure. A structural parameter for soils, *J. Soil Sci.*, 16, 279, 1965.

144. **Currie, J. A.,** Movement of gases in soil respiration, in *Sorption and Transport Processes in Soil*, S.C.I. Monogr. No. 37, Society of Chemical Industry, London, 1970, 152.

145. **Currie, J. A.,** Soil respiration, *Tech. Bull., Min. Agric., Fish. Food*, 29, 461, 1975.

146. **Currie, J. A.,** Rothamsted studies of soil structure. IV. Porosity, gas diffusion and pore complexity in dry soil crumbs, *J. Soil Sci.*, 30, 441, 1979.

147. **Dąbek-Szreniawska, M. and Drążkiewicz, M.,** Problems of microorganisms sorption in soils (Polish), *Problemy Agrofizyki*, Ossolineum, Wrocław, 1977.

148. **Dart, P. J. and Day, J. M.,** Effects of incubation temperature and oxygen tension on nitrogenase activity of legume root nodules, *Plant Soil*, special volume, 167, 1971.

149. **Das, A. C.,** Carbon dioxide evolution from soils. I. A preliminary study, *Technology*, 7, 252, 1970.

150. **Dasberg, S., Enoch, H., and Hillel, D.,** Effect of oxygen and carbon dioxide concentration on the germination of range of grasses, *Agron. J.*, 58, 206, 1966.

151. **Dehérain, P. P. and Demoussy, E.,** The oxidation of organic matter in soil (French), *Ann. Agron.*, 22, 305, 1896.

152. **De Jong, E.,** Soil aeration as affected by slope position and vegetative cover, *Soil Sci.*, 131, 34, 1981.

153. **De Jong, E., Radmann, R. E., and Ripley, E. A.,** A comparison of methods of measure soil respiration, *Soil Sci.*, 127, 300, 1979.

154. **De Jong, E. and Schappert, H. J. V.,** Calculation of soil respiration and activity from $CO_2$ profiles in the soil, *Soil Sci.*, 113, 328, 1972.

155. **Delwiche, C. C. and Rolston, D. E.,** Measurement of small nitrous oxide concentrations by gas chromatography, *Soil Sci. Soc. Am. J.*, 40, 324, 1976.
156. **DeMunk, W. J. and de Rooy, M.,** Influence of ethylene on the development of 5°C precooled "Apeldoorn" tulips during forcing, *Hortic. Sci.*, 6, 40, 1970.
157. **Dixon, R. O. D.,** Hydrogen uptake and exchange by pea root nodules, *Ann. Bot. N. S.*, 31, 179, 1967.
158. **Dixon, R. O. D.,** Hydrogenase in pea root nodule bacteroids, *Arch. Mikrobiol.*, 62, 272, 1968.
159. **Dobson, A. L. and Wilson, H. A.,** Respiration studies on soil treated with some hydrocarbons, *Soil Sci. Soc. Am. Proc.*, 28, 536, 1964.
160. **Doelman, P. and Haanstra, L.,** Effect of lead on soil respiration and dehydrogenase activity, *Soil Biol. Biochem.*, 11, 475, 1979.
161. **Doelman, P. and Haanstra, L.,** Effect of lead on the decomposition of organic matter, *Soil Biol. Biochem.*, 11, 481, 1979.
162. **Domżał, H., Hodara, J., and Słowínska-Jurkiewicz, A.,** Changes of air permeability of soil compacted with wheels of tractors and agricultural machines (Polish, English summary), *Roczn. Glebozn.*, 31, 55, 1980.
163. **Dowdell, R. J.,** Fate of nitrogen applied to agricultural crops with particular reference to denitrification, *Philos. Trans. R. Soc. London, Ser.*, B, 1, 1981.
164. **Dowdell, R. J. and Crees, R.,** Measurement of the nitrous oxide content of the atmosphere, *Lab. Pract.*, 23, 488, 1974.
165. **Dowdell, R. J., Crees, R., Burford, J. R., and Cannell, R. Q.,** Oxygen concentrations in a clay soil after ploughing or direct drilling, *J. Soil Sci.*, 30, 239, 1979.
166. **Dowdell, R. J., Morrison, J., and Hood, A. E. M.,** The fate of fertilizer nitrogen applied to grassland: uptake by plants, immobilisation into soil organic matter and losses by leaching and denitrification, in *Proc. Int. Symp. Eur. Grassland Fed. Role of Nitrogen in Intensive Grassland Production,* Wageningen, 1980, Pudoc, Wageningen, The Netherlands, 1980, 129.
167. **Dowdell, R. J. and Smith, K. A.,** Field studies of the soil atmosphere. II. Occurrence of nitrous oxide, *J. Soil Sci.*, 25, 231, 1974.
168. **Dowdell, R. J., Smith, K. A., Crees, R., and Restall, S. W. F.,** Field studies of ethylene in the soil atmosphere — equipment and preliminary results, *Soil Biol. Biochem.*, 4, 325, 1972.
169. **Drew, M. C.,** Plant responses to anaerobic conditions in soil and solution culture, *Curr. Adv. Plant Sci.*, 36, 1, 1979.
170. **Drew, M. C., Chamel, A., Garrec, J. P., and Fourcy, A.,** Cortical air spaces (aerenchyma) in roots of corn subjected to oxygen stress: structure and influence on uptake and translocation of $^{86}$Rb ions, *Plant Physiol.*, 65, 506, 1980.
171. **Drew, M. C., Jackson, M. B., and Giffard, S.,** Ethylene-promoted adventitious rooting and development of cortical air spaces (aerenchyma) in roots may be adaptive responses to flooding in *Zea mays* L., *Planta*, 147, 83, 1979.
172. **Drew, M. C. and Sisworo, E. J.,** The development of waterlogging damage in young barley plants in relation to plant nutrient status and changes in soil properties, *New Phytol.*, 82, 301, 1979.
173. **Drew, M. C., Sisworo, E. J., and Saker, L. R.,** Alleviation of waterlogging damage to young barley plants by application of nitrate and a synthetic cytokinin, and comparison between the effects of waterlogging, nitrogen deficiency and root excision, *New Phytol.*, 82, 315, 1979.
174. **Drobnik, J.,** The effect of temperature on soil respiration, *Folia Microbiol.*, 7, 132, 1964.
175. **Ebregt, A. and Boldewijn, J. M. A.,** Influence of heavy metals in spruce forest soil and amylase activity, $CO_2$ evolution from starch and soil respiration, *Plant Soil*, 47, 137, 1977.
176. *E.C.A. Working Party on Classification and Survey. Elements of the Legend for the Soil Map of Europe at the Scale of 1:1, 000,000,* Soil Resources Development and Conservations Service, Land and Water Development Division, F.A.O., Rome, 1970.
177. **Edwards, N. T. and Sollins, P.,** Continuous measurement of carbon dioxide evolution from partitioned forest floor components, *Ecology*, 54, 406, 1973.
178. **Effer, W. R. and Ranson, S. L.,** Respiratory metabolism in buckwheat seedlings, *Plant Physiol.*, 42, 1042, 1967.
179. **Ekpete, D. M. and Cornfield, A. H.,** Effect of varying static and changing moisture levels during incubation on the mineralization of carbon in soil, *J. Agric. Sci.*, 64, 205, 1965.
180. **El-Beltagy, A. S. and Hall, M. A.,** Effect of water stress upon endogenous ethylene levels in *Vicia faba*, *New Phytol.*, 73, 47, 1974.
181. **Eliasson, L.,** The inhibitory effect of oxygen on the growth of wheat roots, *Physiol. Plant.*, 11, 572, 1958.
182. **Elkins, Ch. B. and Hoveland, C. S.,** Soil oxygen and temperature effect on tetany potential of three annual forage species, *Agron. J.*, 69, 626, 1977.
183. **Elliott, L. F. and McCalla, T. M.,** The composition of the soil atmosphere beneath a beef cattle feedlot and a cropped field, *Soil Sci. Soc. Am. Proc.*, 36, 68, 1972.

184. **Ellis, R. C.,** The seasonal pattern of nitrogen and carbon mineralization in forest and pasture soils in southern Ontario, *Can. J. Soil Sci.*, 54, 15, 1974.
185. **Engler, R. M., Antie, D. A., and Patrick, W. H.,** Effect of dissolved oxygen on redox potential and nitrate removal in flooded swamp and marsh soils, *J. Environ. Qual.*, 5, 230, 1976.
186. **Engler, R. M. and Patrick, W. H.,** Sulfate reduction and sulfide oxidation in flooded soil as affected by chemical oxidants, *Soil Sci. Soc. Am. Proc.*, 37, 685, 1973.
187. **Engler, R. M. and Patrick, W. H.,** Nitrate removal from floodwater overlying flooded soils and sediments, *J. Environ. Qual.*, 3, 409, 1974.
188. **Enoch, H. and Falkenflug, V.,** An improved membrane system for oxygen probes, *Soil Sci. Soc. Am Proc.*, 32, 445, 1968.
189. *Environmental Biology*, Altman, P. L. and Dittmer, D. S., Eds., Federal American Society for Experimental Biology, Bethesda, Md., 1966.
190. **Epstein, E. and Kohnke, H.,** Soil aeration as effected by organic matter application, *Soil Sci. Soc. Am. Proc.*, 21, 585, 1957.
191. **Erickson, A. E.,** Short-term oxygen deficiencies and plant responses, in *Conf. Proc. Am. Soc. Agric. Eng., Drainage and Efficient Crop Production*, Chicago, 1965, 17.
192. **Erickson, A. E., Fulton, J. M., and Brand, G. M.,** New techniques for relating soil aeration and plant response, in *Trans. 8th Int. Congr. Soil Sci., Bucharest*, 2, 171, 1964.
193. **Erickson, A. E. and Van Doren, D. M.,** The relation of plant growth and yield to soil oxygen availability, in *Trans. 7th Int. Congr. Soil Sci., Madison, Wisc.*, 54, 428, 1960.
194. **Evans, D. D.,** Gas movement, in *Methods of Soil Analysis*, Part 1, Black, C. A., Ed., American Society of Agronomy, Madison, Wisc., 1965, 319.
195. **Evans, D. D. and Kirkham, D.,** Measurement of the air permeability of soils in situ, *Soil Sci. Soc. Am. Proc.*, 14, 65, 1950.
196. **Evans, H. J. and Barber, L. E.,** Biological nitrogen fixation for food and fiber production, *Science*, 197, 332, 1977.
197. **Evans, H. J., Ruiz-Argüeso, T., Jennings, N., and Hanus, J.,** Energy efficiency of symbiotic nitrogen fixation, in *Genetic Engineering for Nitrogen Fixation*, Hollaender, A., Ed., Plenum Press, New York.
198. **Evans, N. T. S. and Ebert, M.,** Radioactive oxygen in the study of gas transport down the root of *Vicia faba*, *J. Exp. Bot.*, 11, 246, 1960.
199. **Fewson, C. A. and Nicholas, D. J. D.,** Utilization of nitrate by micro-organisms, *Nature (London)*, 190, 2, 1961.
200. **Fields, M. L., Der, R., and Hemphill, D. D.,** Influence of DCPA on selected soil microorganisms, *Weeds*, 15, 195, 1967.
201. **Figliolia, A., Izza, C., and Pagliara, S.,** Effect of some nutritive elements on *Avena sativa* root respiration (Italian, English summary), *Note I, Instituto Sperimentale per la Nutrizione delle Piante*, Rome, 1977, 1.
202. **Filip, Z.,** Clay minerals as a factor influencing the biochemical activity of soil microorganisms, *Folia Microbiol.*, 18, 56, 1973.
203. **Filip, Z.,** Interrelationships Between Microorganisms and Clay Minerals and Their Influence on Soil Dynamics (German), Habilitation thesis, Justus Liebieg-University, Giessen, 1975, 173.
204. **Filip, Z., Haider, K., and Martin, J. P.,** Influence of clay minerals on growth and metabolic activity of *Epicoccum nigrum* and *Stachybotrys chartarum*, *Soil Biol. Biochem.*, 4, 135, 1972.
205. **Findlay, W. I. and McKenney, D. J.,** Direct measurement of nitrous oxide flux from soil, *Can. J. Soil Sci.*, 59, 413, 1979.
206. **Fletcher, W. W.,** The effects of herbicides on soil microorganisms, in *Herbicides and the Soil*, Woodford, E. K. and Sagar, G. R., Eds., Blackwell, Scientific, Oxford, 1960, 20.
207. **Flühler, H.,** Oxygen diffusion in soils (German, English summary), *Mitt. Schweiz Anst. Forstl. Versuchswes.*, 49, 125, 1973.
208. **Flühler, H., Ardakani, M. S., Szuszkiewicz, T. E., and Stolzy, L. H.,** Field measured nitrous oxide concentrations, redox potentials, oxygen diffusion rates, and oxygen partial pressures in relation to denitrification, *Soil Sci.*, 122, 107, 1976.
209. **Flühler, H., Stolzy, L. H., and Ardakani, M. S.,** A statistical approach to define soil aeration in respect to denitrification, *Soil Sci.*, 122, 115, 1976.
210. **Focht, D. D.,** The effect of temperature, pH, and aeration on the production of nitrous oxide and gaseous nitrogen — a zero-order kinetic model, *Soil Sci.*, 118, 173, 1974.
211. **Focht, D. D., Stolzy, L. H., and Meek, B. D.,** Sequential reduction of nitrate and nitrous oxide under field conditions as brought about by organic amendments and irrigation management, *Soil Biol. Biochem.*, 11, 37, 1979.
212. **Franken, H.,** The composition of the soil atmosphere in the root region of some arable soils. I. The composition of the soil atmosphere at different depth and the changes in this atmosphere within a vegetation period (German, English summary), *Z. Acker Pflanzenbau*, 128, 258, 1968.

213. **Franken, H.,** The composition of the soil atmosphere in the root region of some arable soils. II. The relationship between the physical properties of the soil, the composition of the soil atmosphere, and root development (German, English summary), *Z. Acker Pflanzenbau,* 130, 110, 1969.
214. **Franz, G.,** The influence of altitude, precipitation and temperature in the investigated region on humus content and microbial activity in soil samples from Nepal, *Pedobiologia,* 16, 136, 1976.
215. **Freytag, A. M. and Lira, E. P.,** Effects of soil-injected ethylene on yields of cotton and sorghum, *Agron., J.,* 64, 524, 1972.
216. **Fujihara, M. P., Gorland, T. R., Wildung, R. E., and Drucker, H.,** Response of microbiota to the presence of heavy metals in soils, *Ann. Meet. Am. Soc. Microbiol.,* 1973, 73, 32.
217. **Fulton, J. M. and Erickson, A. E.,** Relation between soil aeration and ethyl alcohol accumulation in xylem exudate of tomatoes, *Soil Sci. Soc. Am. Proc.,* 28, 610, 1964.
218. **Funke, B. R. and Harris, J. O.,** Early respiratory responses of soil treated by heat or drying, *Plant Soil,* 28, 38, 1968.
219. **Galeva, V. and Kerchev, G.,** Laboratory studies on the aeration status of the leached cinnamonic forest soil (Bulgarian, English summary), *Georgi Dimitrov Agric. Acad., Sofia, Agrophysical Studies,* 1, 77, 1973.
220. **Garcia-Novo, F. and Crawford, R. M. M.,** Soil aeration, nitrate reduction and flooding tolerance in higher plants, *New Phytol.,* 72, 1031, 1973.
221. **Gawlik, J.,** Investigations on the changes of soil air in muck soils in relation to water table variations, *Pol. J. Soil Sci.,* 1, 75, 1968.
222. **Gawlik, J.,** A modified porometer of Loebell and its use in investigations of air properties of soils (Polish, English summary), *Roczn. Glebozn.,* 28, 59, 1977.
223. **Gawlik, J.,** An attempt to determine the upper critical moisture limit of some hydrogenic soils on the basis of the ODR index under laboratory conditions, *Pol. J. Soil Sci.,* 13, 99, 1980.
224. **Gawlik, J., Glínski, J., and Stępniewski, W.,** Oxygen diffusion rate, oxidation-reduction potential and air composition in grassland soils irrigated with sugar plant waste waters (Polish, English summary), *Roczn. Glebozn.,* 27, 27, 1976.
225. **Gawlik, J., Malicki, M., and Stępniewski, W.,** The problem of effective voltage control in measurement of ODR in soil, *Pol. J. Soil Sci.,* 10, 9, 1977.
226. **Geisler, G.,** Morphogenic influence of ($CO_2$ + $HCO_3^-$) on roots, *Plant Physiol.,* 38, 77, 1963.
227. **Geisler, G.,** Interactive effects of $CO_2$ and $O_2$ in soil on root and top growth of barley and peas, *Plant Physiol.,* 42, 305, 1967.
228. **Geisler, G.,** The influence of carbonic acid ($CO_2$ + $HCO_3^-$) in the range of low oxygen concentrations ($O_2$) in soil air on shoot and root growth of cultivated plants (German, English summary), *Z. Acker Pflanzenbau,* 130, 1, 1969.
229. **Geisler, G.,** The influence of oxygen and carbon dioxide concentration on growth of roots and shoots of winter barley, maize and peas (German, English summary), *Bayer. Landwirtsch. Jahrb.,* 46, 259, 1969.
230. **Ghildyal, B. P.,** Nature, physical properties and management of submerged rice soils, Trans. 12th Int. Congr. Soil Sci., New Delhi, India, in *Vertisols and Rice Soils of the Tropics,* Symposium papers II, Shri S. N. Mehta, New Delhi, India, 1982, 121.
231. **Giedrojć, B. and Kowaliński, S.,** The influence of moisture on the $CO_2$ and $O_2$ content in the air of a sandy soil (Polish), *Zesz. Probl. Postepow Nauk Roln.,* 88, 33, 1968.
232. **Gilbert, R. G. and Griebel, G. E.,** Stimulation of soil respiration by volatiles from alfalfa, *Soil Sci. Soc. Am. Proc.,* 33, 270, 1969.
233. **Gill, C. J.,** The flooding tolerance of woody species — a review, *For. Abstr.,* 31, 671, 1970.
234. **Gilmour, J. T. and Kittrick, J. A.,** Solubility and equilibria of zinc in a flooded soil, *Soil Sci. Soc. Am. J.,* 43, 890, 1979.
235. **Gingrich, J. R. and Russell, M. B.,** Effect of soil moisture tension and oxygen concentration on the growth of corn roots, *Agron. J.,* 48, 517, 1956.
236. **Gliński, J., Dobrzański, B., Łabuda, S., and Stępniewski, W.,** Soil oxygen conditions for emergence of tomato seedlings, *Pol. J. Soil Sci.,* 11, 81, 1978.
237. **Gliński, J. and Łabuda, S.,** Oxygen diffusion rate and redox potential as indices of the oxygenation of flax and bean roots during emergence, *Pol. J. Soil Sci.,* 12, 75, 1979.
238. **Gliński, J., Łabuda, S., and Stępniewski, W.,** Oxygen content, redox potential, and ODR as indices characterizing soil oxygen conditions for winter wheat emergence, *Zesz. Probl. Postepow Nauk Roln.,* 220, 125, 1979.
239. **Gliński, J., Przywara, G., and Stępniewski, W.,** The effect of oxygen conditions in the soil on the emergence of sugar beet (Polish, English summary), *Roczn. Glebozn.,* 32, 61, 1981.
240. **Gliński, J. and Stępniewska, Z.,** The relationship between the pH of the soil and changes in the oxygen concentration (Polish, English summary), *Roczn. Glebozn.,* 31, 237, 1980.
241. **Gliński, J., Stępniewska, Z., and Kasiak, A.,** Changes of an enzymatic activity in soils with respect of their water content and oxygen status (Polish, English summary), *Roczn. Glebozn.,* 34, 53, 1983.

242. **Gliński, J. and Stępniewski, W.,** The influence of different substances on the accumulation of high $CO_2$ concentrations in soils, *Pol. J. Soil Sci.*, 6, 63, 1973.
243. **Gliński, J. and Stępniewski, W.,** The development of anaerobic processes in cultivated soils subject to the relation of air to the solid phase of soil, *Trans. 10th Int. Congr. Soil Sci.*, Vol. 1, Moscow, 1974, 262.
244. **Gliński, J. and Stępniewski, W.,** Model investigations of the effect of soil oxygenation on the emergence of cereal plants (Polish, English summary), *Folia Soc. Sci. Lublinensis*, 23, 49, 1981.
245. **Gliński, J., Stępniewski, W., Łabuda, S., and Przywara, G.,** Threshold values of ODR and Eh in soil for the emergence of cultivated plants (Polish, English summary), *Roczn. Glebozn.*, 35, 3, 1984.
246. **Goldbaum, L. R., Domanski, T. J., and Schloegel, E. L.,** The use of a helium ionization detector for the determination of atmospheric gases, *J. Gas Chromatogr.*, 6, 394, 1968.
247. **Gołębiowska, K.,** *Agricultural Microbiology* (Polish), PWRiL, Warsaw, 1975, 99.
248. **Goodlass, G. and Smith, K. A.,** Effect of pH, organic matter content and nitrate on the evolution of ethylene from soils, *Soil Biol. Biochem.*, 10, 193, 1978.
249. **Goodlass, G. and Smith, K. A.,** Effects of organic amendments on evoluation of ethylene and other hydrocarbons from soil, *Soil Biol. Biochem.*, 10, 201, 1978.
250. **Goodlass, G. and Smith, K. A.,** Effects of ethylene on root extension and nodulation of pea (*Pisum sativum* L.) and white clover (*Trifolium repens* L.), *Plant Soil*, 51, 387, 1979.
251. **Goss, J. A.,** Ammonium bicarbonate in plant nutrition, *Soil Sci.*, 89, 296, 1960.
252. **Gotoh, S.,** Reduction processes in waterlogged soils with special reference to transformation of nitrate, manganese, and iron, *Bull. Kyushu Agric. Exp. Stn.*, 16, 669, 1973.
253. **Gotoh, S. and Onikura, Y.,** The production of organic acids in water-logged soils in the presence of organic matter under laboratory conditions, *Bull. Kyushu Agric. Exp. Stn.*, 12, 235, 1967.
254. **Gotoh, S. and Onikura, Y.,** The production of organic acids in the presence of rice straw in relation to the growth and yield of rice plant using pot culture technique, *Bull. Kyushu Agric. Exp. Stn.*, 13, 173, 1970.
255. **Gotoh, S. and Onikura, Y.,** Organic acids in a flooded soil receiving added rice straw and their effect on the growth of rice, *Soil Sci. Plant Nutr.*, 17, 1, 1971.
256. **Gotoh, S. and Patrick, W. H.,** Transformation of manganese in a waterlogged soils as affected by redox potential and pH, *Soil Sci. Soc. Am. Proc.*, 36, 738, 1972.
257. **Gotoh, S. and Patrick, W. H.,** Transformation of iron in a waterlogged soil as influenced by redox potential and pH, *Soil Sci. Soc. Am. Proc.*, 38, 66, 1974.
258. **Gotoh, S. and Yamashita, K.,** Oxidation-reduction potential of a paddy soil in situ with special reference to the production of ferrous iron, manganous manganese and sulfide, *Soil Sci. Plant Nutr.*, 12, 230, 1966.
259. **Grabert, D.,** Measurements of soil respiration in a model experiment on the deeping of the arable layer (German, English summary), *Albrecht-Thaer-Arch.*, 12, 681, 1968.
260. **Grabert, D. and Steinbrener, K.,** Some basic investigations on subsoil improvement of sandy soils. I. The composition of the atmosphere and soil respiration (German, English summary), *Zentralbl. Bakteriol. Parasitenkd. Infektionskr.*, 127, 471, 1972.
261. **Grable, A. R.,** Soil aeration and plant growth, *Adv. Agron.*, 18, 58, 1966.
262. **Grable, A. R. and Danielson, R. E.,** Effect of carbon dioxide, oxygen, and soil moisture suction on germination of corn and soybeans, *Soil Sci. Soc. Am. Proc.*, 29, 12, 1965.
263. **Grable, A. R. and Danielson, R. E.,** Influence of $CO_2$ on growth of corn and soybean seedlings, *Soil Sci. Soc. Am. Proc.*, 29, 233, 1965.
264. **Grable, A. R. and Siemer, E. G.,** Effects of bulk density, aggregate size, and water suction on oxygen diffusion, redox potentials, and elongation of corn roots, *Soil Sci. Soc. Am. Proc.*, 32, 180, 1968.
265. **Gradwell, M. W.,** Laboratory study of the diffusion of oxygen through pasture topsoil, *N. Z. J. Agric. Res.*, 4, 250, 1961.
266. **Gradwell, M. W.,** Soil physical conditions of winter and the growth of ryegrass plants. I. Effects of compaction and puddling, *N. Z. J. Agric. Res.*, 8, 238, 1965.
267. **Gradwell, M. W.,** Soil oxygen and the growth of white clover, *N. Z. J. Agric. Res.*, 12, 615, 1969.
268. **Grechin, I. P. and Ignatiev, N. N.,** Molecular oxygen consumption and carbon dioxide production by soddy-podzolic soils (Russian), *Dokl. Timiryazevsk. Skh. Akad.*, 139, 287, 1968.
269. **Grechin, I. P. and Ignatiev, N. N.,** Adsorption of molecular oxygen and release of carbon dioxide by some soils (Russian, English summary), *Pochvovedenie*, 6, 46, 1969.
270. **Grechin, I. P. and Stratonovich, M. W.,** The dynamics of oxygen and carbon dioxide in the air of loamy sod-podzolic soils (Russian, English summary), *Izv. TSKhA*, 3, 132, 1968.
271. **Green, M. S. and Etherington, J. R.,** Oxidation of ferrous iron by rice (*Oryza sativa* L.) roots: a mechanism for waterlogging tolerance?, *J. Exp. Bot.*, 28, 678, 1977.
272. **Green, R. D. and Fordham, S. J.,** A field method for determining air permeability in soil, in *Soil Physical Conditions and Crop Production*, Tech. Bull. Min. Agric. Fish. Food, No. 29, 1975, 273.
273. **Greenwood, D. J.,** The effect of oxygen concentration on the decomposition of organic materials in soils, *Plant Soil*, 14, 360, 1961.

274. **Greenwood, D. J.,** Nitrification and nitrate dissimilation in soil. II. Effect of oxygen concentration, *Plant Soil,* 17, 378, 1962.
275. **Greenwood, D. J.,** Studies on the transport of oxygen through the stems and roots of vegetable seedling, *New Phytol.*, 66, 337, 1967.
276. **Greenwood, D. J.,** Root growth and oxygen distribution in soil, in *Trans. 9th Int. Congr. Soil. Sci.*, Adelaide, Australia, 1, 823, 1968.
277. **Grenwood, D. J.,** Distribution of carbon dioxide in the aqueous phase of aerobic soil, *J. Soil Sci.*, 21, 314, 1970.
278. **Greenwood, D. J.,** Measurement of soil aeration, in *Soil Physical Conditions and Crop Production,* Tech. Bull. Min. Agric. Fish. Food, No. 29, 1975, 251.
279. **Greenwood, D. J. and Berry, G.,** Aerobic respiration in soil crumbs, *Nature (London),* 195, 161, 1962.
280. **Greenwood, D. J. and Goodman, D.,** Direct measurement of the distribution of oxygen in soil aggregates and in columns of fine soil crumbs, *J. Soil Sci.*, 18, 182, 1967.
281. **Greenwood, D. J. and Goodman, D.,** Studies on the supply of oxygen to the roots of mustard seedlings (*Sinapis alba* L.), *New Phytol.*, 70, 85, 1971.
282. **Griebel, G. E. and Owens, L. D.,** Nature of the transient activation of soil microorganisms by ethanol or acetaldehyde, *Soil Biol. Biochem.*, 4, 1, 1972.
283. **Grineva, G. M.,** The effect and after-effect of anaerobic conditions on the water relations and respirations of plants (Russian, English summary), *Fiziol. Rast.*, 7, 673, 1960.
284. **Grineva, G. M.,** Alcohol formation and excretion by plant roots under anaerobic conditions (Russian, English summary), *Fiziol. Rast.*, 10, 361, 1963.
285. **Grobbelaar, N., Clarke, B., and Hough, M. C.,** The nodulation and nitrogen fixation of isolated roots of *Phaseolus vulgaris* L. III. The effect of carbon dioxide and ethylene, *Plant Soil,* special volume, 215, 1971.
286. **Grossbard, E.,** The effect of repeated field applications of four herbicides on the evolution of carbon dioxide and mineralization of nitrogen in soil, *Weed Res.*, 11, 263, 1971.
287. **Grover, B. L.,** Simplified air permeameters for soil in place, *Soil Sci. Soc. Am. Proc.*, 19, 414, 1955.
288. **Guinn, G.,** Nutritional stress and ethylene evolution by young cotton balls, *Crop Sci.*, 16, 89, 1976.
289. **Hack, H. R. B.,** An application of a method of gas microanalysis to the study of soil air, *Soil Sci.*, 82, 217, 1956.
290. **Haider, K., Filip, Z., and Martin, G. P.,** Influence of montmorillonite on the production of biomass and intermediate metabolites by some microorganisms (German, English summary), *Arch. Microbiol.*, 73, 201, 1970.
291. **Hall, D. M. and Grossbard, E.,** The effects of grazing or cutting a perennial ryegrass and white clover sward on the microflora of the soil, *Soil Biol. Biochem.*, 4, 199, 1972.
292. **Hall, K. C.,** A gas chromatographic method for the determination of oxygen dissolved in water using an electron capture detector, *J. Chromatogr. Sci.*, 16, 311, 1978.
293. **Hall, K. C. and Dowdell, R. J.,** An isothermal gas chromatographic method for the simultaneous estimation of oxygen, nitrous oxide, and carbon dioxide content of gases in the soil, *J. Chromatogr. Sci.*, 19, 107, 1981.
294. **Hammann, R. and Ottow, J. C. G.,** Reductive dissolution of $Fe_2O_3$ by saccharolytic *Clostridia* and *Bacillus polymyxa* under anaerobic conditions, *Z. Pflanzenernaehr. Bodenkd.*, 137, 108, 1974.
295. *Handbook of Physicochemistry* (Polish), Wyd. Naukowo-Techniczne, Warsaw, 1974.
296. **Hanks, R. J. and Thorp, F. C.,** Seedling emergence of wheat as related to soil moisture content, bulk density, oxygen diffusion rate, and crust strength, *Soil Sci. Soc. Am. Proc.*, 20, 307, 1956.
297. **Hansen, G. K.,** Utilization of photosynthates for growth, respiration, and storage in tops and roots of *Lollium multiflorum, Physiol. Plant.*, 42, 5, 1978.
298. **Hansen, G. K. and Jensen, C. R.,** Growth and maintenance respiration in whole plants, tops and roots of *Lollium multiflorum, Physiol. Plant.*, 39, 275, 1977.
299. **Harris, D. G. and van Bavel, C. H. M.,** Growth, yield and water absorption of tobacco plants as affected by the composition of the root atmosphere, *Agron. J.*, 49, 11, 1957.
300. **Harris, D. G. and van Bavel, C. H. M.,** Nutrient uptake and chemical composition of tobacco plants as affected by the composition of the root atmosphere, *Agron. J,* 49, 176, 1957.
301. **Harris, D. G. and van Bavel, C. H. M.,** Root respiration of tobacco, corn, and cotton plants, *Agron. J.*, 49, 182, 1957.
302. **Harter, R. D. and McLean, E. O.,** The effect of moisture level and incubation time on the chemical equilibria of a Toledo clay loam soil, *Agron. J.*, 57, 583, 1965.
303. **Harvey, R. G. and Linscott, J. J.,** Ethylene production in soil containing quackgrass rhizomes and other plant materials, *Soil Sci. Soc. Am. J.*, 42, 721, 1978.
304. **Harvey, E. M. and Rose, R. C.,** The effect of illuminating gas on root systems, *Bot. Gaz.*, 60, 27, 1915.
305. **Hattori, T.,** *Microbiol Life in the Soil, An Introduction,* Marcel Dekker, New York, 1973.

306. **Hattori, T. and Furusaka, H.,** Chemical activities of *Azotobacter agile* adsorbed on a resin, *J. Biochem. (Tokyo),* 50, 312, 1961.

307. **Heinrichs, D. H.,** Flooding tolerance of legumes, *Can. J. Plant Sci.,* 50, 435, 1970.

308. **Heinrichs, D. H.,** Root-zone temperature effects on flooding tolerance of legumes, *Can. J. Plant Sci.,* 52, 985, 1972.

309. **Heisig, W. and Seeboldt, M.,** Influence of straw manuring on the biological activity of a lessive chernozem (German, English summary), *Zentralbl. Bakteriol. Parasitenkd., Infektionskr. Hyg., Abt. 2,* 134, 480, 1979.

310. **Herms, U. and Brümmer, G.,** Solubility of heavy metals in waste deposits and soils as related to pH value, redox conditions and nutrient status (German), *Mitteilungsbl. Dtsch. Bodenkd. Ges.,* 27, 23, 1978.

311. **Herms, U. and Brümmer, G.,** The influence of organic matter on solubility of heavy metals (German), *Mitteilungsbl. Dtsch. Bodenkd. Ges.,* 27, 181, 1978.

312. **Herms, U. and Brümmer, G.,** The influence of redox conditions on solubility of heavy metals in soils and sediments (German), *Mitteilungsbl. Dtsch. Bodenkd. Ges.,* 29, 533, 1979.

313. **Hesse, P. R.,** *A Textbook of Soil Chemical Analysis,* John Murray, London, 1971.

314. **Hirschfelder, J. O., Bird, R. B., and Spotz, E. L.,** Diffusion coefficient in a binary gas mixture, *Chem. Rev.,* 1949, 205.

315. **Hirschfelder, J. O., Curtiss, C. F., and Bird, R. B.,** *Molecular Theory of Gases and Liquids,* John Wiley & Sons, New York, 1954, 1110.

316. **Hoeks, J.,** Changes in composition of soil air near leaks in natural gas mains, *Soil Sci.,* 113, 46, 1972.

317. **Hoeks, J.,** *Effect of Leaking Natural Gas on Soil and Vegetation in Urban Areas,* Pudoc. Centre for Agricultural Publishing and Documentations, Wageningen, The Netherlands, 1972.

318. **Holder, C. B. and Brown, K. W.,** The relationship between oxygen and water uptake by roots of intact bean plants, *Soil Sci. Soc. Am. J.,* 44, 21, 1980.

319. **Holford, I. C. R. and Patrick, W. H.,** Effects of reduction and pH changes on phosphate sorption and mobility in an acid soil, *Soil Sci. Soc. Am. J.,* 43, 292, 1979.

320. **Holthausen, R. S. and Caldwell, M. M.,** Seasonal dynamics of root system respiration in *Atriplex confertifolia, Plant Soil* 55, 307, 1980.

321. **Hopkins, R. M. and Patrick, W. H.,** Combined effects of oxygen concentration and soil compaction on root penetration, *Soil Sci.,* 108, 408, 1969.

322. **Howard, P. J. A.,** Relations between activity of organisms and temperature and the computation of the annual respiration of microorganisms decomposing leaf litter, *IV Int. Coll. Soil Zool., Ann. Zool. Animale Hors,* Serie, 197, 1971.

323. **Howard, P. J. A. and Howard, D. M.,** Respiration of decomposing litter in relation to temperature and moisture. Microbiol decomposition of tree and shrub leaf litter, *Oikos,* 33, 457, 1979.

324. **Huck, M. G.,** Variation in taproot elongation rate as influenced by composition of the soil air, *Agron. J.,* 62, 815, 1970.

325. **Huck, M. G., Hageman, R. H., and Hanson, J. B.,** Diurnal variation in root respiration, *Plant Physiol.,* 37, 371, 1962.

326. **Hughes, T. D., Stone, J. F., Huffine, W. W., and Gingrich, J. R.,** Effect of soil bulk density and soil water pressure on emergence of grass seedlings, *Agron. J.,* 58, 549, 1966.

327. **Huntjens, J. L. M.,** A sensitive method for continuous measurements of the carbon dioxide evolution rate of soil samples, *Plant Soil,* 53, 529, 1979.

328. **Huntjens, J. L. M., Oosterveld-van Vliet, W. M., and Sayed, S. K. Y.,** The decomposition of organic compounds in soil, *Plant Soil,* 61, 227, 1981.

329. **Hutchins, L. M.,** Studies on the oxygen-supplying power of the soil together with quantitative observations on the oxygen-supplying power required for seed germination, *Plant Soil,* 1, 95, 1926.

330. **Ilag, L. and Curtis, R. W.,** Production of ethylene by fungi, *Science,* 159, 1357, 1968.

331. **Ishii, T., Fusao, K., Mizutani, F., Iwasaki, K., and Kadoya, K.,** Ethylene evolution in the rhizosphere of grapevines, and growth disorders as affected by ethylene, *Mem. Coll. Agric. Ehime Univ.,* 2, 97, 1982.

332. **Ishizawa, S.,** Note on nitrate reduction in *Rhizobia, Soil Sci. Plant Nutr.,* 26, 447, 1980.

333. **Ivarson, K. C. and Sowden, F. J.,** Effect of frost action and storage of soil at freezing temperature on the free aminoacids, free sugars and respiratory activity of soil, *Can. J. Soil Sci.,* 50, 191, 1970.

334. **Jackson, M. B. and Campbell, D. J.,** Waterlogging and petiole epinasty in tomato: the role of ethylene and low oxygen, *New Phytol.,* 76, 21, 1976.

335. **Jackson, M. B. and Campbell, D. J.,** Effects of benzyladenine and gibberellic acid on the responses of tomato plants to anaerobic root environments and to ethylene, *New Phytol.,* 82, 331, 1979.

336. **Jackson, M. B., Drew, M. C., and Giffard, S. C.,** Effects of applying ethylene to the root system of *Zea mays* on growth and nutrient concentration in relation to flooding tolerance, *Physiol. Plant.,* 52, 23, 1981.

337. **Jackson, M. B., Gales, K., and Campbell, D. J.,** Effect of waterlogged soil conditions on the production of ethylene and on water relationships in tomato plants, *J. Exp. Bot.,* 29, 183, 1978.

338. **Jackson, M. B. and Goss, M. J.,** Can stress experienced by root systems be alleviated by exogenous growth regulators?, in *Proc. Jt. BCPC and BPGRG Symp. ''Opportunities for Chemical Plant Growth Regulation''*, 1978, 95.
339. **Jackson, W. T.,** The relative importance of factors causing injury to shoots of flooded tomato plants, *Am. J. Bot.*, 43, 637, 1956.
340. **Jacobson, L., Cooper, B. R., and Volz, M. G.,** The interaction of pH and aeration in Cl uptake by barley roots, *Physiol. Plant.*, 25, 432, 1971.
341. **Jager, G.,** Changes in the activity of soil microorganisms influenced by physical factors (drying-remoistening, freezing-thawing), in *Proc. Coll. Dynam. Soil Commun., Progr. Biol.*, Friedr. Vieweg and Sohn GmbH, North-Holland, Amsterdam, 1967, 178.
342. **Jager, G. and Bruins, E. H.,** Effect of repeated drying at different temperatures on soil organic matter decomposition and characteristics and on the soil microflora, *Soil Biol. Biochem.*, 7, 153, 1975.
343. **Jagnow, G.,** Soil respiration, nitrogen mineralization and humus decomposition in East African soils after drying and re-wetting, *Z. Pflanzenernaehr. Bodenkd.*, 131, 56, 1972.
344. **Jain, M. C.,** A simple O.D.R. apparatus for measuring soil aeration status, *J. Indian Soc. Soil Sci.*, 22, 109, 1974.
345. **James, W. O.,** Plant respiration, in *An Introduction to Plant Physiology*, Oxford University Press, 1973, chap. 3.
346. **Janse, A. R. P. and Bolt, G. H.,** The determination of air permeability of soil, *Neth. J. Agric. Sci.*, 8, 124, 1960.
347. **Jastrebov, T.,** A needle for soil air sampling (Russian), *Pochvovedenie*, 4, 89, 1956.
348. **Jenne, E. A.,** Controls on Mn, Fe, Co, Ni, Cu and Zn concentrations in soils and water. The significant role of hydrous Mn and Fe oxides, *Adv. Chem. Ser.*, 73, 337, 1968.
349. **Jensen, C. R., Letey, J., and Stolzy, L. H.,** Labelled oxygen transport through growing corn roots, *Science*, 144, 530, 1964.
350. **Jensen, C. R., Stolzy, L. H., and Letey, J.,** Tracer studies of oxygen diffusion through roots of barley, corn and rice, *Soil Sci.*, 103, 23, 1967.
351. **Jensen, C. R., Van Gundy, S. D., and Stolzy, L. H.,** Recording $CO_2$ in soil-root systems with a potentiometric membrane electrode, *Soil Sci. Soc. Am. Proc.*, 29, 631, 1965.
352. **Jensen, V.,** Effects of lead on biodegradation of hydrocarbons in soil, *Oikos*, 28, 220, 1977.
353. **Jones, L. W. and Bishop, N. I.,** Simultaneous measurement of oxygen and hydrogen exchange from the blue-green alga *Anabaena*, *Plant Physiol.*, 57, 659, 1976.
354. **Jost, W.,** *Diffusion in Solids, Liquids, Gases*, 3rd ed., Academic Press, New York, 1960, 1.
355. **Jugsujinda, A. and Patrick, W. H.,** Growth and nutrient uptake by rice in a flooded soil under controlled aerobic-anaerobic and pH conditions, *Agron. J.*, 69, 705, 1977.
356. **Kaack, K. and Kristensen, K.,** Emergence and seedling growth related to oxygen content and oxygen diffusion rate in different soils, *Agron. J.*, 59, 541, 1967.
357. **Kalir, A. and Poljakoff-Mayber, A.,** Effect of salinity on respiratory pathways in root tips of *Tamarix tetragyna*, *Plant Physiol.*, 57, 167, 1976.
358. **Kanazawa, S., Asami, T., and Takai, Y.,** Characteristics of soil organic matter and soil respiration in subalpine coniferous forest of Mt. Shigayama. III. Soil respiration under field conditions (Japanese, English summary), *J. Sci. Soil Manure*, 47, 549, 1976.
359. **Kanemasu, E. T., Powers, W. L., and Sij, J. W.,** Field chamber measurements of $CO_2$ flux from soil surface, *Soil Sci.*, 118, 233, 1974.
360. **Katayama, T.,** Studies on the intercellular spaces in rice. I., *Proc. Crop Sci. Soc. Jpn.*, 29, 229, 1961.
361. **Kawase, M. J.,** Effect of flooding on ethylene concentration in horticultural plants, *J. Am. Soc. Hortic. Sci.*, 97, 584, 1972.
362. **Kawase, M. J.,** Anaerobic elevation of ethylene concentration in water-logged plants, *Am. J. Bot.*, 65, 736, 1978.
363. **Kays, S. J., Nicklow, C. W., and Simons, D. H.,** Ethylene in relation to the response of roots to physical impedance, *Plant Soil*, 40, 565, 1974.
364. **Kenefick, D. G.,** Formation and elimination of ethanol in sugar beet roots, *Plant Physiol.*, 37, 434, 1962.
365. **Kimball, B. A.,** Soil Atmosphere Composition as Related to Microbial Activity and Diffusion Coefficient, M.S. thesis, Iowa State University, Ames, 1965.
366. **Kimball, B. A. and Lemon, E. R.,** Air turbulence effects upon soil gas exchange, *Soil Sci. Soc. Am. Proc.*, 35, 16, 1971.
367. **Kimura, M., Wada, H., and Takai, Y.,** The studies on the rhizosphere of paddy rice. VI. The effects of anaerobiosis on microbes, *Soil Sci. Plant Nutr.*, 25, 145, 1979.
368. **Kirita, H.,** Studies of soil respiration in warm-temperate evergreen broadleaf forests of southwestern Japan (Japanese, English summary), *Jpn. J. Ecol.*, 2, 230, 1971.
369. **Kirkham, D.,** Field method for determining air permeability of soil in its undisturbed state, *Soil Sci. Soc. Am. Proc.*, 11, 93, 1947.

370. **Kiyosawa, W.,** Studies on the effects of alcohols on membrane water permeability of *Nitella, Protoplasma,* 86, 243, 1975.

371. **Khalid, R. A., Patrick, W. H., and Peterson, F. J.,** Relationship between rice yield and soil phosphorus evaluated under aerobic and anaerobic conditions, *Soil Sci. Plant Nutr.,* 25, 155, 1979.

372. **Klotz, L. J., Stolzy, L. H., De Wolfe, T. A., and Szuszkiewicz, T. E.,** Rate of oxygen supply and distribution of root-rotting fungi in soils, *Soil Sci.,* 99, 200, 1965.

373. **Kobus, J. and Pacewiczowa, T.,** Effect of clay minerals on the biological activity of light soils (Polish, English summary), *Roczn. Glebozn.,* 16, 53, 1966.

374. **Komai, Y. K. and Yamaguchi, M.,** Effects of heavy metals on several microbiological activities in soil. I. Effects on carbon dioxide, *J. Sci. Soil Manure,* 52, 305, 1981.

375. **Konecka-Betley, K.,** Iron problem in soil-forming process (Polish, English summary), *Roczn. Glebozn.,* 19, 51, 1968.

376. **Konings, H. and Jackson, M. B.,** A relationship between rates of ethylene production by roots and the promoting or inhibiting effects of exogenous ethylene and water on root elongation, *Z. Pflanzenphysiol.,* 92, 385, 1979.

377. **Kowalik, P.,** Analysis of influence of water amelioration on soil oxygenation (Polish, English summary), *Zesz. Nauk. Politech. Gdansk.,* 15, 3, 1971.

378. **Kowalik, P.,** Physical theory of depth of gleysation, in *Pseudogley and Gley, Trans. Comm. V and VI ISSS,* Schlichting, E. and Schwertmann, U., Eds., Verlag Chemie, Gmbh, Weinheim, W. Germany, 1973, 453.

379. **Kowalik, P., Barnes, C. J., and Smiles, D. E.,** Oxidation of liquid animal wastes in soil, *Soil Sci. Soc. Am. J.,* 43, 255, 1979.

380. **Kowalik, P. and Obarska-Pempkowiak, H.,** Methods of oxygen diffusion and respiration measurements in the soil irrigated by slurry of pig manure, in *Proc. 2nd Int. Conf., Physical Properties of Agricultural Materials,* Gödöllö, Hungary, Agricultural University, Gödöllö, 1980, 103.

381. **Kowalik, P. and Stępniewski, W.,** The significance of soil aeration for plants, *Zesz. Probl. Postepow Nauk Roln.,* 220, 61, 1979.

382. **Kramer, P. J.,** Causes of injury to plants resulting from flooding of the soil, *Plant Physiol.,* 26, 722, 1951.

383. **Kramer, P. J.,** Permeability in relation to respiration, in *Encyclopedia of Plant Physiology,* Vol. 2, Ruhland, W., Ed., Springer-Verlag, New York, 1956, 358.

384. **Kramer, P. J.,** Physical and physiological aspects of water absorption, in *Encyclopedia of Plant Physiology,* Vol. 3, Ruhland, W. Rd., Springer-Verlag, New York, 1956, 124.

385. **Kramer, P. J.,** *Plant and Soil Water Relationships: A Modern Synthesis,* McGraw-Hill, New York, 1969.

386. **Kramer, P. J. and Jackson, W. T.,** Causes of injury to flooded tobacco plants, *Plant Physiol.,* 29, 241, 1954.

387. **Kratochvil, D. E.,** Determination of the effect of several herbicides on soil microorganisms, *Weeds,* 1, 25, 1951.

388. **Kristensen, K. J.,** Factors affecting measurements of oxygen diffusion rate (ODR) with bare platinum microelectrodes, *Agron. J.,* 58, 351, 1966.

389. **Kristensen, K. J. and Enoch, H.,** Soil air composition and oxygen diffusion rate in soil columns at different heights above a water table, in *Trans. 8th Int. Congr. Soil Sci.,* Bucharest, Romania, 2, 159, 1964.

390. **Kristensen, K. J. and Lemon, E. R.,** Soil aeration and plant root relations. III. Physical aspects of oxygen diffusion in the liquid phase of the soil, *Agron., J.,* 56, 295, 1964.

391. **Kucera, C. L. and Kirkham, D. R.,** Soil respiration studies in tallgrass prairie in Missouri, *Ecology,* 52, 912, 1971.

392. **Kulińska, D.,** Effect of herbicides on oxygen uptake by soil (Polish, English summary), *Roczn. Nauk Roln.,* 93-A-2, 229, 1967.

393. **Kummer, F. A. and Cooper, A. N.,** Soil porosity determinations with the air pressure pycnometer as compared with the tension method, *Agric. Eng.,* 26, 21, 1945.

394. **Kunc, F. and Stotzky, G.,** Effect of clay minerals on heterotrophic microbial activity in soil, *Soil Sci.,* 118, 186, 1974.

395. **Labanauskas, C. K., Letey, J., Stolzy, L. H., and Valoras, N.,** Effects of soil-oxygen and irrigation on the accumulation of macro- and micronutrients in citrus seedlings (*Citrus Sinensis* var. Osbeck), *Soil Sci.,* 101, 378, 1966.

396. **Labanauskas, C. K., Luxmoore, R. J., and Stolzy, L. H.,** Soil temperature and soil aeration effects on protein and free amino acid concentrations in wheat grain, *Plant Soil,* 41, 351, 1974.

397. **Labanauskas, C. K., Stolzy, L. H., and Handy, M. F.,** Protein and nonprotein amino acids in citrus leaves as affected by *Phytophthora* spp. root infestation and soil oxygen content, *J. Am. Soc. Hortic. Sci.,* 97, 433, 1972.

398. **Labanauskas, C. K., Stolzy, L. H., and Handy, M. F.,** Concentrations and total amounts of nutrients in citrus seedlings (*Citrus sinensis* var. Osbeck) and in soil as influenced by differential soil oxygen treatments, *Soil Sci. Soc. Am. Proc.*, 36, 454, 1972.
399. **Labanauskas, C. K., Stolzy, L. H., and Handy, M. F.,** Soil oxygen and *Phytophthora* spp. root infestation effects on the protein and free amino acids in lemon and orange leaves, *J. Am. Soc. Hortic. Sci.*, 99, 497, 1974.
400. **Labanauskas, C. K., Stolzy, L. H., Klotz, L. J., and De Wolfe, T. A.,** Effects of soil temperature and oxygen on the amounts of macronutrients and micronutrients in citrus seedlings (*Citrus sinensis* var. Bessie), *Soil Sci. Soc. Am. Proc.*, 29, 60, 1965.
401. **Labanauskas, C. K., Stolzy, L. H., Klotz, L. J., and De Wolfe, T. A.,** Soil carbon dioxide and mineral accumulation in citrus seedlings (*Citrus sinensis* var. Bessie), *Plant Soil,* 35, 337, 1971.
402. **Labanauskas, C. K., Stolzy, L. H., Klotz, L. J., and De Wolfe, T. A.,** Soil oxygen diffusion rates and mineral accumulations in citrus seedlings (*Citrus sinensis* var. Osbeck), *Soil Sci.*, 11, 386, 1971.
403. **Labanauskas, C. K., Stolzy, L. H., and Luxmoore, R. J.,** Soil temperature and soil aeration effects on concentrations and total amounts of nutrients in "Yecora" wheat grain, *Soil Sci.*, 120, 450, 1975.
404. **Labanauskas, C. K., Stolzy, L. H., Zentmyer, G. A., and Szuszkiewicz, T. E.,** Influence of soil oxygen and soil water on the accumulation of nutrients in avocado seedlings (*Persea americana* Mill.), *Plant Soil,* 29, 391, 1968.
405. **Łabuda, S., Maćkowiak, Cz., and Stępniewski, W.,** Changes of air composition in soil fertilized with slurry, *Pol. J. Soil Sci.*, 10, 3, 1977.
406. **Lahav, N. and Keynan, A.,** The influence of bentonite and attapulgite on the respiration of *Bacillus subtilis, Can. J. Microbiol.*, 8, 565, 1962.
407. **Lai, S. H., Tiedie, J. M., and Erickson, A. E.,** In situ measurement of gas diffusion coefficient in soils, *Soil Sci. Soc. Am. J.*, 40, 3, 1976.
408. **Lamberger, I. and Máté, F.,** Investigations of relationship between biological activity and liming of acid soils (Hungarian, English summary), *Agrokem. Talaj.*, 19, 197, 1970.
409. **Lambers, H.,** Respiration and NADH-oxidation of the roots of flood-tolerant and flood-intolerant *Senecio* species as affected by anaerobiosis, *Physiol. Plant.*, 37, 117, 1976.
410. **Lambers, H. and Smakman, G.,** Respiration of the roots of flood-tolerant and flood-intolerant *Senecio* species: affinity for oxygen and resistance to cyanide, *Physiol. Plant.*, 42, 163, 1978.
411. **Lambers, H. and Steingröver, E.,** Efficiency of root respiration of a flood-tolerant and a flood-intolerant *Senecio* species as affected by low oxygen tension, *Physiol. Plant.*, 42, 179, 1978.
412. **Lambers, H. and Steingröver, E.,** Growth respiration of a flood-tolerant and a flood-intolerant *Senecio* species: correlation between calculated and experimental values, *Physiol. Plant.*, 43, 219, 1978.
413. **Lance, J. C., Whisler, F. D., and Bouwer, H.,** Oxygen utilization in soils flooded with sewage water, *J. Environ. Qual.* 2, 345, 1973.
414. **Lawrie, A. C. and Wheeler, C. T.,** Nitrogen fixation in the root nodules of *Vicia faba* L., in relation to the assimilation of carbon. II. The dark fixation of carbon dioxide, *New Phytol.*, 74, 437, 1975.
415. **Layzell, D. B., Rainbird, R. M., Atkins, C. A., and Pate, J. S.,** Economy of photosynthate use in nitrogen-fixing legume nodules, *Plant Physiol.*, 64, 888, 1979.
416. **Lee, K. K., Holst, R. W., Watanabe, I., and App, A.,** Gas transport through rice, *Soil Sci. Plant Nutr.*, 27, 151, 1981.
417. **Leffelaar, P. A.,** Simulation of partial anaerobiosis in a model soil in respect to denitrification, *Soil Sci.*, 128, 110, 1979.
418. **Lemon, E. R.,** Soil aeration and plant root relations. I. Theory, *Agron. J.*, 54, 167, 1962.
419. **Lemon, E. R. and Erickson, A. E.,** The measurement of oxygen diffusion in the soil with a platinum microelectrode, *Soil Sci. Soc. Am. Proc.*, 16, 160, 1952.
420. **Lemon, E. R. and Erickson, A. E.,** Principles of the platinum microelectrode as a method of characterizing soil aeration, *Soil Sci.*, 79, 383, 1955.
421. **Lemon, E. R. and Wiegand, C. L.,** Soil aeration and plant root relations. II. Root respiration, *Agron J.*, 54, 171, 1962.
422. **Letey, J., Lunt, O. R., Stolzy, L. H., and Szuskiewicz, T. E.,** Plant growth, water use and nutritional response to rhizosphere differentials of oxygen concentration, *Soil Sci. Soc. Am. Proc.*, 25, 183, 1961.
423. **Letey, J., Morgan, W. C., Richards, S. J., and Valoras, N.,** Physical soil amendments, soil compaction, irrigation, and wetting agents in turfgrass management. III. Effects on oxygen diffusion rate and root growth, *Agron. J.*, 58, 531, 1966.
424. **Letey, J. and Stolzy, L. H.,** Measurement of oxygen diffusion rates with the platinum microelectrode. I. Theory and equipment, *Hilgardia,* 35, 545, 1964.
425. **Letey, J., Stolzy, L. H., Blank, G. B., and Lunt, O. R.,** Effect of temperature on oxygen diffusion rates and subsequent shoot growth, root growth and mineral content of two plant species, *Soil Sci.*, 92, 314, 1961.

426. **Letey, J., Stolzy, L. H., Lunt, O. R., and Younger, V. B.,** Growth and nutrient uptake of Newport bluegrass as affected by soil oxygen. *Plant Soil,* 20, 143, 1964.

427. **Letey, J., Stolzy, L. H., Valoras, N., and Szuszkiewicz, T. E.,** Influence of oxygen diffusion rate on sunflower growth at various soil and air temperatures, *Agron. J.,* 54, 316, 1962.

428. **Leyshon, A. J. and Sheard, R. W.,** Influence of short-term flooding on the growth and plant nutrient composition of barley, *Can. J. Soil Sci.,* 54, 463, 1974.

429. **Liberman, M., Kunishi, A., Mapson, L. W., and Wardale, D. A.,** Stimulation of ethylene production in apple tissue slices by methionine, *Plant Physiol.,* 41, 376, 1966.

430. **Liebl, K. H. and Seiler, W.,** CO and $H_2$ destruction at the soil surface, in *Microbiol Production and Utilization of Gases,* Schlegel, H. G., Gottschal, G., and Pfennig, N., Eds., Goltze, Göttingen, W. Germany, 1976, 215.

431. **Ling Chang, Huan Shi-zhen, Tang Long-fei, and Liu Zhung-Zhu,** Studies on nitrogen fixation by association of rice and bacteria, in *Proc. Symp. Paddy Soil China,* Nanjing, China, 1980, Institut Soil Sci., Academia Sinica, Ed., Science Press Beijing-Springer-Verlag, Berlin, 1981, 30.

432. **Lindsay, W. L.,** *Chemical Equilibria in Soils,* John Wiley & Sons, New York, 1979.

433. **Linebarger, R. S., Whisler, F. D., and Lance, J. C.,** A new technique for rapid and continuous measurement of redox potentials, *Soil Sci. Soc. Am. Proc.,* 39, 375, 1975.

434. **Lira, E. P. and Freytag, A. H.,** Effect of ethylene and 2,3,5-triiodobenzoic acid on soybean seedlings grown in hydroponic culture, *Trans. Ill. State Acad. Sci.,* 64, 309, 1971.

435. **Liu Zhi-guang,** Determination of redox potential (Eh) by depolarizing curve method, in *Proc. Symp. Paddy Soils China,* Nanjing, China, 1980, Institut Soil Sci., Academia Sinica, Ed., Science Press Beijing-Springer-Verlag, Berlin, 1981, 258.

436. **Loebell, R.,** Barometer-free air pycnometer (German), *Z.Pflanzenernaehr. Dueng. Bodenkd.,* 60, 172, 1953.

437. **Longsmuir, I. S.,** Respiration rate of bacteria as a function of oxygen concentration, *J. Biochem.,* 57, 81, 1954.

438. **Lundegårdh, H.,** *Carbon Dioxide Turnover in Nature* (German), G. Fisher, Verl., Jena, East Germany, 1924.

439. **Lundegårdh, H. G.,** Carbon dioxide evolution of soil and crop growth, *Soil Sci.,* 23, 417, 1927.

440. **Lunt, O. R., Letey, J., and Clark, S. B.,** Oxygen requirements for root growth in three species of desert shrubs, *Ecology,* 54, 1356, 1973.

441. **Luxmoore, R. J., Sojka, R. E., and Stolzy, L. H.,** Root porosity and growth responses of wheat to aeration and light intensity, *Soil Sci.,* 113, 354, 1972.

442. **Luxmoore, R. J. and Stolzy, L. H.,** Root porosity and growth responses of rice and maize to oxygen supply, *Agron. J.,* 61, 202, 1969.

443. **Luxmoore, R. J. and Stolzy, L. H.,** Oxygen diffusion in the soil-plant system. V. Oxygen concentration and temperature effects on oxygen relations predicted for maize roots, *Agron. J.,* 64, 720, 1972.

444. **Luxmoore, R. J. and Stolzy, L. H.,** Oxygen diffusion in the soil-plant system. VI. A synopsis with commentary, *Agron. J.,* 64, 725, 1972.

445. **Luxmoore, R.J., Stolzy, L. H., and Letey, J.,** Oxygen diffusion in the soil-plant system. I. A model, *Agron. J.,* 62, 317, 1970.

446. **Luxmoore, R.J., Stolzy, L. H., and Letey, J.,** Oxygen diffusion in the soil-plant system. II. Respiration rate, permeability, and porosity of consecutive excised segments of maize and rice roots, *Agron. J.,* 62, 322, 1970.

447. **Luxmoore, R.J., Stolzy, L. H., and Letey, J.,** Oxygen diffusion in the soil-plant system. III. Oxygen concentration profiles, respiration rates, and the significance of plant aeration predicted for maize roots, *Agron. J.,* 62, 325, 1970.

448. **Luxmoore, R. J., Stolzy, L. H., and Letey, J.,** Oxygen diffusion in the soil-plant system. IV. Oxygen concentration profiles, respiration rates and radial oxygen losses predicted for rice roots, *Agron. J.,* 62, 329, 1970.

449. **Lyda, S. D. and Robinson, G.D.,** Soil respiratory activity and organic matter depletion in an arid Nevada soil, *Soil Sci. Soc. Am. Proc.,* 33, 92, 1969.

450. **Lynch, J. M.,** Identification of substrate and isolation of microorganisms responsible for ethylene production in the soil, *Nature (London),* 240, 45, 1972.

451. **Lynch, J. M.,** Ethylene in soil, *Nature (London),* 256, 576, 1975.

452. **Lynch, J. M. and Harper, S. H. T.,** Formation of ethylene by a soil fungus, *J. Gen. Microbiol.,* 80, 187, 1974.

453. **MacFadyen, A.,** Simple methods for measuring and maintaining the proportion of carbon dioxide in air, for use in ecological studies of soil respiration, *Soil Biol. Biochem.,* 2, 9, 1970.

454. **MacFadyen, A.,** Inhibitory effects of carbon dioxide on microbial activity in soil, *Pedobiologia,* 13, 140, 1973.

455. **Machlis, L.,** The respiratory gradient in barley roots, *Am. J. Bot.*, 31, 281, 1944.
456. **Maharajh, D. M. and Walkley, J.,** Lowering of the saturation solubility of oxygen by the presence of another gas, *Nature (London)*, 236, 165, 1972.
457. **Maidl, F. X. and Fischbeck, G.,** A soil probe for measurement of oxygen concentration in soil air (German), *Z. Acker Pflanzenbau*, 150, 489, 1981.
458. **Makarov, B. N. and Patrikeeva, T. A.,** Soil gaseous regime study by the method of gas chromatography (Russian, English summary), *Pochvovedenie*, 9, 126, 1972.
459. **Malicki, M. and Walczak, R.,** A gauge of the redox potential and the oxygen diffusion rate in the soil with an automatic regulation of cathode potential, *Zesz. Probl. Postepow Nauk Roln.*, 220, 457, 1983.
460. **Mapson, L. W.,** Biogenesis of ethylene, *Biol. Rev.*, 44, 155, 1969.
461. **Marshall, K. C.,** Sorptive interactions between soil particles and microorganisms, in *Soil Biochemistry*, Vol. 2, McLaren, A. D. and Skujins, J., Eds., Marcel Dekker, New York, 1971, chap. 14.
462. **Marshall, T. H.,** A relation between permeability and size distribution of pores, *J. Soil Sci.*, 9, 1, 1958.
463. **Marshall, T. J.,** The diffusion of gases through porous media, *J. Soil Sci.*, 10, 79, 1959.
464. **Marszewska-Ziemięcka, J.,** *Microbiology of Soil and Manure* (Polish), PWRiL, Warsaw, 1969.
465. **Matthews, S. and Whitebread, R.,** Factors influencing preemergance mortality in peas, *Plant Pathol.*, 17, 11, 1968.
466. **McBridge, M. B. and Blasiak, J. J.,** Zinc and copper solubility as a function of pH in an acid soil, *Soil Sci. Soc. Am. J.*, 43, 866, 1979.
467. **McCrae, E. R., Hanus, J., and Evans, H. J.,** Properties of the hydrogenase system in *Rhizobium japonicum* bacteroids, *Biochem. Biophys. Res. Commun.*, 80, 384, 1978.
468. **McIntyre, D. S.,** Characterizing soil aeration with a platinum microelectrode. II. Response under controlled soil physical conditions, *Aust. J. Soil Res.*, 4, 103, 1966.
469. **McIntyre, D. S.,** Physical factors affecting operation of the oxygen cathode in unsaturated porous media, *Soil Sci.*, 103, 118, 1967.
470. **McIntyre, D. S.,** The platinum microelectrode method for soil aeration measurement, *Adv. Agron.*, 22, 235, 1970.
471. **Meek, B. D., Donovan, T.J., and Graham, L. E.,** Summertime flooding effects on alfalfa mortality, soil oxygen concentrations, and matric potential in a silty clay loam soil, *Soil Sci. Soc. Am. J.*, 44, 433, 1980.
472. **Meek, B. D. and Grass, L. B.,** Redox potential in irrigated desert soils as an indicator of aeration status, *Soil Sci. Soc. Am. Proc.*, 39, 870, 1975.
473. **Meek, B. D., Owen-Bartlett, E.C., Stolzy, L. H., and Labanauskas, C. K.,** Cotton yield and nutrient uptake in relation to water table depth, *Soil Sci. Soc. Am. J.*, 44, 301, 1980.
474. **Meek, B. D. and Stolzy, L. H.,** Short-term flooding, in *Plant Life in Anaerobic Environments*, Hook, D. D. and Crawford, R. M. M., Eds., Ann Arbor Science, Ann Arbor, Mich., 1978, 351.
475. **Mees, G. C. and Weatherley, P. E.,** The mechanisms of water absorption by roots. II. The role of hydrostatic pressure gradients across the cortex, *Proc. R. Soc. London, Ser. B*, 147, 381, 1957.
476. **Melhuish, F. M., Huck, M. G., and Klepper, B.,** Measurement of aeration, respiration rate and oxygen diffusion in soil during the growth of cotton in a rhizotron, *Aust. J. Soil Res.*, 12, 37, 1974.
477. **Mer, C. L.,** Carbon dioxide and ethanol as factors controlling the growth of etiolated oat seedlings, *Nature (London)*, 191, 260, 1961.
478. *Methodical Guide on Soil Structure Studies* (Russian), Revut, I.B. and Rode, A. A., Eds., Izd-wo "Kolos", Leningrad, 1969, chap. 3.
479. **Michell, M. J.,** An improved method for microrespirometry using gas chromatography, *Soil Biol. Biochem.*, 5, 271, 1973.
480. **Mikkelson, J. P.,** Effects of lead on the microbiological activity in soil, *Tidsskr. Planteavl*, 78, 509, 1974.
481. **Miller, R. D. and Johnson, D. D.,** The effect of soil moisture tension on carbon dioxide evolution, nitrification, and nitrogen mineralization, *Soil Sci. Soc. Am. Proc.*, 28, 644, 1964.
482. **Millington, R.J.,** Diffusion constant and diffusion coefficients, *Science*, 122, 1090, 1955.
483. **Millington, R. J. and Quirk, J. P.,** Transport in porous media, in *Trans. 7th Int. Congr. Soil Sci.*, Madison, Wisc., 1, 97, 1960.
484. **Millington, R. J. and Shearer, R. C.,** Diffusion in aggregated porous media, *Soil Sci.*, 111, 372, 1971.
485. **Minderman, G. and Vulto, J. C.,** Carbon dioxide production by tree roots and microbes, *Pedobiologia*, 13, 337, 1973.
486. **Minderman, G. and Vulto, J. C.,** Comparison of techniques for the measurement of carbon dioxide evolution from soil, *Pedobiologia*, 13, 73, 1980.
487. **Mingeau, M.,** Root porosity and tolerance to flooding (French, English Summary), *Ann. Agron.*, 28, 171, 1977.
488. **Mingo-Castel, A. M., Smith, O. E., and Kumamoto, J.,** Studies on the carbon dioxide promotion and ethylene inhibition of tuberization in potato explants cultured in vitro, *Plant Physiol.*, 57, 480, 1976.

489. **Mitsui, S., Kumazawa, K., and Hishida, T.,** The growth of rice plant with the use of poor drained soil as affected by the accumulation of volatile organic acids. II., *J. Sci. Soil Manure,* 30, 411, 1959.
490. **Mitsui, S., Kumazawa, K., and Mukai, N.,** The growth of rice plant on poorly drained soil as affected by accumulation of volatile organic acids, *J. Sci. Soil Manure,* 30, 345, 1959.
491. **Moneim, A. and Kruglov, Yu. V.,** Effect of the herbicide Atrazine on the biological activity of soil and the microflora of maize rhizospere (Russian), *Pochvovedenie,* 5, 64, 1981.
492. **Moraghan, J. T. and Ayotode, K. A.,** The influence of added organic matter on certain processes occurring in anaerobically incubated soils, in *Trans. 9th Int. Congr. Soil Sci.,* Vol. 4, Adelaide, Australia, 1968, 699.
493. **Moraghan, J. T. and Patrick, W. H.,** Selected metabolic processes in a submerged soil at controlled pH values, in *Trans. 10th Int. Congr. Soil Sci.,* Moscow, Vol. 2, 1974, Izd-vo, "Nauka", Moscow, 1974, 264.
494. **Mosier, A. R. and Mack, L.,** Gas chromatographic system for precise, rapid analysis of nitrous oxide, *Soil Sci. Soc. Am. J.,* 44, 1121, 1980.
495. **Moss, D. N., Musgrave, R. B., and Lemon, E. R.,** Photosynthesis under field conditions. III. Some effects of light, carbon dioxide, temperature and soil moisture on photosynthesis, respiration and transpiration of corn, *Crop Sci.,* 1, 83, 1961.
496. **Munch, J. C., Hillebrand, Th., and Ottow, J. C. G.,** Transformations in the $Fe_o/Fe_d$ ratio of pedogenic iron oxides affected by iron-reducing bacteria, *Can. J. Soil Sci.,* 58, 475, 1978.
497. **Munch, J. C. and Ottow, J. C. G.,** Model experiments on the mechanism of bacterial iron-reduction in waterlogged soils, *Z. Pflanzenernaehr. Bodenkd.,* 140, 549, 1977.
498. **Munch, J. C. and Ottow, J. C. G.,** Preferential reduction of amorphous to cristalline iron oxides by bacterial activity, *Soil Sci.,* 129, 15, 1980.
499. **Munch, J. C. and Ottow, J. C. G.,** Effect of cell contact and iron (III)-oxide form on bacterial iron reduction, *Z. Pflanzenernaehr. Bodenkd.,* 145, 66, 1982.
500. **Murch, R. S. and Paxton, J. D.,** Rhizosphere anaerobiosis and glyceollin accumulation in soybean, *Phytopathol. Z.,* 96, 91, 1979.
501. **Musgrave, R. B. and Moss, D. N.,** Photosynthesis under field conditions. I. A portable, closed system for determining net assimilation and respiration of corn, *Crop Sci.,* 1, 37, 1961.
502. **Nakane, K.,** Dynamics of soil organic matter in different parts of a slope under evergreen oak forest (Japanese, English summary), *Jpn. J. Ecol.,* 25, 206, 1975.
503. **Nakayama, M., Shimura, K., and Ota, Y.,** Physiological action of ethylene in crop plants. IV. Effect of ethylene application to roots on the growth of tomato and soybean plants (Japanese, English summary), *Proc. Crop Sci. Soc. Jpn.,* 42, 493, 1973.
504. **Newton, J. C., Crawford, R.W., and Stemp, R. K.,** A portable mass spectrometer for field and laboratory use, *Lawrence Livermore Lab. Univ. Calif. Rep. U.C.I.D.,* 1623, 1975.
505. **Ng Siew Kee and Bloomfield, C.,** The effect of flooding and aeration on the mobility of certain trace elements in soils, *Plant Soil,* 16, 108, 1962.
506. **Nielsen, W. and Thews, G.,** Electric analoging in solving diffusion problems in physiology (German, English summary), *Pflügers Arch.,* 269, 282, 1959.
507. **Nikolaeva, I. N.,** *Air Conditions of Soddy-Podzolic Soils,* Izd-vo "Kolos", Moscow, 1970.
508. **Nikolski, N. N.,** Carbon dioxide in soddy-podzolic soils and methods of its determination (Russian), *Dokl. TSKA,* 10, 1949.
509. **Nilouskaya, N. T., Kovalenko, V. K., and Laptev, V. V.,** Uptake and evolution of carbon dioxide by plants and microorganisms under artificial environment conditions (Russian, English summary), *Sov. Plant Physiol.,* 17, 567, 1970.
510. **Niranjan Rao, D. and Mikkelsen, D. S.,** Effect of rice straw additions on production of organic acids in a flooded soil, *Plant Soil,* 47, 303, 1977.
511. **Niranjan Rao, D. and Mikkelsen, D.S.,** Effects of $CO_2$, and $CH_4$ and $N_2$ on growth and nutrition of rice seedlings, *Plant Soil,* 47, 313, 1977.
512. **Niranjan Rao, D. and Mikkelsen, D. S.,** Effects of acetic, propionic, and butyric acids on rice seedling growth and nutrition, *Plant Soil,* 47, 323, 1977.
513. **Nitzsch, W.,** Soil pore size and their dependence on soil tillage and moisture (German), *RKTL-Schriften,* 85, Berlin 1938.
514. Non-Symbiotic Nitrogen Fixation and Organic Matter in the Tropics, Symp. papers 1, 12th Int. Congr. Soil Sci., Delhi, India, 1982.
515. **Norris, W. E., Wiegand, C. L., and Johanson, J.,** Effect of carbon dioxide on respiration of excised onion root tips in high oxygen atmospheres, *Soil Sci.,* 88, 144, 1959.
516. **Obermiller, E. L. and Charlier, G. O.,** Gas chromatographic separation of nitrogen, oxygen, carbon monoxide, carbon dioxide, hydrogen sulfide, and sulfur dioxide, *Gas Chromatogr.,* 6, 446, 1968.
517. **Ogawa, H.,** Litter production and carbon cycling in Pasoh Forest, in *Malaysian IBP Synthesis Meeting,* Kuala Lumpur, August, 1974.

518. **Ohmura, T. and Howell, R. W.,** Inhibitory effect of water on oxygen consumption by plant materials, *Plant Physiol.*, 35, 184, 1960.
519. **Ohyama, T. and Kumazawa, K.,** Nitrogen assimilation in soybean nodules. III. Effects of rhizosphere $pO_2$ on the assimilation of $^{15}N_2$ in nodules attached to intact plants, *Soil Sci. Plant Nutr.*, 26, 321, 1980.
520. **Omar, M. H.,** Potential evapotranspiration in a warm arid climate, in *Agroclimatological Methods, Proc. Reading Symp. UNESCO 1968*, Paris 1968, 347.
521. **Ottow, J. C. G.,** Influence of nitrate, chlorate, sulphate, iron oxide forms and growth conditions on the rate of bacterial reduction of iron (German), *Z. Pflanzenernaehr. Bodenkd.*, 124, 238, 1969.
522. **Ottow, J. C. G.,** Selection, characterization and iron reducing capacity of nitrate reductaseless ($nit^-$) mutants of iron reducing bacteria, *Z. Allg., Mikrobiol.*, 10, 55, 1970.
523. **Ottow, J. C. G.,** Bacterial mechanism of iron reduction and gley formation, in *Trans. Comm. V and VI Int. Soc. Soil Sci. Pseudogley and Gley*, Schlichting, E. and Schwertmann, U., Eds., Verlag Chemia GmbH, Weinheim, West Germany, 1973, 29.
524. **Ottow, J. C. G.,** Mechanisms of bacterial iron-reduction in flooded soils, in *Proc. Symp. Paddy Soils China,* Nanjing, China, 1980, Institut Soil Sci., Academia Sinica, Ed., Science Press Beijing-Springer Verlag, Berlin, 1981, 330.
525. **Ottow, J. C. G. and Munch, J. C.,** Mechanisms of reductive transformations in the anaerobic microenvironment of hydromorphic soils, in *Environmental Biogeochemistry and Geomicrobiology*, Vol. 2, Krumbein, W. E., Ed., Ann Arbor Science, Ann Arbor, Mich., 1978, 483.
526. **Ovcharenko, A. D.,** Air-water properties of mountainous meadow and Brown forest soils of Svanetia (Russian, English summary), *Pochvovedenie*, 10, 32, 1981.
527. **Owings, T. R., Iritani, W. M., and Nagel, C. W.,** Respiration rates and sugar accumulation in normal and moisture stressed Russet Burbank potatoes, *Am. Pot. J.*, 55, 211, 1978.
528. **Page, J. B.,** Advantages of the pressure pycnometer for measuring the pore space in soil, *Soil Sci. Soc. Am. Proc.*, 12, 81, 1948.
529. **Pal, S. S., Barik, S., and Sethunathan, N.,** Effects of benomyl on iron and manganese reduction and redox potential in flooded soil, *J. Soil Sci.*, 30, 155, 1979.
530. **Palissa, A. and Koepke, E.,** Biological effects of a synthetic soil conditioner (German, English summary), *Zentralbl. Bakteriol. Parasitenkd. Infektionskr. Hyg.*, 125, 273, 1970.
531. **Pankhurst, C. E. and Sprent, J. I.,** Effects of water stress on the respiratory and nitrogen-fixing activity of soybean root nodules, *J. Exp. Bot.*, 26, 287, 1975.
532. **Papendick, R. I. and Runkles, J. R.,** Transient-state oxygen diffusion in soil. I. The case when rate of oxygen consumption is constant, *Soil Sci.*, 100, 251, 1965.
533. **Papendick, R. I. and Runkles, R.,** Transient-state diffusion in soil. II. A case when rate of oxygen consumption varies with time, *Soil Sci.*, 102, 223, 1966.
534. **Pareilleux, A. and Maignan, C.,** Metabolic activity of *Candida lipolytica* absorbed on bentonites with hydrophobic chains, *Can. J. Microbiol.*, 22, 1065, 1976.
535. **Parkinson, K. J.,** An improved method for measuring soil respiration in the field, *J. Appl. Ecol.*, 18, 221, 1981.
536. **Parr, J. F. and Reuszer, H. W.,** Organic matter decomposition as influenced by oxygen level and method of application to soil, *Soil Sci. Soc. Am. Proc.*, 23, 214, 1959.
537. **Pasricha, N. S. and Ponnamperuma, F. N.,** Influence of salt and alkali on ionic equilibria in submerged soils, *Soil Sci. Soc. Am. J.*, 40, 374, 1976.
538. **Patrick, W. H.,** Extractable iron and phosphorus in submerged soil as affected by redox potential, in *Trans. 8th Int. Congr. Soil Sci.*, Bucharest, Romania, 1964, 605.
539. **Patrick, W. H.,** Critique of "Measurement and prediction of anaerobiosis in soils" by Stolzy, L. M. and Flühler, H., in *Nitrogen in the Environment,* Vol. 1, Nielsen, D. R. and McDonald, J. G., Eds., Academic Press, New York, 1978, 449.
540. **Patrick, W. H.,** The role of inorganic redox systems in controlling reduction in paddy soils, in *Proc. Symp. Paddy Soils China,* Nanjing, China, 1980, Institut Soil Sci. Academia Sinica, Ed., Science Press Beijing-Springer-Verlag, Berlin, 1981, 107.
541. **Patrick, W. H., Delaune, R. D., and Antie, D. A.,** Transformation of added phosphate in flooded soil, in *Trans. 10th Int. Congr. Soil Sci.*, Moscow, 1974, 296.
542. **Patrick, W. H., Delaune, R. D., and Engler, R. M.,** Soil oxygen content and root development of cotton in Mississippi river alluvial soils, *La., Agric. Exp. Stn., Bull.*, 673, 1973.
543. **Patrick, W. H. and Fontenot, W. J.,** Growth and mineral composition of rice at various soil moisture tensions and oxygen levels, *Agron. J.*, 68, 325, 1976.
544. **Patrick, W. H. and Henderson, R. E.,** A method for controlling redox potential in packed soil cores, *Soil Sci. Soc. Am. J.*, 45, 35, 1981.
545. **Patrick, W. H. and Mahapatra, I. C.,** Transformation and availability to rice plant of nitrogen and phosphorus in waterlogged soils, *Adv. Agron.*, 20, 323, 1968.

546. **Patrick, W. H. and Mikkelsen, D. S.,** Plant nutrient behavior in flooded soil, in *Fertilizer Technology and Use,* 2nd ed., Soil Science Society of America, Madison, Wisc., 1971, 187.
547. **Patrick, W. H. and Turner, F. T.,** Effect of redox potential on maganese transformations in waterlogged soils, *Nature (London),* 220, 476, 1968.
548. **Patrick, W. H., Turner, F. T., and Delaune, R. D.,** Soil oxygen content and root development of sugar cane, *La., Agric. Exp. Stn., Bull.,* 641, 1969.
549. **Patrick, W. H., Williams, B. G., and Moraghan, J. T.,** A simple system for controlling redox potential and pH in soil suspensions, *Soil Sci. Soc. Am. Proc.,* 37, 331, 1973.
550. **Patrick, W. H. and Wyatt, R.,** Soil nitrogen loss as a result of alternate submergence and drying, *Soil Sci. Soc. Am. Proc.,* 28, 647, 1964.
551. **Peaslee, D. E. and Moss, D. N.,** Stomatal conductivities in K-deficient leaves of maize (*Zea mays* L.), *Crop Sci.,* 8, 427, 1966.
552. **Penman, L. H.,** Gas and vapour movements in soil. I. The diffusion of vapour through porous solids, *J. Agric. Sci.,* 30, 437, 1940.
553. **Penman, L. H.,** Gas and vapour movements in soil. II. The diffusion of carbon dioxide through porous solids, *J. Agric. Sci.,* 30, 570, 1940.
554. **Pereira, J. S. and Kozlowski, T. T.,** Variations among woody angiosperms in response to flooding, *Physiol. Plant.,* 41, 184, 1977.
555. **Perrier, A., Archer, P., and Planco de Pablos, A.,** Study of real and maximum evapotranspiration of various crops: apparatus and measurements (French, English summary), *Ann. Agron.,* 25, 697, 1974.
556. **Phene, C. J., Campbell, R. B., and Doty, C. W.,** Characterization of soil aeration in situ with automated oxygen diffusion measurements, *Soil Sci.,* 122, 271, 1976.
557. **Pilot, L. and Patrick, W. H.,** Nitrate reduction in soils: effect of soil moisture tension, *Soil Sci.,* 114, 312, 1972.
558. **Pitman, M. G.,** Adaptation of barley roots to low oxygen supply and its relation to potassium and sodium uptake, *Plant Physiol.,* 44, 1233, 1969.
559. *Plant Physiology* (Polish), Zurzycki, J. and Michniewicz, M., Eds., PWRiL, Warsaw, 1977, 263.
560. **Poel, L. W.,** On the use of the katharometer for the estimation of carbon dioxide concentrations in soil, *Ann. Bot.,* 43, 285, 1979.
561. **Pojasov, N. P.,** Soil aeration, in *The Basis of Agrophyscis* (Polish), Joffe, A. F. and Revut, Z. B., Eds., PWRiL, Warsaw, 1967, chap. 5.
562. **Ponnamperuma, F. N.,** The chemistry of submerged soils, *Adv. Agron.,* 24, 29, 1972.
563. **Ponnamperuma, F. N.,** Some aspects of the physical chemistry of paddy soils, in *Proc. Symp. Paddy Soils China,* Nanjing, China, 1980. Institut Soil Sci., Academia Sinica, Ed., Science Press Beijing-Springer-Verlag, Berlin, 1981, 59.
564. **Ponnamperuma, F. N., Loy, T. A., and Tianco, E. M.,** Redox equilibrium in flooded soils. II. The manganese oxide systems, *Soil Sci.,* 108, 48, 1969.
565. **Ponnamperuma, F. N., Martinez, E., and Loy, T.,** Influence of redox potential and partial pressure of carbon dioxide on pH values and the suspension effect of flooded soils, *Soil Sci.,* 101, 421, 1966.
566. **Ponnamperuma, F. N., Loy, T. A., and Tianco, E. M.,** Redox equilibria in flooded soils. II. The manganese oxide systems, *Soil Sci.,* 108, 48, 1969.
567. **Porter, L. K.,** Gaseous products produced by anaerobic reaction of sodium nitrite with oxime compounds and oximes synthesized from organic matter, *Soil Sci. Soc. Am. Proc.,* 33, 696, 1969.
568. **Postgate, J. R.,** Recent advances in the study of the sulfate reducing bacteria, *Bacteriol. Rev.,* 29, 425, 1965.
569. **Poyting, J. M. and Thompson, J. J.,** *Properties of Matter,* 14th ed., Charles Griffin, London, 1947.
570. **Pritchard, D. T.,** Carbon dioxide production in soils under barley, subjected to a range of drought treatments, *J. Sci. Food Agric.,* 30, 547, 1979.
571. **Pritchard, D. T. and Brown, N. J.,** Respiration in cropped and fallow soil, *J. Agric. Sci. Cambridge,* 92, 45, 1979.
572. *Proc. Symp. Paddy Soils China,* Nanjing, China, 1980, Institut Soil Sci., Academia Sinica, Ed., Science Press Beijing-Springer-Verlag, Berlin, 1981.
573. **Prusinkiewicz, Z.,** Microbiological activity of soils as a function of the soil-water potential, *Pol. J. Soil Sci.,* 7, 81, 1974.
574. *Pseudogley and Gley,* Trans. Comm. V and VI Int. Soc. Soil Sci., Schlichting, E. and Schwertmann, U., Eds., Verlag Chemie GmbH, Weinheim, West Germany, 1973.
575. **Quilt, P., Grossbard, E., and Wright, S. J. L.,** Effects of the herbicide barban and its commercial formulation Carbyne on soil micro-organisms, *J. Appl. Bacteriol.,* 46, 431, 1979.
576. **Racz, G. J.,** Release of P in organic soils under aerobic and anaerobic conditions, *Can. J. Soil Sci.,* 59, 337, 1979.
577. **Radford, P. J. and Greenwood, D. J.,** The simulation of gaseous diffusion in soils, *J. Soil Sci.,* 21, 304, 1970.

578. **Radin, J. W. and Loomis, R. S.,** Ethylene and carbon dioxide in the growth and development of cultured radish roots, *Plant Physiol.*, 44, 1584, 1969.
579. **Raison, R. J. and McGarity, J. W.,** Effects of ash, heat, and the ash-heat interaction on biological activity in two contrasting soils. I. Respiration rate, *Plant Soil,* 55, 363, 1980.
580. **Randhawa, N. S. and Katyal, J. C.,** Micronutrients management for submerged rice soils, Trans. 12th Int. Congr. Soil Sci., New Delhi, India, 1982, in *Vertisols and Rice Soils of the Tropics,* Symp. papers II, 192, New Delhi 1982.
581. **Raney, W. A.,** Field measurement of oxygen diffusion through soil, *Soil Sci. Soc. Am. Proc.*, 14, 61, 1949.
582. **Rankin, J. M. and Sumner, M. E.,** Oxygen flux measurement in unsaturated soils, *Soil Sci. Soc. Am. J.*, 42, 869, 1978.
583. **Reardon, E. J., Allison, G. B., and Fritz, P.,** Seasonal chemical and isotopic variations of soil $CO_2$ at Trout Creek, Ontario, *J. Hydrol.*, 43, 355, 1979.
584. **Reddy, C. N. and Patrick, W. H.,** Effect of redox potential on the stability of zinc and copper chelates in flooded soils, *Soil Sci. Soc. Am. J.*, 41, 729, 1977.
585. **Reddy, K. R. and Patrick, W. W.,** Effect of alternate aerobic and anaerobic conditions on redox potential, organic matter decomposition and nitrogen loss in a flooded soil, *Soil Biol. Biochem.*, 7, 87, 1975.
586. **Reddy, K. R. and Patrick, W. H.,** Nitrogen fixation in flooded soil, *Soil Sci.*, 128, 80, 1979.
587. **Reddy, K. R., Rao, P. S. C., and Patrick, W. H.,** Factors influencing oxygen consumption rates in flooded soils, *Soil Sci. Soc. Am. J.*, 44, 741, 1980.
588. **Redmann, R. E.,** Soil respiration in a mixed grassland ecosystem, *Can. J. Soil. Sci.*, 58, 119, 1978.
589. **Reichle, D. E., Dinger, B. E., Edwards, N. T., Harris, W. F., and Sollins, P.,** Carbon flow and storage in a forest ecosystem, in *Carbon and the Biosphere,* Woodwell, G. M. and Pecan, E. V., Eds., U.S. Atomic Energy Commission, 1973, 345.
590. **Reid, C. and Sherwood, T. K.,** *The Properties of Gases and Liquids,* McGraw-Hill, New York, 1966, 646.
591. **Reid, M. S. and Pratt, H. K.,** Effects of ethylene on potato tuber respiration, *Plant Physiol.*, 49, 252, 1972.
592. **Reiners, W. A.,** Carbon dioxide evolution from the floor of three Minnesota forests, *Ecology,* 49, 471, 1968.
593. **Reiners, W. A. and Reiners, N. W.,** Energy and nutrient dynamics of forest floors in three Minnesota forests, *J. Ecol.*, 58, 497, 1970.
594. **Repnevskaya, M. A.,** Release of $CO_2$ from soil in pine woods of the Kola peninsula (Russian, English summary), *Pochvovedenie,* 8, 81, 1967.
595. *Report 1975 ICW,* Wageningen, The Netherlands, Mededeling 153, 1975, 54.
596. **Revut, I. B. and Lebiedieva, W. G.,** Some regularities of gas diffusion processes in soil (Russian), *Dokl. Wsiez. Sielskch. Akad. Nauk W. I. Lenina,* 3, 2, 1967.
597. **Reyes, D. M., Stolzy, L. H., and Labanauskas, C. K.,** Temperature and oxygen effects in soil on nutrient uptake in jojoba seedlings, *Agron. J.*, 69, 647, 1977.
598. **Richter, J.,** A comparative study of soil gas regime in a soil tillage experiment with different soils. I. Field measurements. *Z. Pflanzenernaehr. Bodenkd.*, 137, 135, 1974.
599. **Rickman, R. W.,** Effects of salts on oxygen diffusion rate measurements in unsaturated soils, *Soil Sci. Soc. Am. Proc.*, 32, 618, 1968.
600. **Rickman, R. W., Letey, J., Aubertin, G. M., and Stolzy, L. H.,** Platinum microelectrode poisoning factors, *Soil Sci. Soc. Am. Proc.*, 32, 204, 1968.
601. **Rickman, R. W., Letey, J., and Stolzy, L. H.,** Plant responses to oxygen supply and physical resistance in the root environment, *Soil Sci. Soc. Am. Proc.*, 30, 304, 1966.
602. **Rid, H.,** A field method of measurement of air permeability of soil (German), *Z. Pflanzenernaehr. Dueng. Bodenkd.*, 88, 227, 1960.
603. **Rishbeth, J.,** *Armillaria* on cacao in Sao Tome, *Trop. Agric. (Trinidad)*, 57, 155, 1980.
604. **Ritchie, J. T.,** Soil Aeration Characterization Using Gas Chromatography, Ph.D. dissertation, Iowa State University, Ames, 1964.
605. **Rixon, A. J.,** Oxygen uptake and nitrification at various moisture levels by soils and mats from irrigated pastures, *J. Soil Sci.*, 19, 56, 1968.
606. **Rixon, A. J. and Bridge, B. J.,** Respiratory quotient arising from microbial activity in relation to matric suction and air filled pore space of soil, *Nature (London)*, 218, 961, 1968.
607. **Robertson, G. W. and Bracewell, J. M.,** Rapid representative sampling of soil gases and their determination by mass spectrometry, *J. Soil Sci.*, 30, 681, 1979.
608. **Rolston, D. E., Fried, M., and Goldhamer, D. A.,** Denitrification measurement directly from nitrogen and nitrous oxide gas fluxes, *Soil Sci. Soc. Am. J.*, 40, 259, 1976.
609. **Rolston, D. E., Hoffman, D. L., and Toy, D. W.,** Field measurement of denitrification. I. Flux of $N_2$ and $N_2O$, *Soil Sci. Soc. Am. J.*, 42, 863, 1978.

610. **Romell, L. G.,** Luftvaxlingen i marken som ekologisk faktor (Norwegian), *Medd. Statens Skogsförsöksves.*, 19, 125, 1922.
611. **Rosene, H. F.,** Effect of cyanide on rate of exudation in excised onion roots, *Ann. J. Bot.*, 31, 127, 1944.
612. **Ross, D. J.,** Some factors influencing the estimation of dehydrogenase activities of some soils under pasture, *Soil Biol. Biochem.*, 3, 97, 1971.
613. **Ross, D. J.,** Effects of freezing and thawing of some grassland topsoils on oxygen uptake and dehydrogenase activities, *Soil Biol. Biochem.*, 4, 115, 1974.
614. **Ross, D. J. and Cairns, A.,** Influence of temperature on biochemical processes in some soils from tussock grasslands, *N.Z.J. Sci.*, 21, 581, 1978.
615. **Ross, D. J., Cairns, A., and Speir, T. W.,** Effect of irrigation with municipal water or sewage effluent on the biology of soil cores. IV. Respiratory and enzyme activities, *N.Z. J. Agric. Res.*, 21, 411, 1978.
616. **Ross, D.J., Campbell, I. B., and Bridger, B. A.,** Biochemical activities of organic soils from subantarctic tussock grasslands on Campbell Island. I. Oxygen uptakes and nitrogen mineralization, *N.Z. J. Sci.*, 22, 161, 1979.
617. **Ross, D. J. and Roberts, H. S.,** A study of activities of enzymes hydrolyzing sucrose and starch and of oxygen uptake in a sequence of soils under tussock grassland, *J. Soil Sci.*, 19, 186, 1968.
618. **Ruiz-Argüso, T., Hanus, J., and Evans, H. J.,** Hydrogen production and uptake by pea nodules as affected by strains of *Rhizobium leguminosarum*, *Arch. Mikrobiol.*, 116, 113, 1978.
619. **Runkles, J. R.,** Diffusion Sorption and Depth Distribution of Oxygen in Soils, Ph. D. dissertation, Iowa State University, Ames, 1956.
620. **Runkles, J. R., Scott, A. D., and Nakayama, F. S.,** Oxygen sorption by moist soils and vermiculite, *Soil Sci. Soc. Am. Proc.*, 22, 15, 1958.
621. **Russell, E. J. and Appleyard, A.,** The atmosphere of the soil: its composition and the cause of variation, *J. Agric. Sci.*, 7, 1, 1915.
622. **Russell, M. B.,** A simplified air pycnometer for field use, *Soil Sci. Soc. Am. Proc.*, 14, 73, 1949.
623. **Russell, M. B.,** Soil aeration and plant growth, in *Soil Physical Conditions and Plant Growth*, Show, B. T., Ed., Academic Press, New York, 1952, 253.
624. **Russell, W. E.,** *Soil Conditions and Plant Growth*, Willan Clowes, London, 1973.
625. **Rust, R. H., Klute, A., and Gieseking, J. E.,** Diffusion-porosity measurements using a non-steady state system, *Soil Sci.*, 84, 453, 1956.
626. **Ryden, J. C., Lund, L. J., and Focht, D. D.,** Direct in-field measurement of nitrous oxide flux from soils, *Soil Sci. Soc. Am. J.*, 42, 731, 1978.
627. **Ryden, J. C., Lund, L. J., and Focht, D. D.,** Direct measurement of denitrification loss from soil. I. Laboratory evaluation of acetylene inhibition of nitrous oxide reduction, *Soil Sci. Soc. Am. J.*, 43, 104, 1979.
628. **Ryden, J. C., Lund, L. J., Letey, J., and Focht, D. D.,** Direct measurements of denitrification loss from soil. II. Development and application of field methods, *Soil Sci. Soc. Am. J.*, 43, 110, 1979.
629. **Saini, G. R.,** Relationship between soil oxygen diffusion rate and yield of oats in a coastal alluvial soil at critical salinity level, *Agron. J.*, 65, 841, 1973.
630. **Saini, G. R.,** Relationship between potato yield and oxygen diffusion rate of subsoil, *Agron. J.*, 68, 823, 1976.
631. **Salisbury, F. B. and Ross, C.,** *Plant Physiology*, Wodsworth, Belmont, Calif., 1969, chap. 6.
632. **Salonius, P. O.,** Effects of mixing and various temperature regimes on the respiration of fresh and air-dried coniferous raw humus materials, *Soil Biol. Biochem.*, 10, 479, 1978.
633. **Salonius, P. O. and Mahendrappa, M. K.,** Respiration and nitrogen immobilization in forest soil treated with sulfur and urea, *Soil Sci.*, 127, 358, 1979.
634. **Sauerbeck, D. and Johnen, B.,** The turnover of plant roots during the growth period, its influence on "soil respiration" (German, English summary), *Z. Pflanzenernaehr. Dueng. Bodenkd.*, 3, 309, 1976.
635. **Savant, N. K. and Ellis, R.,** Changes in redox potential and phosphorus availability in submerged soil, *Soil Sci.*, 98, 388, 1964.
636. **Savickaya, N. N.,** Influence of waterlogging on barley at different stages of development (Russian), *Dokl. Akad. Nauk. SSSR*, 128, 850, 1959.
637. **Schaffer, G.,** Soil respiration under the influence of long term differentiated soil fertilization (German, English summary), *Z. Pflanzenernaehr. Dueng. Bodenkd.*, 67, 219, 1954.
638. **Schinner, F. A., Niederbacher, R., and Neuwinger, I.,** Influence of compound fertilizer and cupric sulfate on soil enzymes and $CO_2$-evolution, *Plant Soil*, 57, 85, 1980.
639. **Schlichting, E., Schweikle, V., and Pfaff, B.,** The evaluation of long lasting $O_2$-deficienty in soils from nutrient analyses, *Z. Pflanzenernaehr. Bodenkd.*, 142, 537, 1979.
640. **Scholander, P. F.,** Analyzer for accurate estimation of respiratory gases in one-half cubic centimeter samples, *J. Biol. Chem.*, 167, 235, 1947.
641. **Scholander, P. F. and Evans, H. J.,** Microanalysis of fractions of a cubic millimeter of gas, *J. Biol. Chem.*, 169, 551, 1947.

642. **Schubert, K. R. and Evans, H. J.,** Hydrogen evolution: a major factor affecting the efficiency of nitrogen fixation in nodulated symbionts, *Proc. Natl. Acad. Sci. U.S.A.*, 73, 1207, 1976.
643. **Schulze, E. D.,** Soil respiration of tropical vegetation types, *Ecology*, 48, 652, 1967.
644. **Schumacher, T. E. and Smucker, A. J. M.,** Mechanical impedance effects on oxygen uptake and porosity of drybean roots, *Agron. J.*, 73, 51, 1981.
645. **Scott, T. W. and Erickson, A. E.,** Effect of aeration and mechanical impedance on the root development of alfalfa, sugar beets and tomatoes, *Agron. J.*, 56, 575, 1964.
646. **Scotter, D. R., Thurtell, G.W., and Tanner, C. B.,** Measuring oxygen uptake by the roots of intact plants under controlled conditions, *Soil Sci.*, 104, 374, 1967.
647. **Seiler, W.,** The influence of the biosphere on the atmospheric CO and $H_2$ cycles, in *Environmental Biogeochemistry and Geomicrobiology*, Vol. 3, Krumbein, W. E., Ed., Ann Arbor Science, Ann Arbor, Mich., 1978, 773.
648. **Seiler, W., Liebl, K. H., Stöhr, W. T., and Zakosek, H.,** CO and $H_2$ evolution in soils, *Z. Pflanzenernaehr. Bodenkd.*, 140, 257, 1977.
649. **Shearer, R.C., Millington, R. J., and Quirk, J. P.,** Oxygen diffusion through sands in relation to capillary hysteresis. I. Calibration of oxygen cathode for use in diffusion studies, *Soil Sci.*, 101, 361, 1966.
650. **Shearer, R. C., Millington, R. J., and Quirk, J. P.,** Oxygen diffusion through sands in relation to capillary hysteresis. II. Quasisteady-state diffusion of oxygen through partly saturated sands, *Soil Sci.*, 11, 432, 1966.
651. **Shurikova, V. I.,** Sorption of $O_2$ and evolution of $CO_2$ in dark grey forest soils of different erosion degree (Russian, English summary), *Pochvovedenie*, 1, 82, 1978.
652. **Siegel, S. M. and Rosen, L. A.,** Effect at reduced oxygen tension on germination and seedling growth, *Physiol. Plant.*, 15, 437, 1962.
653. **Sikes, D. A. and Pettit, R. D.,** Soil temperature, oxygen, and water level effects on sand shinnery oak, *Soil Sci.*, 130, 344, 1980.
654. **Silberbush, M., Gornat, B., and Goldberg, D.,** Effect of irrigation from a point source (trickling) on oxygen flux and on root extension in the soil, *Plant Soil*, 52, 507, 1979.
655. **Sillanpää, M.,** *Micronutrient and the Nutrient Status of Soils*, FAO Soils Bulletin, 48, Rome, 1982, chap. 2.
656. **Simon-Sylvestre, G. and Fournier, J. C.,** Effects of pesticides on the soil microflora, *Adv. Agron.*, 31, 1, 1979.
657. **Simpson, F. B., Maier, R. J., and Evans, H. J.,** Hydrogen stimulated $CO_2$ fixation and coordinate induction of hydrogenase and rhibulosebiphosphate carboxylase in $H_2$-uptake positive strain of *Rhizobium japonicum*, *Arch. Microbiol.*, 123, 1, 1979.
658. **Sims, J. L. and Patrick, W. H.,** The distribution of micronutrient cations in soil under conditions of varying redox potential and pH, *Soil Sci. Soc. Am. J.*, 42, 258, 1978.
659. **Singh, R. and Ghildyal, B. P.,** Influence of soil edaphic factors and their critical limits on seedlings emergence of corn (*Zea mays* L.), *Plant Soil*, 47, 125, 1977.
660. **Skerman, V. B. D. and MacRae, I. C.,** The influence of oxygen on the reduction of nitrate by adapted cells of *Pseudomonas denitrificans*, *Can. J. Microbiol.*, 3, 505, 1957.
661. **Slowik, K., Labanauskas, C. K., Stolzy, L. H., and Zentmyer, G. A.,** Influence of rootstocks, soil oxygen, and soil moisture on the uptake and translocation of nutrients in young avocado plants, *J. Am. Soc. Hortic. Sci.*, 104, 172, 1979.
662. **Smith, A. M.,** Ethylene production by bacteria in reduced microsites in soil and some implications to agriculture, *Soil Biol. Biochem.*, 8, 293, 1976.
663. **Smith, A. M. and Cook, R. J.,** Implications of ethylene production by soil bacteria for biological balance of soil, *Nature (London)*, 222, 769, 1974.
664. **Smith, G. J. and Chalk, P. M.,** Determination of nitrogenous gases evolved from soils in closed systems, *Analyst (London)*, 104, 538, 1979.
665. **Smith, G. J. and Chalk, P. M.,** Gaseous, nitrogen evolution during nitrification of ammonia fertilizer and nitrite transformations in soils, *Soil Sci. Soc. Am. J.*, 44, 277, 1980.
666. **Smith, J. E. and Fletcher, W. W.,** 3:5-Dihalogeno-4-hydroxybenzonitriles and soil micro-organisms, *Hortic. Res.*, 3, 60, 1964.
667. **Smith, J. H., Gilbert, R. G., and Miller, J. B.,** Redox potentials in a cropped potato processing waste water disposal field with a deep water table, *J. Environ. Qual.*, 7, 571, 1978.
668. **Smith, K. A.,** Soil aeration, *Soil Sci.*, 123, 284, 1977.
669. **Smith, K. A.,** Critique of "Application of gaseous-diffusion theory to measurement of denitrification", in *Nitrogen in the Environment*, Vol. 1, Nielson, D. R. and McDonald, J. G., Eds., Academic Press, New York, 1978, 337.
670. **Smith, K. A.,** A model of the extent of anaerobic zones in aggregated soils, and its potential application to estimates of denitrification, *J. Soil Sci.*, 31, 263, 1980.

671. **Smith, K. A. and Dowdell, R. J.,** Gas chromatographic analysis of the soil atmosphere. Automatic analysis of gas samples for $O_2$, $N_2$, Ar, $CO_2$, $N_2O$, and $C_1$-$C_4$ hydrocarbons, *J. Chromatogr. Sci.*, 11, 655, 1973.
672. **Smith, K. A. and Dowdell, R. J.,** Field studies of the soil atmosphere. I. Relationships between ethylene, oxygen, soil moisture content, and temperature, *J. Soil Sci.*, 25, 217, 1974.
673. **Smith, K. A. and Hall, K. C.,** Gas chromatographic measurement of dissolved oxygen and trace levels of nitrous oxide using a helium ionization detector, Letcombe Laboratory, *Annu. Rep.*, 1973, 53.
674. **Smith, K. A. and Restall, S. W. F.,** The occurrence of ethylene in anaerobic soil, *J. Soil Sci.*, 22, 430, 1971.
675. **Smith, K. A. and Robertson, P. D.,** Effect of ethylene on root extension of cereals, *Nature (London)*, 234, 148, 1971.
676. **Smith, K. A. and Russell, R. S.,** Occurrence of ethylene, and its significance, in anaerobic soil, *Nature (London)*, 222, 769, 1969.
677. **Smith, M. S. and Weeraratna, C. S.,** A study of the effect of Simazine on soil microbiol activity and available nitrogen, in *Trans. 10th ISSS Congr.*, Moscow, Izd.-No "Nauka", Moscow, 1974, 173.
678. **Smucker, A. J. M.,** Interactions of soil oxygen and water stresses upon growth, disease, and production of navy beans, *Report of the Bean Improvement Cooperative*, 77, 1975.
679. **Smucker, A. J. M., Schumacher, B. D., Kenezek, B. D., and Erickson, A. E.,** Oxygen requirements for the growth and development of navy beans, in *Agronomy Abstracts*, American Society of Agronomy, Madison, Wisc., 1976, 123.
680. Soil Classification, A comprehensive system, 7th Approximation, U.S. Department of Agriculture, 1980.
681. **Sojka, R. E., Joseph, H. A., and Stolzy, L. H.,** Wheat response to short-term heat stress and to soil oxygen stress at three stages of growth, *Agron. J.*, 64, 450, 1972.
682. **Sojka, R. E. and Stolzy, L. H.,** Soil-oxygen effects on stomatal response, *Soil Sci.*, 130, 350, 1980.
683. **Sojka, R. E., Stolzy, L. H., and Kaufmann, M. R.,** Wheat growth related to rhizosphere temperature and oxygen levels, *Agron. J.*, 67, 591, 1975.
684. **Sokolova, T. A. and Polteva, R. N.,** The study of iron-manganese concretions from a strongly podzolic soil profile, in *Trans. 9th Int. Congr. Soil Sci.*, Vol. 4, Adelaide, Australia, 1968, 459.
685. **Solomos, T. and Laties, G. G.,** The mechanisms of ethylene and cyanide action in triggering the rise in respiration in potato tubers, *Plant Physiol.*, 55, 73, 1975.
686. **Solovyev, E. V.,** Construction of a mathematical model for the process of $CO_2$ outgo from chernozem samples (Russian, English summary), *Pochvovedenie*, 5, 130, 1978.
687. **Sommers, L. E. and Lindsay, W. L.,** Effect of pH and redox on predicted heavy metal-chelate equilibria in soils, *Soil Sci. Soc. Am. J.*, 43, 39, 1979.
688. **Spalding, B. P.,** Effects of divalent metal chlorides on respiration and extractable enzymatic activities of Douglas-fir needle litter, *J. Environ. Qual.*, 8, 105, 1979.
689. **Steinbrenner, E. C.,** A portable air penetrometer for forest soil, *Soil Sci. Soc. Am. Proc.*, 23, 478, 1959.
690. **Stelzer, R. and Läuchli, A.,** Salt- and flooding tolerance of *Puccinellia peisonis*. IV. Root respiration and the role of aerenchyma in providing atmospheric oxygen to the roots, *Z. Pflanzenphysiol.*, 97, 171, 1980.
691. **Stemmet, M. C., De Bruyn, J. A., and Zeeman, P. B.,** The uptake of carbon dioxide by plant roots, *Plant Soil*, 17, 357, 1962.
692. **Stępniewska, Z.,** Changes of Redox Potential, pH and Reduced Forms of Manganese and Iron in Soil at Different Oxygen and Moisture Contents (Polish), Ph.D. dissertation, Agricultural University, Lublin, Poland, 1979.
693. **Stępniewski, W.,** Influence of Mineral Fertilization and Watering of Arable Soil on its Oxygenation (Polish), Ph.D. dissertation, Agricultural University, Lublin, Poland, 1975.
694. **Stępniewski, W.,** The effect of irrigation and increased mineral fertilization on the aeration status of a cultivated soil, *Zesz. Probl. Postepow Nauk Roln.*, 197, 203, 1977.
695. **Stepniewski, W.,** The usefulness of the tip-point microelectrode in the measurement of ODR in soil, *Roczn. Glebozn.*, 30, 15, 1979.
696. **Stępniewski, W.,** Gas diffusion in a silty brown soil, *Zesz. Probl. Postepow Nauk Roln.*, 220, 559, 1980.
697. **Stępniewski, W.,** Oxygen diffusion and strength as related to soil compaction. I. ODR, *Pol. J. Soil Sci.*, 13, 3, 1980.
698. **Stępniewski, W.,** Oxygen diffusion and strength as related to soil compaction. II. Oxygen diffusion coefficient, *Pol. J. Soil Sci.*, 14, 3, 1981.
699. **Stępniewski, W.,** Oxygen diffusion and strength as related to soil compaction. III. Soil strength and water retention, *Pol. J. Soil Sci.*, 14, 15, 1981.
700. **Stępniewski, W.,** Oxygen diffusion and strength as related to soil compaction. IV. Choice of optimum parameters, *Pol. J. Soil Sci.*, 14, 3, 1981.
701. **Stevens, J. R. and Cornforth, I. S.,** The effect of pig slurry applied to a soil surface on the composition of the soil atmosphere, *J. Sci. Ed. Agric.*, 25, 1263, 1974.
702. **Stevenson, I. L.,** Some observation on the microbial activity in remoistened air dried soils, *Plant Soil*, 7, 170, 1957.

703. **Stewart, E. R. and Freebairn, H. T.,** Ethylene, seed germination, and epinasty, *Plant Physiol.*, 44, 955, 1969.
704. **Stolwijk, J. A. A. and Thimann, K. V.,** On the uptake of carbon dioxide and bicarbonate by roots, and its influence on growth, *Plant Physiol.*, 32, 513, 1957.
705. **Stolzy, L. H.,** Soil atmosphere, in *The Plant Root and its Environment*, Carson, E. W., Ed., University Press of Virginia, Charlottesville, 1974, 335.
706. **Stolzy, L. H. and Flühler, H.,** Measurement and prediction of anaerobiosis in soils, in *Nitrogen in the Environment*, Vol. 1. Nielsen, D. R. and McDonald, A. G., Eds., Academic Press, New York, 1978, 363.
707. **Stolzy, L. H., Focht, D. D., and Flühler, J.,** Indicators of soil aeration status, *Flora*, 171, 236, 1981.
708. **Stolzy, L. H., Labanauskas, C. K., Klotz, L. J., and De Wolfe, T. A.,** Nutritional responses and root rot of *Citrus limon* and *Citrus sinensis* under high and low soil oxygen supplies in the presence and absence of *Phytophthora* spp., *Soil Sci.*, 119, 136, 1975.
709. **Stolzy, L. H. and Letey, J.,** Characterizing soil oxygen conditions with a platinum microelectrode, *Adv. Agron.*, 16, 249, 1964.
710. **Stolzy, L. H., Letey, J., Klotz, L., and De Wolfe, T. A.,** Soil aeration and root-rotting fungi as factors in decay of citrus feeder roots, *Soil Sci.*, 99, 403, 1965.
711. **Stolzy, L. H., Letey, J., Szuszkiewicz, T. E., and Lunt, O. R.,** Root growth and diffusion rates as functions of oxygen concentration, *Soil Sci. Soc. Am. Proc.*, 25, 463, 1961.
712. **Stolzy, L. H., Taylor, O. C., Dugger, W. M., and Mersereau, D.,** Physiological changes in an ozone susceptibility of the tomato plant after short periods of inadequate oxygen diffusion to the roots, *Soil Sci. Soc. Am. Proc.*, 28, 305, 1964.
713. **Stolzy, L. H., Taylor, O. C., Letey, J., and Szuszkiewicz, T. E.,** Influence of soil-oxygen diffusion rates on susceptibility of tomato plants to air-borne oxidants, *Soil Sci.*, 91, 15, 1961.
714. **Stolzy, L. H., Van Gundy, S. D., Labanauskas, C. K., and Szuskiewicz, T. E.,** Response of *Tylenhulus semipenetrans* infected citrus seedlings to soil aeration and temperature, *Soil Sci.*, 96, 292, 1963.
715. **Stotzky, G.,** Microbial respiration, in *Methods of Soil Analysis*. II. Chemical and Microbiological Properties, Black, C. A. Ed., American Society of Agronomy, Madison, Wisc., 1965.
716. **Stotzky, G.,** Influence of clay minerals on microorganisms. II. Effect of various clay species, homoionic clays, and other particles on bacteria, *Can. J. Microbiol.*, 12, 831, 1966.
717. **Stotzky, G.,** Influence of clay minerals on microorganisms. III. Effect of particle size, cation exchange capacity, and surface area on bacteria, *Can. J. Microbiol.*, 12, 1235, 1966.
718. **Stotzky, G. and Rem, L. T.,** Influence of clay minerals on microorganisms. I. Montmorillonite and kaolinite on bacteria, *Can. J. Microbiol.*, 12, 547, 1966.
719. **Stotzky, G. and Rem, L. T.,** Influence of clay minerals on microorganisms. IV. Montmorillonite and kaolinite on fungi, *Can. J. Microbiol.*, 13, 1535, 1967.
720. **Stumm, W. and Morgan, J.,** *Aquatic Chemistry: Introduction Emphasizing Chemical Equilibria in Natural Waters*, Wiley-Interscience, New York, 1970.
721. **Summerfield, R. J. and Rieley, J. O.,** Growth of *Narthecium ossifragum* in relation to the dissolved oxygen concentration of the rooting substrate, short communication, *Plant Soil.*, 41, 701, 1974.
722. **Tackett, J. L.,** Theory and application of gas chromatography in soil aeration research, *Soil Sci. Soc. Am. Proc.*, 33, 346, 1968.
723. **Tackett, J. L. and Pearson, R. W.,** Oxygen requirements of cotton seedlings roots for penetration of compacted soil cores, *Soil Sci. Soc. Am. Proc.*, 28, 600, 1964.
724. **Tackett, J. L. and Pearson, R. W.,** Effect of carbon dioxide on cotton seedling root penetration of compacted soil cores, *Soil Sci. Soc. Am. Proc.*, 28, 741, 1964.
725. **Tadano, T., Kirimoto, K., Aoyama, J., and Tanaka, A.,** Comparison of tolerance to high moisture conditions in the soil among crop plants: studies on the comparative plant nutrition (Japanese), *J. Sci. Soc. Manure*, 50, 261, 1979.
726. **Takai, Y. and Kamura, T.,** The mechanisms of reduction in waterlogged paddy soil, *Folia Microbiol.*, 11, 304, 1966.
727. **Tamm, E. and Krzysch, G.,** The effect of soil temperature and moisture on carbon-di-oxide production in a sandy loam (German, English summary), *Z. Acker Pflanzenbau.*, 117, 359, 1963.
728. **Tanner, C. B. and Wengel, R. W.,** An air permeameter for field and laboratory use, *Soil Sci. Soc. Am. Proc.*, 21, 663, 1957.
729. **Tate, R. L.,** Effect of flooding on microbial activities in organic soils: carbon metabolism, *Soil Sci.*, 128, 267, 1979.
730. **Taylor, D. I.,** Influence of oxygen tension on respiration, fermentation and growth of wheat and rice, *Am. J. Bot.*, 29, 721, 1942.
731. **Taylor, S. A.,** Oxygen diffusion in porous media as a measure of soil aeration, *Soil Sci. Soc. Am. Proc.*, 14, 55, 1949.

732. **Terry, R. E.,** Nitrogen mineralization in Florida Histosols, *Soil Sci. Soc. Am. J.*, 44, 747, 1980.
733. **Tesarova, M. and Gloser, J.,** Total $CO_2$ output from alluvial soils with two types of grassland communities, *Pedobiologia*, 16, 364, 1976.
734. **Thomas, R. E., Schwartz, W. A., and Bendixen, T. W.,** Pore gas composition under sewage spreading, *Soil Sci. Am. Proc.*, 32, 419, 1968.
735. **Tjepkema, J.,** The role of oxygen diffusion from the shoots and nodule roots in nitrogen fixation by root nodules of *Myrica gale*, *Can. J. Bot.*, 56, 1365, 1978.
736. Trans. 12th Int. Congr. Soil Sci., New Delhi, India, 1982, in *Vertisols and Rice Soils of the Tropics*, Symp. papers II, New Delhi 1982.
737. **Troeh, F. R., Jabro, J. D., and Kirkham, D.,** Gaseous diffusion equation for porous materials, *Geoderma*, 27, 239, 1982.
738. **Trought, M. C. T. and Drew, M. C.,** The development of water-logging damage in wheat seedlings (*Triticum aestivum* L.). I. Shoot and root growth in relation to changes in the concentrations of dissolved gases and solutes in the soil solution, *Plant Soil*, 54, 55, 1980.
739. **Trought, M. C. T. and Drew, M. C.,** The development of waterlogging damage in wheat seedlings (*Triticum aestivum* L.).II. Accumulation and redistribution of nutrients by the shoot, *Plant Soil*, 56, 187, 1980.
740. **Trybis, W. P. and Aviedeev, L. B.,** Gas exchange dynamics in a peat soil during incubation (Russian, English summary), *Pochvovedenie*, 7, 101, 1978.
741. **Tu, C. M.,** A screening technique for assessing effects of pesticides on population and activities of non-target soil microorganisms, *Bull. Environ. Contam. Toxicol.*, 20, 212, 1978.
742. **Tu, C. M.,** Effects of insecticides on population of microflora, nitrification and respiration in soil, *Soil Sci. Plant Anal.*, 9, 629, 1978.
743. **Turner, F. T. and Patrick, W. H.,** Chemical changes in waterlogged soils as a result of oxygen depletion, in *Trans. 9th Int. Congr. Soil Sci.*, Adelaide, Australia, 1968, 55.
744. **Turski, R., Domżał, H., and Słowińska-Jurkiewicz, A.,** Air permeability as an index of the physical state of soil (Polish, English summary), *Roczn. Glebozn.*, 29, 3, 1978.
745. **Tyler, G.,** Heavy metal pollution and soil enzymatic activity, *Plant Soil*, 41, 303, 1974.
746. **Uhling, F., Maidl, F. X., and Fischbeck, G.,** A stationary method useful for gas analysis of soil air (German), *Z. Acker Pflanzenbau*, 150, 474, 1981.
747. **Unger, P. W. and Danielson, R. E.,** Influence of oxygen and carbon dioxide on germination and seedling development of corn (*Zea mays* L), *Agron. J.*, 57, 56, 1965.
748. **Valoras, N. and Letey, J.,** Soil oxygen and water relationships to rice growth, *Soil Sci.*, 101, 210, 1966.
749. **Van Amerongen, G. J.,** The permeability of different rubbers to gases and its relation to diffusivity and solubility, *J. Appl. Phys.*, 17, 972, 1946.
750. **Van Bavel, C. H. M.,** A soil aeration theory based on diffusion, *Soil Sci.*, 73, 33, 1951.
751. **Van Bavel, C. H. M.,** Gaseous diffusion and porosity in porous media, *Soil Sci.*, 73, 91, 1952.
752. **Van Bavel, C. H. M.,** Composition of soil atmosphere, in *Methods of Soil Analysis*. I. Black, C. A., Ed., American Society of Agronomy, Madison, Wisc., 1965, 315.
753. **Van Cleemput, O., Patrick, W. H., and McIlhenny, R. C.,** Formation of chemical and biological denitrification products in flooded soil at controlled pH and redox potential, *Soil Biol. Biochem.*, 7, 329, 1975.
754. **Van Cleemput, O., Patrick, W. H., and McIlhenny, R. C.,** Nitrate decomposition in flooded soil under different pH and redox potential conditions, *Soil Sci. Soc. Am. J.*, 40, 55, 1976.
755. **Van Cleve, K., Coyne, P. I., Goodwin, E., Johnson, C., and Kelley, M.,** A comparison of four methods for measuring respiration in organic material, *Soil Biol. Biochem.*, 11, 237, 1979.
756. **Van der Heide, H., Boer-Bolt, B. M., and van Raalte, M. H.,** The effect of a low oxygen content of the medium on the roots of barley seedlings, *Acta Bot. Neerl.*, 12, 231, 1963.
757. **Van Diest, A.,** Effect of soil aeration and compaction upon yield, nutrient uptake and variability in a greenhouse fertility experiment, *Agron. J.*, 54, 515, 1962.
758. **Van Doren, D. M. and Erickson, A. E.,** Factors affecting the platinum microelectrode method for measuring the rate of oxygen diffusion through the soil solution, *Soil Sci.*, 102, 23, 1966.
759. **Van Groenewoud, H.,** Methods and apparatus for measuring air permeability of the soil, *Soil Sci.*, 106, 275, 1968.
760. **Van Gundy, S. D., McElroy, F. D., Cooper, A. F., and Stolzy, L. H.,** Influence of soil temperature, irrigation and aeration on *Hemicycliophora arenaria*, *Soil Sci.*, 106, 270, 1968.
761. **Van Gundy, S. D. and Stolzy, L. H.,** The relationship of oxygen diffusion rates to the survival, movement and reproduction of *Hemicycliophora arenaria*, *Nematologica*, 9, 605, 1963.
762. **Van Overbeek, J.,** Water uptake by excised root systems of the tomato due to non-osmotic forces, *Ann. J. Bot.*, 29, 677, 1942.

763. **Van Praag, H. J.,** Integral studies of pseudo-gley soils under the natural vegetation of Hautes Fagnes Plateau (Belgium). II. Theoretique and experimental considerations on the ODR measurements with the platinum microelectrode (French), *Catena,* 8, 325, 1981.
764. **Van Raalte, M. H.,** On the oxygen supply of rice roots, *Ann. Bot., Gard. Buitenzorg,* 51, 43, 1941.
765. **Van Raalte, M. H.,** On the oxidation of the environment by the roots of rice (*Oriza sativa* L.), *Hortus botanicus, Bogotiensis,* Java, Syokubutu-Iho, 1, 15, 1943.
766. **Van Schreven, D. A., Lindenberg, D. J., and Koridon, A.,** Effect of several herbicides on bacterial populations and activity and the persistence of these herbicides in soil, *Plant Soil,* 33, 513, 1970.
767. **Varade, S. B., Stolzy, L. H., and Letey, J.,** Influence of temperature, light intensity, and aeration on growth and root porosity of wheat, *Triticum aestivum, Agron. J.,* 62, 505, 1970.
768. **Velich, J. and Honz, J.,** Microbial activity in gleyed meadow soil and pseudogley soil under permanent meadow, (Czechoslovakian, English summary), *Rostl. Vyroba,* 17, 1203, 1971.
769. **Vilain, M.,** Soil aeration (French), *Ann. Agron.,* 14, 967, 1963.
770. **Visser, J., Locher, J. Th., and Brouwer, R.,** Effects of aeration and mineral supply on growth and mineral content of shoots and roots of apple trees (var. ''Golden Delicious'' on M IX), *Neth. J. Agric. Sci.,* 19, 125, 1971.
771. **Visser, W. C.,** Pore space determination as a field method, *Soil Sci.,* 44, 467, 1937.
772. **Vlamis, D. and Davis, A. R.,** Germination growth and respiration of rice barley seedlings at low oxygen pressures, *Plant Physiol.,* 18, 685, 1943.
773. **Waddington, D. V. and Baker, J. H.,** Influence of soil aeration on the growth and chemical composition of three grass species, *Agron. J.,* 57, 253, 1965.
774. **Wallihan, E. F., Garber, N. J., Sharpless, R. G., and Printy, W. L.,** Effect of soil oxygen deficit on iron nutrition of orange seedlings, *Plant Physiol.,* 36, 425, 1961.
775. **Wanner, H.,** Oxygen diffusion as a limiting factor of plant root respiration, *Vierteljahrsschr. Naturforsch. Ges. Zuerich,* 90, 98, 1945.
776. **Wanner, H.,** Soil respiration, litter fall and productivity of tropical rain forest, *J. Ecol.,* 58, 543, 1970.
777. **Watanabe, I.,** Biological nitrogen fixation associated with wetland rice, in *Proc. Symp. Paddy Soils China,* Nanjing, China, 1980, Institut Soil Sci., Academia Sinica, Ed., Science Press Beijing-Springer-Verlag, Berlin, 1981, 348.
778. **Watson, E. R., Lapins, P., and Barron, R. J. W.,** Effect of waterlogging on the growth, grain and straw yield of wheat, barley and oats, *Aust. J. Exp. Agric. Anim. Husb.,* 16, 116, 1976.
779. **Weaver, R. W.,** A simple, inexpensive apparatus for simultaneous collection of $CO_2$ evolved from numerous soils, *Soil Sci. Soc. Am. Proc.,* 38, 853, 1974.
780. **Weeraratna, C. S.,** Effect of 2-chloro-4, 6-bis (ethylamino)-S-triazine (Simazine) on some soil microbial processes, *Zentralbl. Bakteriol. Parasitenkd., Infectionskr. Hyg., Abt. 2,* 134, 115, 1979.
781. **Weller, K. R., Stenhouse, N. S., and Watts, M.,** Diffusion of gases in porous solids. I and II. *Can. J. Chem.,* 52, 2684 and 2692, 1974.
782. **Welp, G., Herms, U., and Brümmer, G.,** Influence of soil reaction, redox condtions and organic matter on the phosphate content of soil solutions, *Z. Pflanzenernaehr. Bodenkd.,* 146, 38, 1983.
783. **Wengel, R. W.,** Emergence of corn in relation to soil oxygen diffusion rates, *Agron. J.,* 58, 69, 1966.
784. **Wesseling, J.,** Some solutions of the steady state diffusion of carbon dioxide through soils, *Neth. J. Agric. Sci.,* 10, 109, 1962.
785. **Wesseling, J. and van Wijk, W. R.,** Land drainage in relation to soils and crops. I. Soil physical conditions in relation to drain depth, in *Drainage of Agricultural Land,* Luthin, J. D., Ed., American Society of Agronomy, Madison, Wisc., 1957, 461.
786. **Whiteside, J. S. and Alexander, M.,** Measurements of microbiological effects of herbicides, *Weeds,* 8, 204, 1960.
787. **Wiedenroth, E. M.,** Relations between photosynthesis and root metabolism of cereal seedlings influenced by root anaerobiosis, *Phytosynthetica,* 15, 575, 1981.
788. **Wiegand, C. L. and Lemon, E. R.,** A field study of some plant-soil relation in aeration, *Soil Sci. Soc. Am. Proc.,* 22, 216, 1958.
789. **Wiersma, D. and Mortland, M. M.,** Response of sugar beets to peroxide fertilization and its relationship to oxygen diffusion, *Soil Sci.,* 75, 355, 1953.
790. **Wiersum, L. K.,** Potential subsoil utilization by roots, *Plant Soil,* 27, 383, 1967.
791. **Wiersum, L. K.,** Presumed aerotropic growth of roots of certain species, *Naturwissenschaften,* 54, 203, 1967.
792. **Wiersum, L. K.,** Tulip root behaviour and aeration requirements, *Acta Hortic.,* 23, 318, 1971.
793. **Wildung, R. E., Garland, T. R., and Buschbom, R. L.,** The interdependent effects of soil temperature and water content on soil respiration rate and plant root decomposition in arid grassland soils, *Soil Biol. Biochem.,* 7, 373, 1975.

794. **Willett, I. R.,** Redox potentials and pH of two Murrumbidgee irrigation area rice soils during flooding for rice cultivation, CSIRO Division of Soils, *Div. Rep. No. 31,* Bibliography, Commonwealth Scientific and Industrial Research Organization, Adelaide, Australia, 1978.

795. **Willey, C. R.,** Effect of short periods of anaerobic and near-anaerobic conditions on water uptake by tobacco roots, *Agron. J.*, 62, 224, 1970.

796. **Willey, C. R. and Tanner, C. B.,** Membrane-covered electrode for measurement of oxygen concentration in soil, *Soil Sci. Soc. Am. Proc.*, 27, 511, 1963.

797. **Willey, C. R. and Tanner, C. B.,** A steady-state method for laboratory measurements of the oxygen diffusivity of porous media, *Soil Sci. Soc. Am. Proc.*, 28, 733, 1964.

798. **Williams, S. T., Shameemullah, M., Watson, E. T., and Mayfield, C. I.,** Studies on the ecology of actinomycetes in soil. IV. The influence of moisture tension on growth and survival, *Soil Biol. Biochem.*, 4, 215, 1972.

799. **Williamson, R .E.,** The effect of root aeration on plant growth, *Soil Sci. Soc. Am. Proc.*, 28, 86, 1964.

800. **Williamson, R. E.,** Effect of soil gas composition and flooding on growth of *Nicotiana tabacum* L., *Agron. J.*, 62, 80, 1970.

801. **Williamson, R. E. and Splinter, W. E.,** Effect of gaseous composition of root environment upon root development and growth of *Nicotiana tabacum* L., *Agron. J.*, 60, 365, 1968.

802. **Williamson, R. E. and Splinter, W. E.,** Effects of light intensity temperature, and root gaseous environment on growth of *Nicotiana tabacum* L., *Agron, J.*, 61, 285, 1969.

803. **Wilson, J. M. and Griffin, D. M.,** Water potential and the respiration of microorganisms in the soils, *Soil Biol. Biochem.*, 7, 199, 1975.

804. **Wilson, J. M and Griffin, D. M.,** Respiration and radial growth of soil fungi at two osmotic potentials, *Soil Biol. Biochem.*, 7, 269, 1975.

805. **Wójcikowska-Kapusta, A.,** The Content of Some Trace Elements in Soils of the Lublin Coal Mine Region with Particular Reference to Their Accumulation and Solubility (Polish), Ph.D. dissertation, Agricultural University, Lublin, Poland, 1982.

806. **Wood, M. B.,** An application of gas chromatography to measure concentrations of ethane, propane, and ethylene found in industrial soil gases, *J. Chromatogr. Sci.*, 18, 307, 1980.

807. **Wood, J. T. and Greenwood, D. J.,** Distribution of carbon dioxide and oxygen in the gas phase of aerobic soils, *J. Soil Sci.*, 22, 281, 1971.

808. **Yamane, I.,** Metabolism in muck paddy soil. II. Determination of gases evolved from paddy field. Estimation of decomposable organic matter, *Soil Plant Food*, 4, 25, 1958.

809. **Yamane, I.,** Eh-pH diagrams of iron systems in relation to flooded rice soils, *Rep. Inst. Agric. Res. Tohoku Univ.*, 21, 39, 1970.

810. **Yamane, I.,** Ep-pH diagrams of manganese systems in relation to flooded soils, *Rep. Inst. Agric. Res. Tohoku Univ.*, 24, 1, 1973.

811. **Yamane, I. and Okazaki, M.,** Chemical properties of submerged rice soils, in *Trans. 12th Int. Congr. Soil Sci.*, New Delhi, India, 1982, Symp. Papers II, 143.

812. **Yamane, I. and Sato, K.,** Decomposition of organic acids and gas formation in flooded soils, *Soil Sci. Plant Nutr.*, 9, 32, 1963.

813. **Yamane, I. and Sato.,** Decomposition of glucose and gas formation in flooded soil, *Soil Sci. Plant Nutr.* 10, 127, 1964.

814. **Yamane, I. and Sato, K.,** Effect of temperature on the decomposition of organic substances in flooded soil, *Soil Sci. Plant Nutr.*, 13, 94, 1967.

815. **Yamane, I. and Sato, K.,** Initial rapid drop of oxidation-reduction potential in submerged air-dried soils, *Soil Sci. Plant Nutr.*, 14, 68, 1968.

816. **Yasuda, T.,** Studies on soil air. III. Measurement of oxygen diffusion rates with the platinum microelectrode and its application to plant growth, *J. Sci. Soil Manure*, 43, 223, 1973.

817. **Yentur, S. and Leopold, A. C.,** Respiratory transition during seed germination, *Plant Physiol.*, 57, 274, 1976.

818. **Yoneda, T.,** Studies on the rate of decay of wood litter, on the forest floor. II. Dry weight loss and $CO_2$ evolution of decaying wood, *Jpn. J. Ecol.*, 25, 132, 1975.

819. **Yoneda, T. and Kirita, H.,** Soil respiration, in *Biological Production in a Warm-Temperate Evergreen Oak Forest of Japan,* Kira, T., Ono, Y., and Hosokawa, T., Eds., Japanese Comm. for the Int. Biol. Progr. Univ. of Tokyo Press, 7 IBP Synthesis, Tokyo, 1978, 18, 239.

820. **Yoshida, T. and Suzuki, M.,** Formation and degradation of ethylene in submerged rice soils, *Soil Sci. Plant Nutr.*, 21, 129, 1975.

821. **Yoshida, T., Takai, Y., and Del Rosario, D. C.,** Molecular nitrogen content in a submerged rice field, *Plant Soil,* 42, 653, 1975.

822. **Yoshinari, T. and Knowles, R.,** Acetylene inhibition of nitrous oxide reduction by denitrifying bacteria, *Biochem. Biophys. Res. Commun.*, 69, 705, 1976.

823. **Young, A. and Harris, P. J.,** Preliminary investigation into the biodegrability of polyvinyl alcohol and polyvinyl acetate, *Meded. Fac. Landbouwwet. Rijksuniv. Gent,* 41/1, 253, 1976.
824. **Yu, P. T., Stolzy, L. H., and Letey, J.,** Survival of plants under prolonged flooded conditions, *Agron. J.,* 61, 844, 1969.
825. **Yu Tian-ren,** Oxidation-reduction properties of paddy soils, in *Proc. Symp. Paddy Soils China,* Nanjing, China, 1980. Institut Soil Sci., Academia Sinica, Ed., Science Press Beijing-Springer-Verlag, Berlin, 1981, 95.
826. **Zablotowicz, R. M., Eskew, D. L., and Focht, D. D.,** Dentrification in *Rhizobium, Can. J. Microbiol.,* 24, 757, 1978.
827. **Zimmerman, J. F. and Berry, L. J.,** Studies of onion root respiration. III. Amperometric titration as a method for the measurement of respiratory overshoot, *Biochem. Biophys. Acta,* 3, 615, 1949.
828. **Zvyagintsev, D. G.,** *Interaction on Microorganisms with Solid Surface* (Russian), Moskov University, Moskow, 1973.

# INDEX

## C



## D

## E



## F

## N



## O

## P

## R

## S

## Y



## Z